AF361808

Unconventional Drone-Based Surveying

Unconventional Drone-Based Surveying

Editors

Arianna Pesci
Giordano Teza
Massimo Fabris

Basel • Beijing • Wuhan • Barcelona • Belgrade • Novi Sad • Cluj • Manchester

Editors

Arianna Pesci
Istituto Nazionale di
Geofisica e Vulcanologia
Sezione di Bologna
Bologna
Italy

Giordano Teza
Department of Physics
and Astronomy
University of Bologna
Bologna
Italy

Massimo Fabris
Department of Civil,
Environmental and
Architectural Engineering
University of Padova
Padova
Italy

Editorial Office
MDPI
St. Alban-Anlage 66
4052 Basel, Switzerland

This is a reprint of articles from the Special Issue published online in the open access journal *Drones* (ISSN 2504-446X) (available at: www.mdpi.com/journal/drones/special_issues/unconventional_drone_surveying).

For citation purposes, cite each article independently as indicated on the article page online and as indicated below:

Lastname, A.A.; Lastname, B.B. Article Title. *Journal Name* **Year**, *Volume Number*, Page Range.

ISBN 978-3-0365-8727-1 (Hbk)
ISBN 978-3-0365-8726-4 (PDF)
doi.org/10.3390/books978-3-0365-8726-4

Contents

Preface . . . . **vii**

Arianna Pesci, Giordano Teza and Massimo Fabris
Editorial of Special Issue "Unconventional Drone-Based Surveying"
Reprinted from: *Drones* **2023**, *7*, 175, doi:10.3390/drones7030175 . . . . **1**

Yeong-Ju Go and Jong-Soo Choi
An Acoustic Source Localization Method Using a Drone-Mounted Phased Microphone Array
Reprinted from: *Drones* **2021**, *5*, 75, doi:10.3390/drones5030075 . . . . **8**

Andrey V. Savkin, Satish Chandra Verma and Stuart Anstee
Optimal Navigation of an Unmanned Surface Vehicle and anAutonomous Underwater Vehicle
Collaborating for Reliable Acoustic Communication with Collision Avoidance
Reprinted from: *Drones* **2022**, *6*, 27, doi:10.3390/drones6010027 . . . . **26**

Xiaoyu Liu, Xugang Lian, Wenfu Yang, Fan Wang, Yu Han and Yafei Zhang
Accuracy Assessment of a UAV Direct Georeferencing Method and Impact of the Configuration
of Ground Control Points
Reprinted from: *Drones* **2022**, *6*, 30, doi:10.3390/drones6020030 . . . . **48**

**Alexander Shelekhov, Alexey Afanasiev, Evgeniya Shelekhova, Alexey Kobzev, Alexey
Tel'minov and Alexander Molchunov et al.**
Low-Altitude Sensing of Urban Atmospheric Turbulence with UAV
Reprinted from: *Drones* **2022**, *6*, 61, doi:10.3390/drones6030061 . . . . **63**

Sunan Huang, Rodney Swee Huat Teo and William Wai Lun Leong
Multi-Camera Networks for Coverage Control of Drones
Reprinted from: *Drones* **2022**, *6*, 67, doi:10.3390/drones6030067 . . . . **85**

Eric Villeneuve, Abdallah Samad, Christophe Volat, Mathieu Béland and Maxime Lapalme
An Experimental Apparatus for Icing Tests of Low Altitude Hovering Drones
Reprinted from: *Drones* **2022**, *6*, 68, doi:10.3390/drones6030068 . . . . **108**

Manaram Gnanasekera and Jay Katupitiya
A Time-Efficient Method to Avoid Collisions for CollisionCones: An Implementation for UAVs
Navigating inDynamic Environments
Reprinted from: *Drones* **2022**, *6*, 106, doi:10.3390/drones6050106 . . . . **127**

Matteo Cutugno, Umberto Robustelli and Giovanni Pugliano
Structure-from-Motion 3D Reconstruction of the Historical Overpass Ponte della Cerra: A
Comparison between MicMac® Open Source Software and Metashape®
Reprinted from: *Drones* **2022**, *6*, 242, doi:10.3390/drones6090242 . . . . **153**

**Antonio L. Diaz, Andrew E. Ortega, Henry Tingle, Andres Pulido, Orlando Cordero and
Marisa Nelson et al.**
The Bathy-Drone: An Autonomous Uncrewed Drone-Tethered Sonar System
Reprinted from: *Drones* **2022**, *6*, 294, doi:10.3390/drones6100294 . . . . **176**

**Rolando Salas López, Renzo E. Terrones Murga, Jhonsy O. Silva-López, Nilton B.
Rojas-Briceño, Darwin Gómez Fernández and Manuel Oliva-Cruz et al.**
Accuracy Assessment of Direct Georeferencing for Photogrammetric Applications Based on
UAS-GNSS for High Andean Urban Environments
Reprinted from: *Drones* **2022**, *6*, 388, doi:10.3390/drones6120388 . . . . **195**

Massimo Fabris, Pietro Fontana Granotto and Michele Monego
Expeditious Low-Cost SfM Photogrammetry and a TLS Survey for the Structural Analysis of
Illasi Castle (Italy)
Reprinted from: *Drones* **2023**, 7, 101, doi:10.3390/drones7020101 . . . . **210**

Preface

Nowadays, unmanned aerial vehicles (UAVs), as well as unmanned surface vehicles (USVs) or unmanned underwater vehicles (UUVs), later on simply called drones, have reached a sufficient degree of maturity to allow their use for various purposes. Drones are used in remote sensing, in particular in geological mapping and architectural surveying. New applications continually appear; among them, disaster management, precision agriculture, weather forecast, wildlife monitoring, search and rescue, law enforcement, shipping and delivery, and also entertainment, with different type of sensors.

This increasing diffusion of drones has various reasons. On the one hand, technologically advanced low-cost drones, generally equipped with GNSS receivers and, often, also equipped with compact inertial platforms, are available today. On the other hand, nowadays, besides professional and prosumer cameras for structure-from-motion photogrammetry (SfM), other relatively affordable sensors like light infrared cameras or compact LiDARs are available. Last but not least, the drone-based survey can be performed by planning the navigation consistently with the specific application. Furthermore, the aspect of the use of drone swarms, or in any case of operations in environments where there are numerous drones in motion simultaneously, is always better studied.

In some applications, the use of drones is now completely standardized. However, there is significant room for growth both to improve the accuracy and resolution of survey products, optimizing flight plans (in general, navigation plans) according to them, and to identify new applications. These systems are suitable both for routine surveys, even predefined at scheduled times, and for surveys in emergency conditions. However, it must not be forgotten that the simplicity of the survey and the possibility of greatly automating the survey and data processing activities in any case require an adequate evaluation of the quality of the results.

Arianna Pesci, Giordano Teza, and Massimo Fabris
Editors

Editorial

Editorial of Special Issue "Unconventional Drone-Based Surveying"

Arianna Pesci [1,*], Giordano Teza [2] and Massimo Fabris [3]

[1] Istituto Nazionale di Geofisica e Vulcanologia, Sezione di Bologna, Via Creti 12, 40128 Bologna, Italy
[2] Department of Physics and Astronomy, Alma Mater Studiorum University of Bologna, Viale Berti Pichat 6/2, 40127 Bologna, Italy
[3] Department of Civil, Environmental and Architectural Engineering—ICEA, University of Padova, 35131 Padova, Italy
[*] Correspondence: arianna.pesci@ingv.it

Citation: Pesci, A.; Teza, G.; Fabris, M. Editorial of Special Issue "Unconventional Drone-Based Surveying". *Drones* **2023**, *7*, 175. https://doi.org/10.3390/drones7030175

Academic Editor: Diego González-Aguilera

Received: 28 February 2023
Accepted: 1 March 2023
Published: 3 March 2023

1. Introduction

Nowadays, Unmanned Aerial Vehicles (UAVs), as well as Unmanned Surface Vehicles (USVs) or also Unmanned Underwater Vehicles (UUVs), later on simply called drones, have reached a sufficient degree of maturity to allow their use for various purposes. Drones are used in remote sensing, as shown by [1] and references therein, where regulatory and safety facts are also discussed, in particular in geological mapping [2] and architectural surveying [3]. New applications continually appear; among them, disaster management, precision agriculture, weather forecast, wildlife monitoring, search and rescue, law enforcement, shipping and delivery, and also entertainment, with different type of sensors.

This increasing diffusion of drones has various reasons. On the one hand, technologically advanced low-cost drones, generally equipped with GNSS receivers and, often, also equipped with compact inertial platforms, are available today. On the other hand, nowadays, besides professional and prosumer cameras for Structure-from-Motion photogrammetry (SfM), other relatively affordable sensors like light infrared cameras [4] or compact LiDARs [5] are available. Last but not least, the drone-based survey can be performed by planning the navigation consistently with the specific application. Furthermore, the aspect of the use of drone swarms, or in any case of operations in environments where there are numerous drones in motion simultaneously, is always better studied [6].

In some applications the use of drones is now completely standardized. However, there is significant room for growth both to improve the accuracy and resolution of survey products, optimizing flight plans (in general, navigation plans) according to them, and to identify new applications [7]. These systems are suitable both for routine surveys, even predefined at scheduled times, and for surveys in emergency conditions. However, it must not be forgotten that the simplicity of the survey and the possibility of greatly automating the survey and data processing activities in any case require an adequate evaluation of the quality of the results, as shown, e.g., by [8] in the case of evaluation of surface deformations due to induced liquefaction.

For all these reasons, the special issue of Drones "Unconventional Drone-Based Surveying", to which this Editorial is dedicated, was born. The articles published in the framework of the Special Issue are briefly described in Section 2, while the main lessons learned are discussed in Section 3, with emphasis on multidisciplinarity, reliability, exportability of the techniques used and, obviously, quality of the obtainable results. Insights about future researches are also provided.

2. Overview of Contributions

The papers included in this Special Issue cover a wide spectrum of applications related to the unconventional surveying using drones in different areas of the world and characterized by multi-source data. The accepted works are briefly presented here in order of publishing.

Go and Choi [9] implemented a system to detect the location of acoustic sources generated on the ground using a drone equipped with an array of microphones. In order to reduce the noise generated by the drone, and better identify the acoustic signals of interest, the authors successfully applied spectral subtraction method. Moreover, they implemented a method to locate the acoustic sources by fusing flight information from the navigation system of the drone. The validation of the proposed method was performed using a drone mounted with a 32-channel microphone array and the verification of the sound source location detection method was limited to the explosion sound generated from the firecrackers: in the experiment conducted at the Chungnam National University stadium (Korea), the authors detected the acoustic source with a ground distance of 151.5 m between the drone and the explosion location, and errors of 8.8° and 10.3° for the horizontal and vertical angles respectively.

Savkin et al. [10] analyzed safe navigation of an USV in proximity to a UUV. In this context, the authors aimed to maximize the amount of data successfully transmitted between the vehicles using underwater acoustic communications, while minimizing the probability of collisions between them. For these reasons, they implemented a real-time navigation algorithm belonging to the class of sliding mode control laws. The authors verified that the developed algorithm is asymptotically optimal, because as the interval subdivision parameters tend to infinity, the value of the cost function that describe the amount of successfully transmitted data converges to the global maximum. In addition, they checked the effectiveness of the proposed method with MATLAB simulations.

Liu et al. [11] evaluated accuracy of the direct georeferencing GNSS-assisted UAV method linked to the number and distribution of ground control points (GCPs) in the study area. After collected data in Xishan, Taiyuan, Shanxi Province (China) using a FEIMA D2000 multi-rotor UAV, the authors developed several photogrammetric projects varying the number and distribution of GCPs (measured by means of a GNSS real-time kinematic—RTK receiver) used in the bundle adjustment (BA) process of the Agisoft Photoscan software. The evaluation of the obtained results was performed using the root mean square error (RMSE) of ground-measured checkpoints (CPs) and the Multiscale Model to Model Cloud Comparison (M3C2) distance. In this context, in the direct georeferencing the authors obtained 0.087 m and 0.041 m in the vertical and horizontal RMSE, respectively. Finally, they suggested the use of at least 1 GCP in the BA, while when the resolution of the GCPs is greater than $10/12$ GCP/km^2 horizontal and vertical errors decrease slowly.

Shelekhov et al. [12] investigated the effects of a turbulent atmosphere on a quadcopter for low-altitude surveys on morphologically complex urban areas. The authors provided the equations for fluctuations of the longitudinal and lateral wind speed; the fluctuations are proportional to the variations of the pitch and roll angles. In the experiment conducted in Academgorodok, Tomsk (Russian Federation) they used a DJI Phantom 4 Pro quadcopter in the hover mode and combined with AMK-03 ultrasonic weather stations. Turbulence spectra with high spatial resolution in the atmosphere on complex areas, under different weather conditions and seasons were obtained. The authors concluded that, on the one hand, the quadcopter is a promising tool for solving problems related to the drone movement control under unfavorable weather conditions and that, on the other hand, such a kind of UAV can be used to provide climatic data in urban environments.

Huang et al. [13] presented a solution related to the coverage control of a multi-UAV system with downward facing cameras: for this issue the authors extended a coverage control algorithm to include a new sensor model and two constrains, the view angle and the collision avoidance. They provided the theoretical analysis to solve the issue of the view angle in the design of the coverage problem linked to the collision avoidance. Subsequently, the authors performed computer simulation tests using the MATLAB software; results showed automatic coverage with cameras of the area of interest (AOI) by multi-UAV

However, in their proposal the i-th UAV can be outside to the AOI and cannot manage invisible and blacked areas.

Villeneuve et al. [14] developed an experimental test rig to study the effects of icing on the rotor of a Bell APT70 drone rotor in take-off/hover flying mode. In the Anti-Icing Materials International Laboratory—AMIL (Canada), a 9-m-high cold chamber which allow the study of the ground effect and rotor proximity to the icing nozzle array at different rotor heights (h) was used by the authors for the test rig. Results of preliminary experimental comparison of aerodynamic parameters between rotor heights at $h = 2$ m and $h = 4$ m showed no differences. Further icing tests were performed by the authors at lower height taking advantage by the experimental convenience and increased distance from the nozzles. The studies related to the effects of ice on UAV performances are relevant in extreme environments, and the development of ice protection systems for drone applications in these situations is needed.

Gnanasekera and Katupitiya [15] studied a method to avoid collisions of aerial drones. They proposed an algorithm based on collision cone approach to avoid a collision in a time-efficient way (PHS—Purely Heading Solution). In this context, possible scenarios were analyzed including three methods to avoid a potential collision with dynamic obstacles: heading change, speed change and combined heading and speed change. The method proposed by the authors was mathematically demonstrated to be the most time-efficient, compared with other available works, and validated with simulation conducted in a MATLAB-based testing platform. Experiments were conducted using a matrice 600 Pro hexacopter in an open field in Menangle, New South Wales (Australia). Some differences between simulation and experimental results were observed by the authors. These differences were due to the dynamics of the drone and external resistances, e.g., the wind.

Cutugno et al. [16] assessed the quality of sparse point cloud extracted from UAV and using both free-and-open-source software (MicMac®) and commercial software package (Agisoft Metashape®). To this aim, they conducted a survey test over Conte della Cerra street, a historical overpass in the Vomero neighborhood of Naples (Italy) using a DJI model Mavic 2 Pro UAV. The authors extracted the 3D data using the two software and the photogrammetric results were evaluated analyzing the image residuals, the GCPs and CPs statistic errors, the relative accuracy assessment, and the Cloud-to-Cloud distance comparison. They obtained comparable results highlighting that the MicMac® open source software can produce reliable results and similar with those generated by a commercial software package.

Diaz et al. [17] developed a new drone-based system for bathymetric surveys: the "Bathy-drone" is composed by a multi-rotor drone that, by means of a tether, drags a small vessel on the water surface; the latter is equipped with a commercial sonar unit that has down-scan, side-scan, and chirp capabilities and is linked to a GPS-reference onboard. The authors conducted extensive tests over a 5 acre pond located at the University of Florida Plant Science and Education Unit in Citra (Florida) using a DJI Matrice 600 drone. To evaluate the performances of the system, they acquired the ground-truth data of the pond by using a Trimble RTK GNSS system on a pole. Results of the comparison provided a deeper average measurement of 21.6 cm of the Bathy-drone compared with the ground-truth, justifying the integration of RTK and inertial measurement unit (IMU) corrections.

Salas López et al. [18] analyzed the accuracy in positioning of the direct georeferencing approach together with the use of GCPs: they evaluated the repeatability and reproducibility of Digital Surface Model and ortho-mosaic photogrammetric products of a commercial multi-rotor system equipped with a GNSS receiver in urban environments. The authors conducted a test on an urban area of the Chachapoyas city (Peru) using a DJI Phantom 4 RTK UAS and acquiring 10 GCPs and 4 Validation Points with a Trimble R10 GNSS receiver worked in PPK (Post-Processing Kinematic) mode. They obtained viable solution of the direct georeferencing for applications in urban areas. In addition, PPK solution using at least 1 GCP significantly improved the RMSE when compared with the use of 5 or also 10 GCPs without PPK.

Fabris et al. [19] investigated expeditious survey procedures using low-cost photogrammetric sensors for the structural analysis of degraded historical buildings. The authors performed the 3D survey of Illasi Castle (northern Italy) using a Parrot Anafi low-cost drone, a single-lens reflex (SLR) camera, and a smartphone. They generated four different 3D models by means of the SfM technique from the acquired data using the Agisoft Metashape software. To evaluate accuracies and performances of the fast and low-cost photogrammetric approach, the authors compared the obtained models with a high-resolution and high-precision terrestrial laser scanning survey performed using the Leica ScanStation P20 instrument: they obtained standard deviation values of the point cloud differences of about 2–3 cm for the model generated integrating the images acquired with the drone and SLR camera, and double values when the images acquired with the smartphone were used. Subsequently, the best 3D model was used in the finite element (FE) analysis that provided the safety assessment of the historical building.

3. Lessons Learned and New Research Perspectives

The articles published in the framework of this Special Issue concern several topics (they are presented here, unlike in the previous chapter, in order of affinity of the research topics): surveys with non-optical sensors, in particular acoustic [9] and sonar [17]; use of USV and UUV, including maximization of exchanged information via underwater acoustic communication channel and minimization of collision risk [10]; use of a swarm of drones or one or more drones in an environment where other drones are flying, including maximization of obtained information and minimization of collision risk [13], whereas [15] shows a method for avoiding collisions of independent UAVs; use of a UAV physically connected to a vessel equipped with a sonar [17]; solving of issues related to flying in hard environment, in particular turbulent atmosphere [12] and icing [14]; direct georeferencing with different approaches [11,18]; comparison between open source and commercial SfM packages [16]; low-cost cultural heritage applications including numerical modeling [19].

3.1. Non-Optical Sensors

The compact camera is the most common type of sensor carried by a UAV; the data are subsequently processed by means of SfM, even if this poses some problems discussed e.g., by [8]. However, today other types of compact sensors are also available and can be successfully used. If the aim is the localization of an acoustic source, the presence on board of engines and drive systems poses problems due to noise that can be solved by means of an array of microphones and by subtracting the modeled drone-related acoustic noise, as shown by [9]. The main lesson learned is that sensor architecture and noise estimation and subtraction are both necessary to obtain relevant results affected by acceptable uncertainties. New research lines may concern on the one hand the optimization of the sensor array design, and on the other hand the definition of time-varying noise models (therefore time frequency analysis) to take into account the different motion conditions of the drone in different phases of the flight. In perspective, this could be implemented by a manufacturer of UAVs specifically designed to identify acoustic sources.

The bathymetric system articulated into a vessel towed by a multirotor UAV by means of a tether, presented in [17], on the one hand it is an interesting result and on the other it can lead to interesting developments. The system is conceived for bridge infrastructure inspection and marine geomatics and the developments are consistent with this. Since obstacles such as bridges, pilings, and trees are commonly near bodies of water, the introduction of an active vessel control with a servo-actuated rudder is necessary. This mean that vessel control and multirotor UAS control could be semi-independent, with the need to solve the problems associated with the presence of the tether.

3.2. Multiple Drones, Swarms of Drones and Navigation in Crowded Environment

A standard survey typically involves a single drone that performs the measurement from several viewpoints progressively reached during the navigation. However, in the case of multimodal observations or in case of need for simultaneous observations from different points of view, several drones may be needed, not necessarily of the same type (e.g., some UAVs, UAV(s) and USV(s), USV(s) and UUV(s), or other combinations). Furthermore, it is increasingly probable that a drone may have to operate together with similar systems but not managed by the same operator or in any case by the same control system, with the need to dynamically adapt to the scenario in order to avoid collisions.

The experiments shown in [10] demonstrate the safe navigation of an USV in proximity to a UUV with maximization of data transmitted between the vehicles using underwater acoustic communications and minimization of the probability of collisions between them. The developments will concern two main research lines: (i) solution of the issues related to USV operating in proximity to a surfaced UUV, mainly due to loss of both Wi-Fi radio-frequency communications and acoustic connectivity because of wave motions and surface currents that could lead to sporadic submersion of the UUV; (ii) study of scenarios where a USV operates together with several UUSs, or where there are teams of several USVs and UUSs, in particular with the introduction, among other things, of decision systems that can identify which UUS must be placed in acoustic communication with a given USV at a given time. Both research lines require the development of controllers suitable to the specific applications.

The use of a swarm of UAVs is advantageous when compared to a single one. In some case, a single UAV could take an excessive long time to carry out a task or could not be able to accomplish it effectively. The research challenge is then to develop the appropriate cooperation logic so that the UAVs work together to complete missions effectively and efficiently. A possible solution of the challenges related to the use of several multirotor UAVs with downward facing cameras is proposed by [13], where the key factor is that coverage control involves the collision avoidance. In this way, the whole interesting area can be covered with cameras without collisions, as demonstrated by simulations. Future researches will focus on: (i) evaluation of the deformation of reality that may arise in the image due to the differentiation of the tilt angle or rotation; (ii) handling invisible and obscured areas, to have a truly 3D coverage system; (iii) implementing the method as a real multi-UAV system.

The research activity carried out by [15] also deals with prevention of collisions between UAVs, even in this case the focus is on operation of a single UAV in an environment potentially crowded by other independent UAVs. The main learned lesson, obtained by both simulations and real flight tests, is that the better way to avoid a potential collision with dynamic obstacles is changing the heading. This is because changing in speed and changing in both speed and heading are characterized by slower response time and, therefore, are less effective. Future developments will be mainly aimed at enhancing performance in 3D environments and facing the impact of the wind resistance by using an additional controller.

3.3. Flight in Hard Environment

A drone is generally used in environmental conditions suitable for observation, for example by avoiding operations in windy days or in days in any case unsuitable for flying or, in general, for navigation. However, if a drone (or set of drones) is used in monitoring and/or in emergency conditions, it may be necessary to operate in difficult weather conditions. It could also be necessary to monitor atmospheric turbulence, in which case, by definition, the conditions under which the flight is performed are potentially difficult. The latter case is addressed in [12], whose lesson learned is that a quadcopter can operate in hovering in a turbulent atmosphere providing the turbulence spectra at low-altitude. Future researches will focus on use of quadcopter UAV swarms in order to provide profiles of atmospheric turbulence in both the vertical and horizontal planes.

Another interesting case of study of UAV behavior in hard environment is shown by [14], which propose an innovative experimental test rig to study the effects of icing on the rotor of a drone in take-off/hover flying mode. The key results are the design of test rig and the fact that the ground effect does not take place if the UAV rotor height is 2 m. Further developments will allow the study of different active and passive solutions to the icing problematic. In this way, the impact of ice on UAV performance can be investigated and ice protection systems for drone applications can be developed.

3.4. Direct Georeferencing

Direct georeferencing (DG) essentially consists in the use of the localization data provided by the GNSS receiver integrated in the drone to georeferencing the final photogrammetric model with the use of a low number of GCPs acquired with topographic techniques. This is easy to say but complex to do, given that the accuracy of a receiver is not sufficient and that at least the use of differential GNSS is necessary. Possible solutions of some issues related to DG are provided by [11,18]. Both the research teams highlight the importance of use of RTK positioning with a base station, which nowadays is accessible not only to the researcher, but also to the practitioner. In particular, [11] finds that a density of 10–12 GCPs/km^2 allows an accuracy of some centimeters for the final result and that a decisive increase in density does not lead to a significant improvement in accuracy. In [18], the fact that the use of a single GCP with PPK mode allows better results than several GCPs without GNSS data in PPK mode is highlighted. This last result is very interesting where it is required the surveying of areas that are difficult to access. Future developments will concern on the one hand the use of lenses which, while compatible with light UAVs, have satisfactory characteristics (or which, thanks to pre-calibration, can in any case allow good photogrammetric modeling), and on the other hand the improvement of RTK positioning.

3.5. Low Cost Surveying and Modeling

In the past, surveys were often very expensive (think, for example, of the cost of a laser scanning instrument, or of a photogrammetric survey carried out with a helicopter). Today, photogrammetric surveys can be carried out by means of drones of adequate performance. However, the surveys require the use of support techniques (for example, GNSS) and the data processing still requires the use of software packages that can provide the desired products (3D models or further quantitative analyses).

In particular, [16] shows that the use of an Open Source package such as MicMac allows to obtain results completely comparable to those obtainable using commercial packages of a higher level (and therefore cost). It should also considered that a key element of Open Source software is the possibility of having additional free tools intended for the solution of specific problems and which can make such a king of software even richer than the commercial one. Therefore, further developments will be aimed at defining correct procedures for 3D model generation and quality assessment employing ultra-low-cost equipment, as well as at disseminating the corresponding implementations.

Sometimes, for example in the case of buildings damaged by natural events (earthquakes, landslides, etc.), it may be necessary to carry out expeditious surveys with relatively limited technical and financial resources. The study shown in [19] demonstrates that low-cost sensors and expeditious surveys can be used to provide 3D photogrammetric models with accuracy adequate for the subsequent Finite Element modeling (several centimeters), including the comparison between the observed crack pattern and the obtained principal stress distribution. This is interesting for assessing the conditions of historic buildings, especially if disasters affecting large areas occur.

3.6. Conclusive Remarks

The articles published in the framework of the Special Issue "Unconventional Drone-Based Surveying" show that the drone (not necessarily flying, but also operating on the surface or underwater) can be used in different contexts and equipped with different sensors. The environment in which the drone can acquire data may not be the classic one of a survey implemented in optimal conditions, but one in which various systems (interconnected or even independent of each other) can operate simultaneously, even in unfavorable weather conditions. This because the surveying could be required to be carried out in a short time for monitoring purposes.

All the articles refer to ongoing studies, to be continued in order to provide useful tools and data for both the researcher and the practitioner. The Editors of this Special Issue hope that these articles will spark discussion and inspire new research lines.

Author Contributions: A.P., G.T. and M.F. contributed equally to all aspects of this Editorial. All authors have read and agreed to the published version of the manuscript.

Funding: This research received no external funding.

Acknowledgments: The guest Editors would like to sincerely thank all the Authors who contributed to this Special Issue and the Reviewers who dedicated their time and provided the Authors with valuable and constructive recommendations. They would also like to thank the Editorial team of Remote Sensing for their support.

Conflicts of Interest: The authors declare no conflict of interest.

References

1. Colomina, I.; Molina, P. Unmanned aerial systems for photogrammetry and remote sensing: A review. *ISPRS J. Photogramm. Remote Sens.* **2014**, *92*, 79–97. [CrossRef]
2. Nex, F.; Remondino, F. UAV for 3D mapping applications: A review. *Appl. Geomat.* **2014**, *6*, 1–15. [CrossRef]
3. Videras Rodríguez, M.; Melgar, S.G.; Cordero, A.S.; Márquez, J.M.A. A critical review of Unmanned Aerial Vehicles (UAVs) use in architecture and urbanism: Scientometric and bibliometric analysis. *Appl. Sci.* **2021**, *11*, 9966. [CrossRef]
4. Burke, C.; Wich, S.; Kusin, K.; McAree, O.; Harrison, M.E.; Ripoll, B.; Ermiasi, Y.; Mulero-Pázmány, M.; Longmore, S. Thermal-drones as a safe and reliable method for detecting subterranean peat fires. *Drones* **2019**, *3*, 23. [CrossRef]
5. Liao, J.; Zhou, J.; Yang, W. Comparing LiDAR and SfM digital surface models for three land cover types. *Open Geosci.* **2021**, *13*, 497–504. [CrossRef]
6. Asaamoning, G.; Mendes, P.; Rosário, D.; Cerqueira, E. Drone swarms as networked control systems by integration of networking and computing. *Sensors* **2021**, *21*, 2642. [CrossRef] [PubMed]
7. Monego, M.; Achilli, V.; Fabris, M.; Menin, A. 3-D survey of rocky structures: The dolomitic spire of the gusela del Vescovà. *Commun. Comput. Inf. Sci.* **2020**, *1246*, 211–228. [CrossRef]
8. Pesci, A.; Teza, G.; Loddo, F.; Rollins, K.; Andersen, P.; Minarelli, L.; Amoroso, S. Remote sensing of induced liquefaction: TLS and SfM for a full scale blast test. *J. Surv. Eng.* **2022**, *148*, 04021026. [CrossRef]
9. Go, Y.-J.; Choi, J.-S. An acoustic source localization method using a drone-mounted phased microphone array. *Drones* **2021**, *5*, 75. [CrossRef]
10. Savkin, A.V.; Verma, S.C.; Anstee, S. Optimal navigation of an unmanned surface vehicle and an autonomous underwater vehicle collaborating for reliable acoustic communication with collision avoidance. *Drones* **2022**, *6*, 27. [CrossRef]
11. Liu, X.; Lian, X.; Yang, W.; Wang, F.; Han, Y.; Zhang, Y. Accuracy assessment of a UAV direct georeferencing method and impact of the configuration of ground control points. *Drones* **2022**, *6*, 30. [CrossRef]
12. Shelekhov, A.; Afanasiev, A.; Shelekhova, E.; Kobzev, A.; Tel'minov, A.; Molchunov, A.; Poplevina, O. Low-altitude sensing of urban atmospheric turbulence with UAV. *Drones* **2022**, *6*, 61. [CrossRef]
13. Huang, S.; Teo, R.S.H.; Leong, W.W.L. Multi-camera networks for coverage control of drones. *Drones* **2022**, *6*, 67. [CrossRef]
14. Villeneuve, E.; Samad, A.; Volat, C.; Béland, M.; Lapalme, M. An experimental apparatus for icing tests of low altitude hovering drones. *Drones* **2022**, *6*, 68. [CrossRef]
15. Gnanasekera, M.; Katupitiya, J. A Time-efficient method to avoid collisions for collision cones: An implementation for UAVs navigating in dynamic environments. *Drones* **2022**, *6*, 106. [CrossRef]
16. Cutugno, M.; Robustelli, U.; Pugliano, G. Structure-from-motion 3D reconstruction of the historical overpass ponte della cerra: A comparison between MicMac® open source software and Metashape®. *Drones* **2022**, *6*, 242. [CrossRef]
17. Diaz, A.L.; Ortega, A.E.; Tingle, H.; Pulido, A.; Cordero, O.; Nelson, M.; Cocoves, N.E.; Shin, J.; Carthy, R.R.; Wilkinson, B.E.; et al. The Bathy-Drone: An autonomous uncrewed drone-tethered sonar system. *Drones* **2022**, *6*, 294. [CrossRef]
18. Salas López, R.; Terrones Murga, R.E.; Silva-López, J.O.; Rojas-Briceño, N.B.; Gómez Fernández, D.; Oliva-Cruz, M.; Taddia, Y. Accuracy assessment of direct georeferencing for photogrammetric applications based on UAS-GNSS for high andean urban environments. *Drones* **2022**, *6*, 388. [CrossRef]
19. Fabris, M.; Fontana Granotto, P.; Monego, M. Expeditious low-cost SfM photogrammetry and a TLS survey for the structural analysis of Illasi Castle (Italy). *Drones* **2023**, *7*, 101. [CrossRef]

Article

An Acoustic Source Localization Method Using a Drone-Mounted Phased Microphone Array

Yeong-Ju Go and Jong-Soo Choi *

Department of Aerospace Engineering, Chungnam National University, 99 Daehak-ro, Yuseong-gu, Daejeon 34134, Korea; yjgo@cnu.ac.kr
* Correspondence: jchoi@cnu.ac.kr

Abstract: Currently, the detection of targets using drone-mounted imaging equipment is a very useful technique and is being utilized in many areas. In this study, we focus on acoustic signal detection with a drone detecting targets where sounds occur, unlike image-based detection. We implement a system in which a drone detects acoustic sources above the ground by applying a phase difference microphone array technique. Localization methods of acoustic sources are based on beamforming methods. The background and self-induced noise that is generated when a drone flies reduces the signal-to-noise ratio for detecting acoustic signals of interest, making it difficult to analyze signal characteristics. Furthermore, the strongly correlated noise, generated when a propeller rotates, acts as a factor that degrades the noise source direction of arrival estimation performance of the beamforming method. Spectral reduction methods have been effective in reducing noise by adjusting to specific frequencies in acoustically very harsh situations where drones are always exposed to their own noise. Since the direction of arrival of acoustic sources estimated from the beamforming method is based on the drone's body frame coordinate system, we implement a method to estimate acoustic sources above the ground by fusing flight information output from the drone's flight navigation system. The proposed method for estimating acoustic sources above the ground is experimentally validated by a drone equipped with a 32-channel time-synchronized MEMS microphone array. Additionally, the verification of the sound source location detection method was limited to the explosion sound generated from the fireworks. We confirm that the acoustic source location can be detected with an error performance of approximately 10 degrees of azimuth and elevation at the ground distance of about 150 m between the drone and the explosion location.

Keywords: acoustic source localization; phased microphone array; spectral subtraction; beamforming

Citation: Go, Y.-J.; Choi, J.-S. An Acoustic Source Localization Method Using a Drone-Mounted Phased Microphone Array. *Drones* **2021**, *5*, 75. https://doi.org/10.3390/drones5030075

Academic Editors: Arianna Pesci, Giordano Teza, Massimo Fabris and Dimitrios Zarpalas

Received: 23 June 2021
Accepted: 4 August 2021
Published: 6 August 2021

Publisher's Note: MDPI stays neutral with regard to jurisdictional claims in published maps and institutional affiliations.

1. Introduction

One way to cope with disasters and security situations is to use drones to search for where events occur [1,2], which can provide visual information by taking images from the sky. In general, drone imaging has been used for military purposes, and it is also the most commercially used method for video production. The usability of drones has already been demonstrated, and as the operating area gradually gets closer to humans, there is a demand for new detection methods beyond existing mission equipment that only provides images. One new method is to detect sounds.

Acoustic-based detection methods in disaster and security situations can complement the limitations of image-based target detection methods. In addition, acoustic-based detection methods are needed in situations where image identification is needed at night in bad weather, and over complex terrain with features that can be difficult to identify [3,4]. This technology extends detection methods by adding human hearing-like sensations to drones. In particular, it is considered to be a necessary technology for detecting the location of explosions in disaster situations such as fires or detecting distress signals for lifesaving situations.

The technology of detecting targets using acoustic signals has been previously studied in a variety of fields [5–7], from implementing a robot that detects sounds by mimicking human ears to noise source instrumentation systems for the analysis of mechanical noise. Prior to the development of radar, this technique was used to detect the invasion of enemy aircraft by sound to defend against anti-aircraft. In recent years, it has also been applied to systems that track where guns are fired. Recently, it has become an essential technology for low noise research and has been used to detect and quantify acoustic sources of noise generated by transport vehicle systems such as automobiles and aviation. This technique is based on the principle of detecting acoustic sources by reconstructing acoustic pressure measured by the concept of phase difference using microphone arrays.

The most significant problem associated with detecting external noise by mounting microphones on a drone is the noise of the drone when it operates. The main noise generated by a drone is aeroacoustics noise generated mainly by rotating rotors [8]. Depending on the mechanism of occurrence, a significant factor that affects the tonal or broadband frequency band is related to the blade passing frequency [9,10]. A typical drone has more than four rotors, which creates a more complex noise field environment [11,12]. Furthermore, the operating noise of a drone in close proximity to a microphone can cause the sound to be masked and indistinguishable. Therefore, robust denoising techniques are essential for identifying external sounds in highly noisy environments. It is ideal for a drone to reduce its own noise in terms of operability, but this is difficult to implement immediately with the current technology; therefore, signal processing that reduces noise from signals acquired from microphones is realistic. Spectrum subtraction is a method of eliminating noise and obtaining clean sound to improve the signal-to-noise ratio in a noisy environment. This is one of the methods of background noise suppression and has been studied for the purpose of clarifying voices in the field of voice recognition [13,14]. In general, background noise is removed by using statistical characteristics of background noise, but applicable methods are also being studied in non-stationary noise environments [15–17]. In addition, there are cases that are effective in improving the signal-to-noise ratio in sound detection such as impact sound [18].

Beamforming methods are well known as methods for estimating the direction of arrival of acoustic sources [19–22]. Beamforming is the principle of calculating beam power by using multiple microphones to correct the phase difference of signals according to the geometric positional relationships among the microphones and estimating the direction of arrival of the source from that intensity [23]. The direction of arrival estimation performance of the beamforming method is determined by the number of microphones, the form of the array, and an improved algorithm. Beamforming methods can also reduce the strength of uncorrelated signals by phase difference correction and increase the strength of the signal of interest to enhance the signal-to-noise ratio by reducing noise. Generally, the greater the number of microphones and the higher the caliber, the better the performance; however, limited number and size microphone arrays are used considering real-time processing and hardware performance.

Recently, studies have been conducted to detect sound sources by installing microphone arrays in drones. Although the use of microphone arrays to detect or distinguish sound sources is common, there are various results depending on the hardware that makes up the array and the array signal processing algorithms. In [24,25], the authors studied how to attach microphone arrays to drones and embedded systems for signal acquisition. In [26–28], the authors presented an arrival angle detection algorithm for sound sources based on beamforming. In addition, valid results for the detection of sound sources were shown through verification experiments on near-field sound sources in a well-refined indoor environment. There are also sound source detection studies on the actual operation of drones in outdoor environments. In [29,30], detection studies were conducted on whistles and voices. These studies confirmed the performance of sound source detection in terms of the signal-to-noise ratio of sound source and background noise, and showed that it is possible to detect nearby sound sources located about 20 m away from drones.

In this study, a method is proposed for detecting sound sources generated above the ground using a drone-mounted phased microphone array. Especially, we tried to effectively remove noise caused by rotors from the microphone signal and to confirm the expectation of the accuracy of the sound source localization in detection mission that occurs during flight. Each of the techniques needed to detect the sound source suggested a major perspective that must be addressed. The goal is to confirm the expected performance of the location detection of sound sources by actually implementing the process of connecting techniques. Since microphones are so close to the drone, the sound of interest is distorted by noise generated when the drone operates, and the spectral subtraction method is used to improve the signal-to-noise ratio. The application of the general spectral subtraction method applies the average model in the entire frequency band. However, in order to effectively remove the drone noise that is an obstacle to us, we separated the spectral band with clear spectral characteristics and applied the subtraction method through different models. The separated bands are divided into bands due to the main appearance of BPF and other noises, including turbulent flow, and are designed to be effective in reducing drone noise.

We distinguished impact sound through spectral subtraction. Spectral reduction has been shown to be effective in restoring acoustic signals of impact sounds. Using this denoised signal, we detected the direction of arrival for ground impact sounds by beamforming methods based on microphone arrays. We applied a method of representing the direction of arrival as absolute coordinates for the ground source, measured on the basis of the microphone array coordinates mounted on a drone, and a data fusion method that can detect the location of sound sources by correcting the changing posture in real time using drone flight information. The entire algorithm for detecting sound sources experimentally identifies detection performance using a 32-channel microphone array. To verify the proposed methods, a clear detection sound source was needed. We focused on identifying the localization error for point sound sources with a clear signal such as impulse. In the verification experiment, the localization performance was confirmed by limiting the impact sound using a firecracker.

In Section 2, we describe how to detect the location of a ground acoustic source. We describe a spectral reduction method to improve the signal-to-noise ratio for acoustic sources of interest, a beamforming method to estimate the direction of arrival of acoustic sources, and a method to represent the direction of arrival of acoustic sources detected by drones with geographical information above the ground. In Section 3, we describe the experiments to verify the location detection performance for ground acoustic sources. We describe the microphone array system and experimental environment mounted on a drone. In Section 4, we address the analysis of data measured through the experiments and the estimation performance of ground acoustic sources.

2. Materials and Methods

The method for detecting ground acoustic sources can be divided into three stages: (1) the measured signal in a microphone array uses spectral subtraction to improve the signal-to-noise ratio, (2) the beamforming method uses time domain data to estimate the angle of arrival for the acoustic source of interest, (3) the flight data of the drone are fused to estimate the location of the source on the ground.

2.1. Spectral Subtraction

Spectral subtraction is a method of restoring a signal by subtracting estimates of noise from a noisy signal [31]. In a noisy environment, the measured signals are represented by the sum of the signals of interest, $s(t)$, and noise, $n(t)$. Here, the signal of interest is the signal we want to measure, and noise is the signal we want to remove, not the signal of interest:

$$y(t) = s(t) + n(t) \tag{1}$$

Spectral subtraction is performed by all computations in the frequency domain. Thus, signals in the time domain are extracted sequentially from block to block through window functions and analyzed as frequency signals through Fourier transformations.

Noise spectra are typically estimated from time intervals where only steady noise exists without signals of interest. This assumes the amplitude of the estimated spectrum as a noise model and computes the difference from the amplitude of the measured signal spectrum to eliminate the averaged noise. Amplitude in the subtracted spectrum applies the phase of the measured signal, and the noise-reduced time domain signal, $s\prime(t)$, is extracted through the inverse Fourier transform (Figure 1). An important view of the spectral subtraction method applied in this study is that, independently, noise reduction for each microphone signal preserves the phase of the signal of interest, so that the phase relationship between the signals of each microphone is not distorted. Preserving the phase is directly related to estimation error, as the method of estimating the location of acoustic sources by array signal processing is based on the phase relationship of the signals measured on each microphone. The noise model assumes that the signal is consistently and continuously affecting the interval at which it is measured. As it is difficult to estimate sufficient averaged spectra when noise properties change over time, in such cases, a sufficiently averaged spectrum can be obtained by applying a smoothing filter.

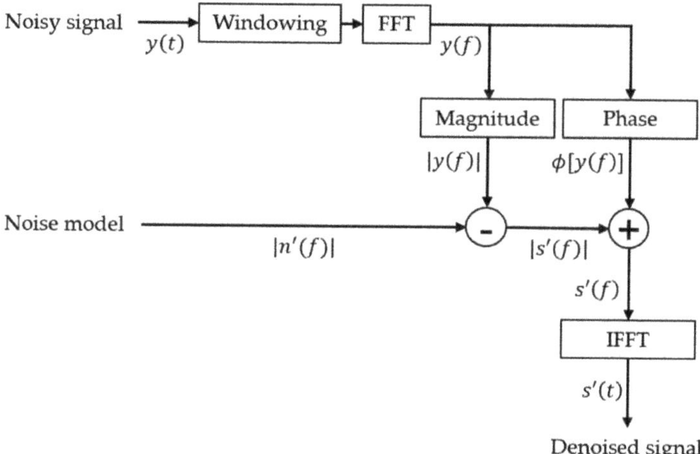

Figure 1. Block diagram of the spectral subtraction method.

2.2. Beamforming

A delay-and-sum beamforming calculated in the time domain was applied. In the time domain, beam power, b, can be expressed by the arrival vector of acoustic waves defined in the three-dimensional space geometry, x, and time, t, as follows:

$$b(x,t) = \frac{1}{M} \sum_{i=0}^{M-1} w_i p_i (t - \tau_i) \tag{2}$$

where p is the acoustic pressure measured on the microphone and relative time delay, τ is the relative time delay caused by the position of each microphone, w is the weight and does not give a differentiated weight, M represents the number of stagnant microphones used, and i represents the microphone index. Beam power is calculated for each direction of arrival of a hypothetical acoustic source. The magnitude of the calculated beam powers is determined for the direction of arrival of the acoustic source by making a relative comparison in the entire virtual source space and estimating the direction of arrival with the maximum value. The phase of the sound pressure signal measured between the reference point of the microphone, P_{ref}, and the position between each microphone, P_i, is caused by a time delay, τ_i; τ_i is analytically related by speed of sound, c_0, and delay distance, dl_i, as in Equation (3) [32,33], and dl_i is calculated from the position vector r_m defined as P_{ref} and P_i and the incident direction of the plane wave, i.e., the normal vector of the plane wave, n (Figure 2). The normal vector of the plane wave is the relation between the azimuth angle, Φ, and elevation angle, Θ, which can be represented in the spherical coordinate system. The speed of sound was corrected by atmospheric temperature, T [34], as follows:

$$\tau_i = dl_i / c_0 \tag{3}$$

$$dl_i = n \cdot r_m / |n| \tag{4}$$

$$r_m = P_i - P_{ref} \tag{5}$$

$$n = (\cos \Phi \sin \Theta, \ \sin \Phi \sin \Theta, \ \cos(\Theta) \ for \ 0 \leq \Theta \leq 2\pi, \ 0 \leq \Phi \leq \pi/2 \tag{6}$$

$$c_0 = 331.3 \sqrt{(1 + T/273.15)} \tag{7}$$

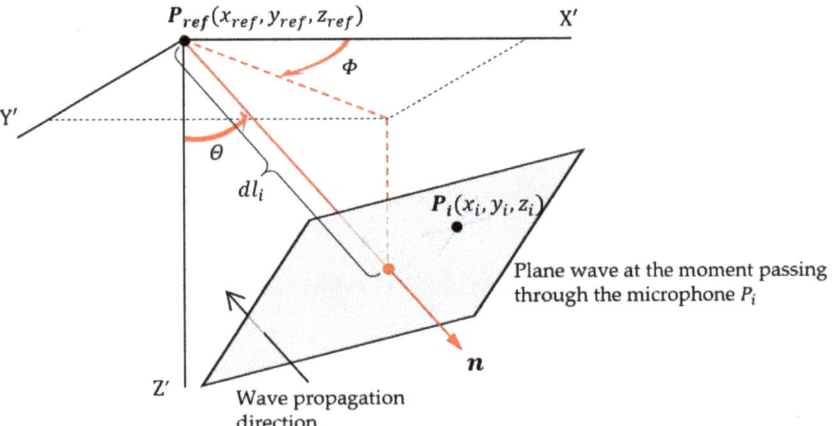

Figure 2. The relationship between the incident direction of the plane wave and the location of the microphone.

A virtual source for calculating beam power was assumed to be a ground source, and based on the drone's headings, the azimuth angle, Φ, was detected from 0 to 360° and the elevation angle, Θ, from 0 to 90°. The definition of detection angles defined between the drone's heading vector and the direction of arrival of the detected acoustic source is shown in Figure 3. The beam power calculated from the lower hemisphere range of the drone explored the maximum beam power and estimated its orientation as the angle of arrival at which the acoustic source enters.

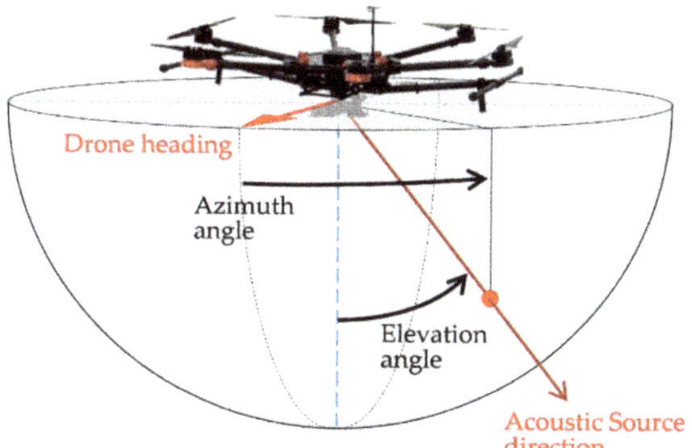

Figure 3. DOA estimation coordinate system of the microphone array based on drone heading direction.

2.3. Acoustic Source Localization on the Ground

The location detection of acoustic sources generated on the ground is determined by matching the state information the drone is flying with the direction of arrival of acoustic sources estimated from the drone-mounted microphone array. The direction of arrival of an acoustic source is estimated to be the azimuth and elevation angles with the maximum beam power, based on the heading direction of the drone's body frame coordinate system. The body frame coordinate is calculated by the roll, pitch, yaw angle, and heading vector based on the magnetic northward and navigation frame coordinate system from the GPS [35–37]. The drone's posture and position changing in real time is updated to 30 Hz on the data processing board with signals from the microphone. The drone-mounted microphone array estimates the angle of arrival based on the drone's heading direction. The conversion of the heading direction to the navigation frame coordinate system coordinates converts the estimated angle of arrival to the same criterion to geometrically derive the point of intersection with the ground with geometric information above the ground, and estimates this intersection as an acoustic source (Figure 4).

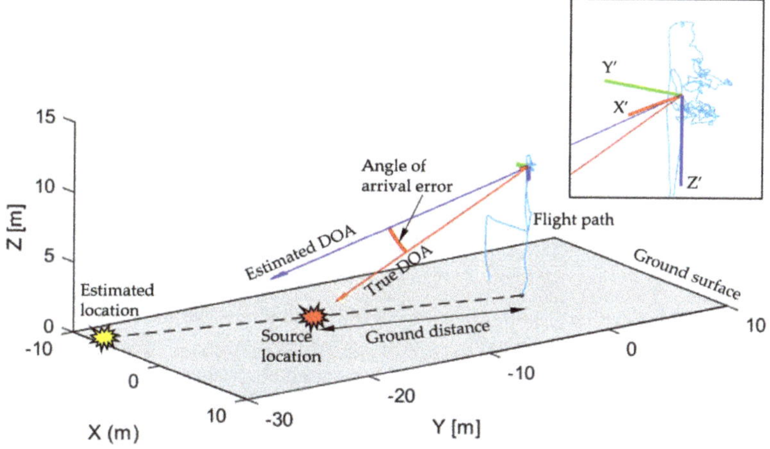

Figure 4. The concept of estimating the location of the acoustic source on the ground based on the DOA measured by the microphone array.

2.4. Phased Microphone Array Mounted on the Drone

The microphone array system consisted of 4 microphone modules with support mounts, and a signal processing board. One microphone module had a radius of 0.025 m, and 8 microphones were arranged in a circular equidistant array (Figure 5). Each microphone module was attached to the underside of the hemispherical dome to avoid the flow-induced pressure caused by the propeller wake. As shown in Figure 6, four microphone modules were attached to each end of the X-shaped mount with a radius of 0.54 m, and thus an array of 32 microphones was implemented. The 32-channel microphone array is a prototype built for the purpose of studying signal acquisition and location detection for acoustic sources of interest outside the drone, designed to verify the performance of several signal processing algorithms. The x-shaped support allows for adjustments of angle and height, as well as the array shape. All the acoustic pressure signals measured on each microphone are time synchronized and can be collected via a separate data acquisition board and stored in external memory in accordance with the acoustic event or trigger switch from the external controller. For verification experiments, signals were manually recorded using the trigger function of an external controller in time for acoustic source generation, and acoustic signals stored on memory cards were post-processed. The microphone array specifications are listed in Table 1.

Figure 5. Microphone module consisting of 8 microphones.

Figure 6. Microphone array system mounted on the drone.

Table 1. Microphone array specifications.

Parameter	Specification
Number of channels	32
Sensitivity	−26 dBFS
Signal-to-noise ratio	64 dB(A)
Maximum pressure level	122.5 dB
Sampling rate	25.6 kS/s
Frequency range	25 Hz~12.8 kHz

3. Experiment

Experiments to verify the acoustic source localization performance were conducted in a stadium at Chungnam National University. The experimental environment is shown in Figure 7, indicating the locations of the acoustic sources, i.e., firecrackers, and where the drone was hovering in place. To quantify the locations of the sources, a line in the soccer stadium was referred to and the absolute location was measured using GPS. The Universal Transverse Mercator Coordinate System (UTM) using the WGS84 ellipsoid was used to derive plane coordinates based on the central position (Location 4) of the soccer stadium. The locations of the acoustic sources on this plane coordinate and the location of the drone are defined as absolute coordinates, and the results of estimating the location of the acoustic source based on the drone's GPS information can also be expressed.

Figure 7. Experimental site. Firecracker locations (red circles) and drone location (blue arrow) in hovering position.

The acoustic sources used for the experiment were commercial firecrackers, which use gunpowder to generate short impulse signals after a certain period of time when ignited. The acoustic pressure signal from the explosion of the firecracker used in the experiment is shown in Figure 8. This signal was collected separately without a drone and measured at a distance of about 20 m away from the location of the firecracker near the center of Figure 6. That is, the signal was completely unrelated to the operation of the drone and was intended to observe pure acoustic signal features. Strong impulse signals, such as explosions, are easy to distinguish in domains because of their strong amplitude and clear features. In the practical problem of finding acoustic sources, impulses are only one example of the detection targets, but they are important underlying experiments in the goal of assessing the detection performance that this study wants to identify. Additionally, it is possible to identify whether spectral subtraction methods work well for a specific frequency or frequency band that wants to denoise. The impulse signal has the constant response characteristic across all frequencies. In order to restore the energy of pure impulse signals as much as possible in a state of noise mixing, it is important to subtract only a specific frequency for the noise to be removed. The impulse is one of the useful targets for providing basic clues to determine the acoustic frequency of the drone to be removed. It was confirmed that the explosion signal was characterized by a short and strong shock sound in an instant. The firecrackers were fixed just above the ground. The location estimation results analyzed in the Results section were performed for approximately 20 min during 1 sortie flight. The firecrackers were detonated three times at each point in the order of location numbers from L1 to L6. The interval between each three repeated explosions was about 10 s. A straight line from the hovering position to the detonating location is described in Table 1, up to 151.5 m. During this flight, the drone maintained a hovering position at an altitude of about 150 m. The drone's position was near the right-hand corner north of the soccer stadium, heading southwards, and it maintained a stable hovering position due to the stabilizer mode.

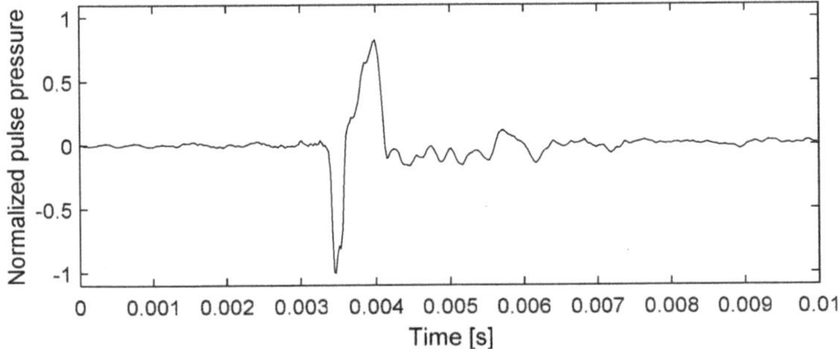

Figure 8. Acoustic pressure signal from the explosion of a firecracker.

4. Results and Discussion

Figure 9 shows the acoustic pressure signals measured by the drone-mounted microphone system, in a hovering position. It is a spectrum that averages acoustic pressure signals measured over 3 s using a total of 32 microphones. The frequency decomposition was 1 Hz, and a 5 Hz high pass filter was applied. This spectrum is meaningful for identifying noise characteristics that actually operate underneath the drone. Noise can be observed together with atmospheric flow and background noise present in the atmosphere when the actual drone is maneuvering. In particular, it is a very poor measurement environment in which fluctuating pressure caused by the propellers' wakes at the bottom of the drone directly affects the microphones. Analysis of the mean spectrum is the basis for determining the frequency bands and amplitude that are necessary to determine the noise generated by drones when flying, and that should be deducted when applying spectral subtraction.

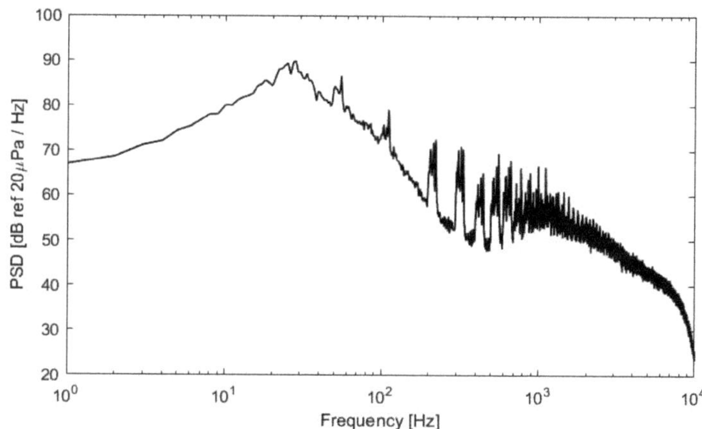

Figure 9. Averaged hovering noise of the drone measured by the mounted microphones.

The drone operates with six rotors, each with two blades. The average motor rotation at hovering was about 3000 rpm. In order to maintain the drone's position during hovering, the number of motor rotations changed by about 200 rpm. Changes in motor rotation are directly related to the blade passing frequency and affect the harmonic components that appear as tonal component features. In the average spectrum, six motors remained hovering for three seconds, outputting slightly different rotations, resulting in peak frequencies scattered for each harmonic. Accurately predicting the blade passing frequency and its amplitude should be able to measure each motor's rotation speed and rotation phase, as it is efficient to match the same number of rotations in each motor from the point of view of drone control.

With respect to the average motor rotation speed, the shaft rate frequency characteristic was shown at about 50 Hz, the first harmonic of blade passing frequency at 100 Hz, and harmonic frequencies for its multiples were observed. Harmonics are observed for rotational frequencies up to 2 kHz, and the tonal frequencies shown here were caused by propellers and motors. Broadband noise above 800 Hz is characterized by turbulence around the propellers. Strong broadband components were observed in bands below 200 Hz. It was observed that this frequency band was mainly caused by atmospheric flow and propeller wake at fluctuating pressures that directly affected the microphones, resulting in high amplitude.

Spectral reduction methods were treated differently for each of the three frequency band intervals in the frequency domain. The first band is the low frequency band below 200 Hz, mainly affected by atmospheric flow and rotor wake. Here, the fluctuating pressure components caused by the flow around the microphones have the greatest effect on the acoustic pressure of the entire frequency characteristics. In the signal we were trying to detect, this band was completely eliminated because it contributed less and was generally associated with high noise amplitude. The second band is from 200 to 1000 Hz, which features the blade passing frequency of the rotor. Blade passing frequency is directly affected by rotor rotation speed, as shown in Figure 9. Since six rotors were operating at different rotations for maneuvering during flight, peak frequencies continued to change over time even in hover flight. Since this peak frequency is difficult to detect accurately in real time, the excess of amplitude as compared with the spectral reduction model was deducted in the same amount as the model. The third band is over 1000 Hz, containing motor noise, which makes it difficult to specify the frequency characteristics of this section. As the processing method is not clear, we simply subtracted the amount of amplitude the subtraction model had from the measured amplitude. The subtraction of amplitude is calculated for each frequency in the spectrum and is in Pascal units, not dB scales. The analysis of impulse sources requires short-time spectral analysis to process signals in real time. Short-time spectral analysis does not show a clear spectrum curve as compared with the averaged spectrum. Small peak amplitudes were removed through the spectrum smoothing filter to derive amplitude envelope similar to the mean properties and applied as a subtraction model. For the subtraction model, the negative pressure data used signals from the negative pressure signal window of interest, which took into account real-time processing. The signals for the subtraction model were used one second before the window frame of the acoustic pressure signal of interest, considering real-time processing. The frequency and acoustic pressure level models to be subtracted calculate the spectrum of acoustic pressure data measured one second in advance from real-time measured pressure signals and derive the mean characteristics through a smoothing filter. It was not appropriate to use a model to deduct long-time averaged signals because the impact of the number of propeller rotations and wake changed rapidly as the drone was maneuvered.

Figures 10 and 11 show the data that indicate before and after the application of the spectral subtraction method and are measured on microphone Channel 0 for the first of three explosions at Location 1. Explosive sounds represent strong impulse signal characteristics, and the ideal impulse signal has a flat amplitude in the whole band in the frequency domain. The explosive sound of a firecracker affects amplitude over the entire band of frequencies because it produces impulse waves due to rapid pressure changes. Therefore, in order to effectively preserve the amplitude of the impulse signal, spectral subtraction must be performed with the correct focus on the noise frequency. Figure 11 shows the result of applying spectral subtraction to extract frequencies for explosives. By comparing the spectra, it can be observed that the low frequency band below 200 Hz has been eliminated and that the strong tonal component between 100 and 1000 Hz has been reduced. The 32-channel signal, which passed the spectral reduction method for each channel, was restored back to the time domain signal by a 0.5 s, 50% overlap window function. This process made it possible to specify the time the impulse signal and the time the signal occurred in the time domain data.

The results of spectral subtraction on the impulse signals of firecrackers can be observed more clearly in Figure 12. We compared the subtraction model obtained through the smoothing filter, the measured signal, and the spectrum after applying the subtraction method. Different subtractions can be distinguished in bands below 200 Hz, between 200 Hz and 1000 Hz, and above 1000 Hz. From this setting, we were able to effectively extract the acoustic pressure of the impulse signal of the firecracker. Defining frequency band discrimination was empirically distinguished here by already measured drone acoustic data, but in the future, it is considered that automatic detection will be possible if the blade passing frequency is detected through the peak finder of the spectrum or if a learned real-time adaptive discrimination filter is used. In addition, the process of spectral subtraction applied in this study can be applied to clarify other acoustic sound sources such as voices. However, this study only confirmed the analysis of the impulse signal and it is necessary to verify the performance of other sounds separately. The spectral subtraction method, implemented separately for each frequency band, was used to eliminate frequencies affected by the drone's own noise and found to be effective in enhancing the signal-to-noise ratio of acoustic pressure. Especially, it is effective in situations where it is difficult to accurately estimate the harmonic components of blade passing frequency that constantly change during flight. However, if a tonal sound source such as a whistle is detected in a band with a smoothing filter to remove the harmonics of blade passing frequency, it may be considered as blade passing frequency, which may cause a decrease in the signal-to-noise ratio. If this situation is to be detected, it is necessary to consider other ways to only subtract blade passing frequencies.

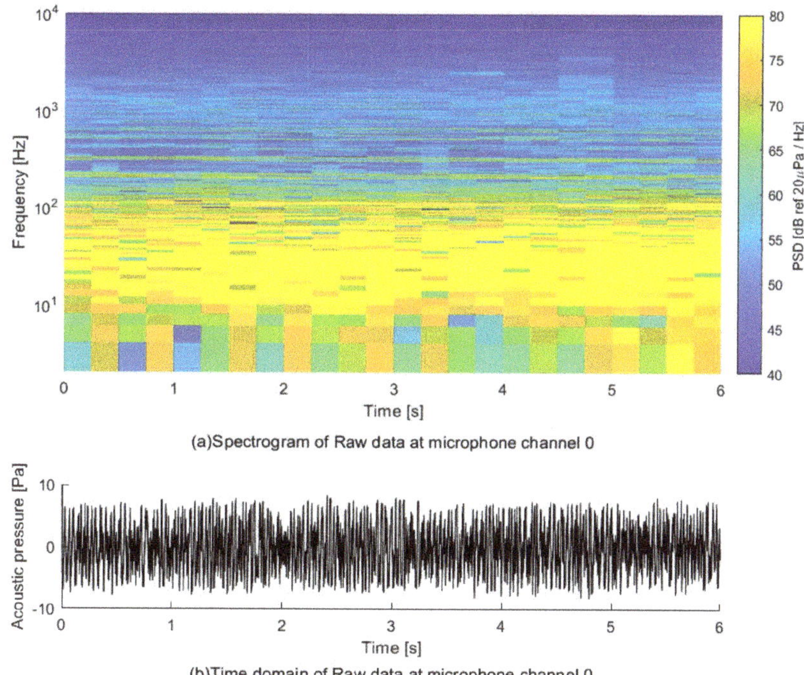

(a)Spectrogram of Raw data at microphone channel 0

(b)Time domain of Raw data at microphone channel 0

Figure 10. Measured acoustic signal at Location 1 for the first explosion.

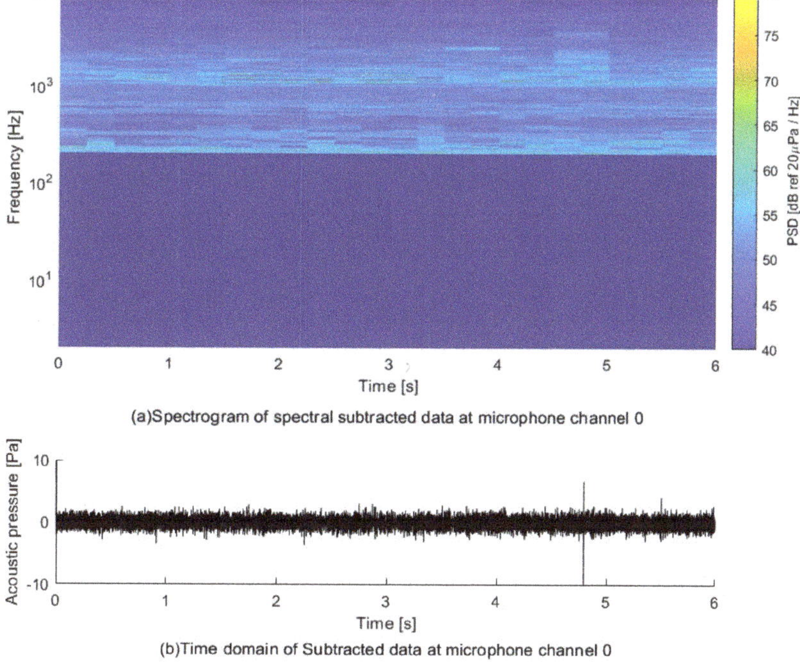

(a)Spectrogram of spectral subtracted data at microphone channel 0

(b)Time domain of Subtracted data at microphone channel 0

Figure 11. Spectral subtracted acoustic signal at Location 1 for the first explosion.

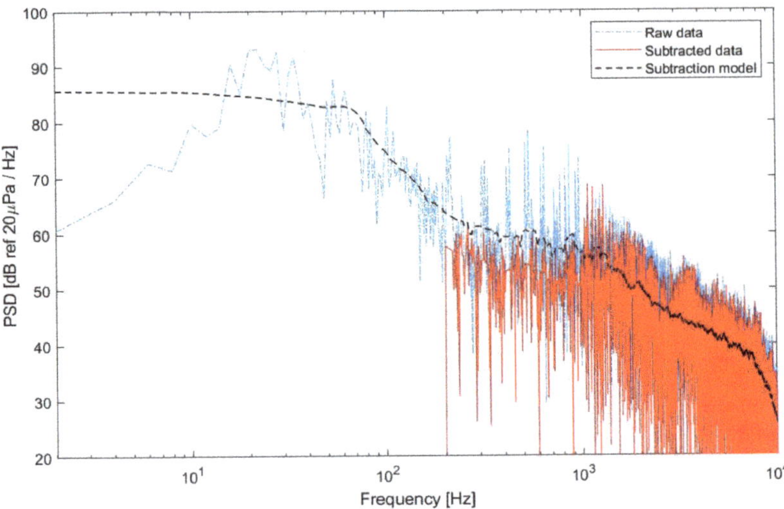

Figure 12. Comparison of the application results of spectral subtraction method to impulse signal (spectra for 4.8 s frames in the spectrogram of Figures 10 and 11).

Time-domain signals with improved signal-to-noise ratio by spectral subtraction methods become input signals that calculate beam power for detecting arrival angles.

Although the beamforming method has the effect of improving the signal-to-noise ratio according to the correlation by phase difference calibration, it is difficult to detect target acoustic sources due to reduced beamforming performance if there is a strong correlation in the proximity field, such as propeller noise. Figure 13 shows the beam power calculated using signals measured from 32 microphones mounted on the drone. The beam power represents the magnitude relative to the lower directions of the drone based on the center point of the drone.

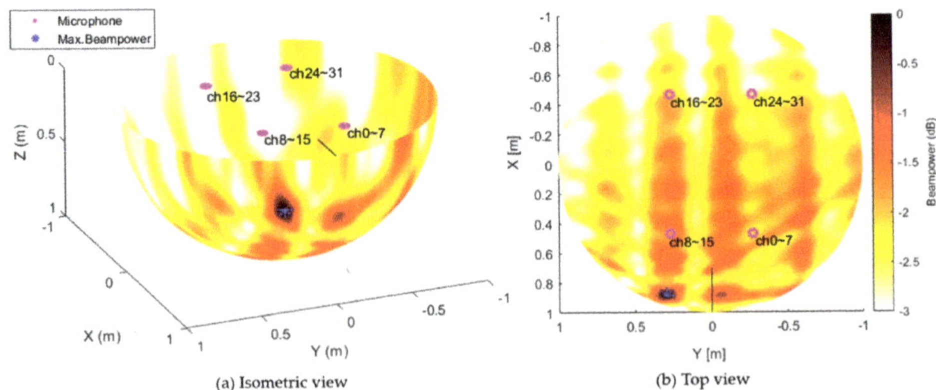

Figure 13. Beam power results at Location 1 for the first explosion.

In general, -3 dB is judged as the effective range as compared with the calculated maximum beam power. Since the 32-channel microphone array used in this study was not considered to be optimized for beamforming methods, sidelobes were also prominent but we found that the direction of arrival calculated with maximum beam power was the direction of arrival for the actual acoustic source. Beamforming methods are also computable in frequency domains, where signals that are characterized across broadband, such as impulse signals, are useful for finding features that are calculated in the time domain. Improving beam power performance requires the optimization of the number of available microphones, the maximum implementable aperture size, and the microphone spacing that matches the frequency characteristics of the target acoustic source. Generally, the greater the number of microphones and the larger the aperture size, the better the performance tends to be, but additional consideration is needed to minimize the operation and acoustic measurement interference of drones to be mounted on drones.

The direction of arrival of acoustic waves was estimated by dividing the maximum beam power into the azimuth and elevation angles based on the heading of the flying drone. The drone's headings are related to the body coordinate system, and the roll, pitch, and yaw angle are output from the drone's flight control system in the reference direction. The pose angle of the drone and the estimated arrival angle of the acoustic wave were corrected relative to the ground. Figure 14 shows the record of flight posture and positional data collected from the time of takeoff until the drone lands after the completion of the measurement and maintains hovering during experiments measuring the sound of the explosion. The flight data showed the amount of change based on the average value of the hovering time interval. The time data also showed when the first explosion was detected at each explosion location from Locations 1 to 6. The remaining firecrackers exploded sequentially, seconds apart after the first explosion.

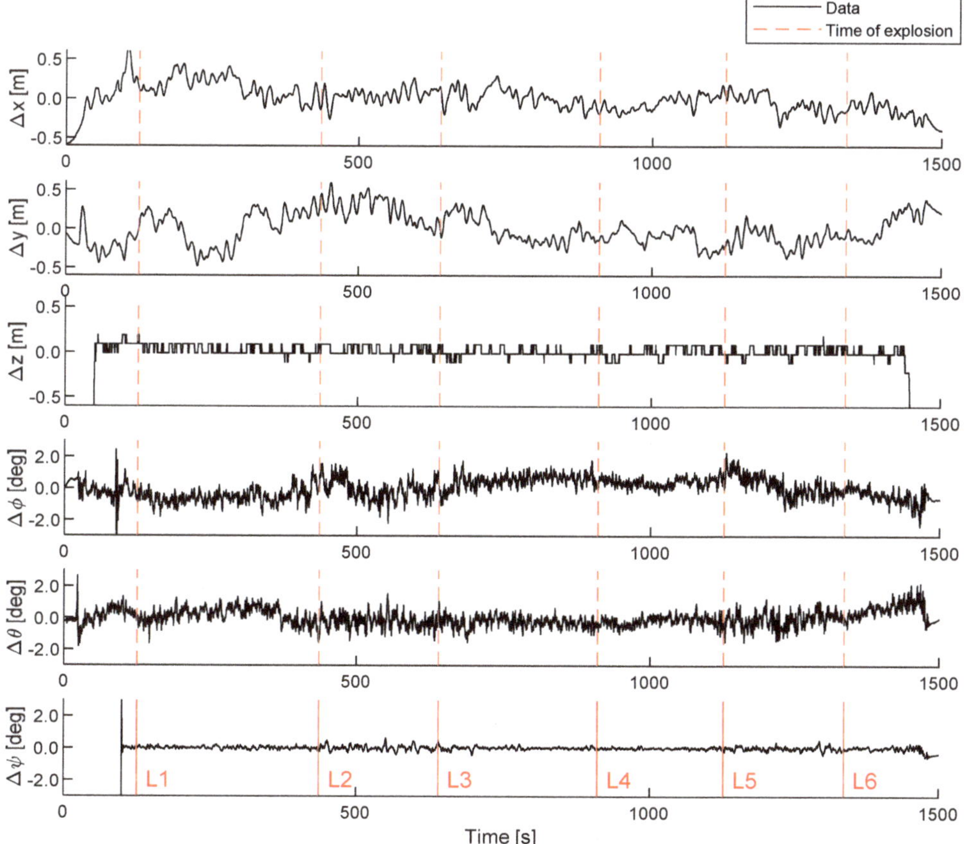

Figure 14. The amount of change in position and tilt angle during flight.

The estimated angle of arrival at the time an acoustic source was detected and the position and posture data of the drone could be corrected to specify the source location on the ground. Figure 15 shows the results of determining the estimated accuracy of the acoustic sources above the ground by the error of the horizontal and vertical angles. In this graph, the angle error of 0 degrees implies the true direction for the actual position based on GPS localization, and the measurement error represents the directional error angle that occurred from the true direction. In this experiment, acoustic source detection was performed up to a maximum ground distance of 151.5 m. The mean estimation error for three repeated impact sounds confirmed the detection performance of 8.8 degrees for the horizontal angle error and 10.3 degrees for the vertical angle error (Table 2). The estimated direction error for the entire location tended to increase both horizontally and vertically as the ground distance increased; the vertical angle error tended to be slightly higher than the horizontal angle error. The estimation error was the most sensitive during the drone's positional information correction process. In other words, we experimentally confirm that the location and posture of the drone at the time of detecting an acoustic source of interest must be applied to enable accurate localization.

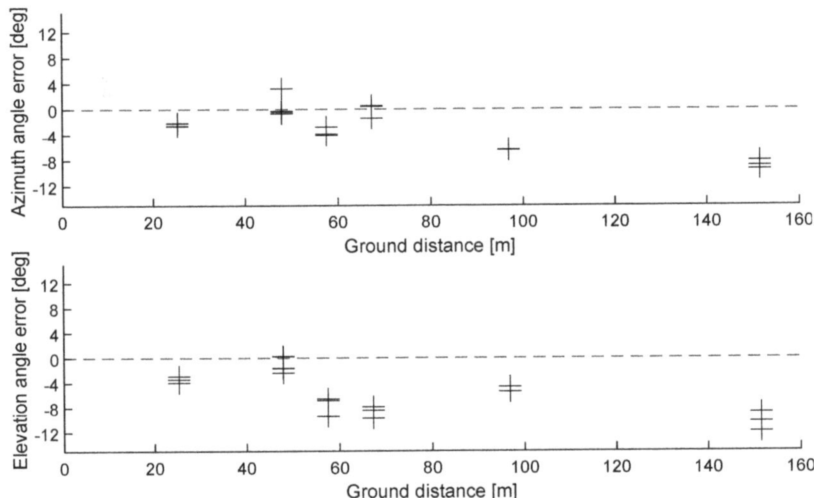

Figure 15. Estimated angle of arrival error for azimuth and elevation angles.

Table 2. Estimated angle of arrival error.

Ground Distance (m)	Explosion Location	Mean Azimuth Angle Error (Degree)	Mean Elevation Angle Error (Degree)
25.3	L6	2.42	3.42
48.0	L1	1.40	1.42
57.6	L4	3.55	7.61
67.3	L5	0.80	8.67
96.8	L2	6.40	5.07
151.5	L3	8.81	10.39

5. Conclusions

Using an array of microphones mounted on a drone, we described how to detect the location of acoustic sources generated on the ground and confirmed the performance of the system through experiments. The spectral subtraction method was applied to enhance the signal-to-noise ratio of the acoustic source of interest and was effective in eliminating noise generated by the drone. It was able to remove the fluctuating pressure and turbulence noise affected by the blade passing frequency of drone propellers and the wake, and to preserve the impact sound that was intended to be detected. Furthermore, we confirmed that the direction of arrival could be estimated when applied to beamforming methods that detected the direction of arrival of acoustic sources using phase difference relationships, and that the phase difference relationship could be effectively restored even with spectral subtraction methods. It is important to highlight that the spectral subtraction method implemented in this study was able to improve the signal-to-noise ratio, and also to effectively preserve the phase for each microphone. A drone mounted with a 32-channel microphone array was used to detect acoustic sources, and flight data could be used to detect the locations on the ground. We implemented a valid acoustic source detection method by fusing flight data at the same time as when the acoustic source was detected. The detection performance of acoustic sources with a ground distance of 151.5 m was confirmed with a horizontal angle error of 8.8 degrees and a vertical angle error of 10.3 degrees.

In this study, we proposed and implemented the necessary elements to implement a method for exploring external sound sources of interest detected by drones. Although each element of the technique is not entirely state-of-the-art, it is meaningful to specify and implement what must be performed when detecting acoustic sources using a drone, and to describe the process of connecting them. Language recognition is a detection target that should immediately become of interest. It can be used to detect people or to strengthen the voice of a person targeted from the air. Drones, especially for lifesaving activities, can be a very useful technology. There have already been several studies on language recognition technology using deep learning, and in the near future, applications of this technology are expected. Deep learning can also be applied to noise cancellation technology for the drone's self-noise. From a hardware perspective, there should also be additional studies on optimal arrays that can effectively fit drones with fewer microphones to improve location detection performance.

Author Contributions: Conceptualization, Y.-J.G. and J.-S.C.; methodology, Y.-J.G.; software, Y.-J.G.; validation, Y.-J.G. and J.-S.C.; formal analysis, Y.-J.G.; investigation, Y.-J.G.; resources, Y.-J.G.; data curation, Y.-J.G.; writing—original draft preparation, Y.-J.G.; writing—review and editing, Y.-J.G. and J.-S.C.; visualization, Y.-J.G.; supervision, J.-S.C.; project administration, J.-S.C.; funding acquisition, J.-S.C. Both authors have read and agreed to the published version of the manuscript.

Funding: This work was supported by the research fund of Chungnam National University.

Institutional Review Board Statement: Not applicable.

Informed Consent Statement: Not applicable.

Data Availability Statement: Not applicable.

Conflicts of Interest: The authors declare no conflict of interest.

References

1. Erdelj, M.; Natalizio, E.; Chowdhury, K.R.; Akyildi, I.F. Help from the Sky: Leveraging UAVs for Disaster Management. *IEEE Pervasive Comput.* **2017**, *16*, 24–32. [CrossRef]
2. Vattapparamban, E.; Güvenç, I.; Yurekli, A.I.; Akkaya, K.; Uluağaç, S. Drones for smart cities: Issues in cybersecurity privacy and public safety. In Proceedings of the International Wireless Communications and Mobile Computing Conference (IWCMC), Paphos, Cyprus, 5–9 September 2016; pp. 216–221. [CrossRef]
3. Ciaburro, G.; Iannace, G. Improving Smart Cities Safety Using Sound Events Detection Based on Deep Neural Network Algorithms. *Informatics* **2020**, *7*, 23. [CrossRef]
4. Ciaburro, G. Sound Event Detection in Underground Parking Garage Using Convolutional Neural Network. *Big Data Cogn. Comput.* **2020**, *4*, 20. [CrossRef]
5. Rascon, C.; Ivan, M. Localization of sound sources in robotics: A review. *Robot. Auton. Syst.* **2017**, *96*, 184–210. [CrossRef]
6. Liaquat, M.U.; Munawar, H.S.; Rahman, A.; Qadir, Z.; Kouzani, A.Z.; Mahmud, M.A.P. Localization of Sound Sources: A Systematic Review. *Energies* **2021**, *14*, 3910. [CrossRef]
7. Lee, J.; Go, Y.; Kim, S.; Choi, J. Flight Path Measurement of Drones Using Microphone Array and Performance Improvement Method Using Unscented Kalman Filter. *J. Korean Soc. Aeronaut. Space Sci.* **2018**, *46*, 975–985.

8. Hubbard, H.H. *Aeroacoustics of Flight Vehicles: Theory and Practice*; The Acoustical Society of America: New York, NY, USA, 1995; Volume 1, pp. 1–205.
9. Brooks, T.F.; Marcolini, M.A.; Pope, D.S. Main Rotor Broadband Noise Study in the DNW. *J. Am. Helicopter Soc.* **1989**, *34*, 3–12. [CrossRef]
10. Brooks, T.F.; Jolly, J.R., Jr.; Marcolini, M.A. Helicopter Main Rotor Noise: Determination of Source Contribution using Scaled Model Data. *NASA Technical Paper 2825*, August 1988.
11. Intaratep, N.; Alexander, W.N.; Devenport, W.J.; Grace, S.M.; Dropkin, A. Experimental Study of Quadcopter Acoustics and Performance at Static Thrust Conditions. In Proceedings of the 22nd AIAA/CEAS Aeroacoustics Conference, Lyon, France, 30 May–1 June 2016.
12. Djurek, I.; Petosic, A.; Grubesa, S.; Suhanek, M. Analysis of a Quadcopter's Acoustic Signature in Different Flight Regimes. *IEEE Access* **2020**, *8*, 10662–10670. [CrossRef]
13. Boll, S. Suppression of acoustic noise in speech using spectral subtraction. *IEEE Trans. Acoust. Speech Signal Process.* **1979**, *27*, 113–120. [CrossRef]
14. Berouti, M.; Schwartz, R.; Makhoul, J. Enhancement of speech corrupted by acoustic noise. In Proceedings of the IEEE International Conference on Acoustics, Speech, and Signal Processing, Washington, DC, USA, 2–4 April 1979; pp. 208–211. [CrossRef]
15. Bharti, S.S.; Gupta, M.; Agarwal, S. A new spectral subtraction method for speech enhancement using adaptive noise estimation. In Proceedings of the 2016 3rd International Conference on Recent Advances in Information Technology (RAIT), Dhanbad, India, 3–5 March 2016; pp. 128–132. [CrossRef]
16. Yamashita, K.; Shimamura, T. Nonstationary noise estimation using low-frequency regions for spectral subtraction. *IEEE Signal Process. Lett.* **2005**, *12*, 465–468. [CrossRef]
17. Cohen, I.; Berdugo, B. Speech enhancement for non-stationary noise environments. *Signal Process.* **2001**, *81*, 2403–2418. [CrossRef]
18. Ramos, A.; Sverre, H.; Gudvangen, S.; Otterlei, R. A Spectral Subtraction Based Algorithm for Real-time Noise Cancellation with Application to Gunshot Acoustics. *Int. J. Electron. Telecommun.* **2013**, *59*, 93–98. [CrossRef]
19. Allen, C.S.; Blake, W.K.; Dougherty, R.P.; Lynch, D.; Soderman, P.T.; Underbrink, J.R. *Aeroacoustic Measurements*; Springer: Berlin/Heidelberg, Germany, 2002; pp. 62–215.
20. Chiariotti, P.; Martarelli, M.; Castellini, P. Acoustic beamforming for noise source localization—Reviews, methodology and applications. *Mech. Syst. Signal Process.* **2019**, *120*, 422–448. [CrossRef]
21. Michel, U. History of acoustic beamforming. In Proceedings of the 1st Berlin Beamforming Conference—1st BeBeC, Berlin, Germany, 22–23 November 2006.
22. Merino-Martínez, R.; Sijtsma, P.; Snellen, M.; Ahlefeldt, T.; Antoni, J.; Bahr, C.J.; Blacodon, D.; Ernst, D.; Finez, A.; Funke, S.; et al. A review of acoustic imaging methods using phased microphone arrays. *CEAS Aeronaut. J.* **2019**, *10*, 197–230. [CrossRef]
23. Van Trees, H.L. *Optimum Array Processing: Part IV of Detection, Estimation, and Modulation Theory*; Wiley-Interscience: New York, NY, USA, 2002; pp. 231–331.
24. Clayton, M.; Wang, L.; McPherson, A.P.; Cavallaro, A. An embedded multichannel sound acquisition system for drone audition. *arXiv* **2021**, arXiv:abs/2101.06795.
25. Ruiz-Espitia, O.; Martínez-Carranza, J.; Rascón, C. AIRA-UAS: An Evaluation Corpus for Audio Processing in Unmanned Aerial System. In Proceedings of the International Conference on Unmanned Aircraft Systems (ICUAS), Dallas Marriott City Center Dallas, TX, USA, 12–15 June 2018; pp. 836–845.
26. Blanchard, T.; Thomas, J.H.; Raoof, K. Acoustic localization estimation of an Unmanned Aerial Vehicle using microphone array. *J. Acoust. Soc. Am.* **2020**, *148*, 1456. [CrossRef] [PubMed]
27. Strauss, M.; Mordel, P.; Miguet, V.; Deleforge, A. DREGON: Dataset and Methods for UAV-Embedded Sound Source Localization. In Proceedings of the 2018 IEEE/RSJ International Conference on Intelligent Robots and Systems(IROS), Madrid, Spain, 1–5 October 2018; pp. 1–8. [CrossRef]
28. Misra, P.; Kumar, A.A.; Mohapatra, P.; Balamuralidhar, P. DroneEARS: Robust Acoustic Source Localization with Aerial Drones. In Proceedings of the 2018 IEEE International Conference on Robotics and Automation (ICRA), Brisbane, Australia, 21–25 May 2018; pp. 80–85. [CrossRef]
29. Salvati, D.; Drioli, C.; Ferrin, G.; Foresti, G.L. Acoustic Source Localization from Multirotor UAVs. *IEEE Trans. Ind. Electron.* **2020**, *67*, 8618–8628. [CrossRef]
30. Hoshiba, K.; Washizaki, K.; Wakabayashi, M.; Ishiki, T.; Kumon, M.; Bando, Y.; Gabriel, D.; Nakadai, K.; Okuno, H.G. Design of UAV-Embedded Microphone Array System for Sound Source Localization in Outdoor Environments. *Sensors* **2017**, *17*, 2535 [CrossRef] [PubMed]
31. Vaseghi, S.V. *Advanced Signal Processing and Digital Noise Reduction*; Springer: Berlin/Heidelberg, Germany, 1996; pp. 242–260.
32. Johnson, D.H.; Dudgeon, D.E. *Array Signal Processing, Concepts and Techniques*; PTR Prentice Hall: Englewood Cliffs, NJ, USA, 1993.
33. Frost, O.L., III. An algorithm for linear constrained adaptive array processing. *IEEE* **2009**, *60*, 925–935.
34. Bies, D.A.; Hansen, C.H. *Engineering Noise Control—Theory and Practice*, 4th ed.; CRC Press: New York, NJ, USA, 2009; pp. 18–19 ISBN 978-0-415-48707-8.

35. Hoffmann, G.M.; Huang, H.; Waslander, S.L.; Tomlin, C.J. Quadrotor helicopter flight dynamics and control: Theory and experiment. In Proceedings of the AIAA Guidance, Navigation and Control Conference and Exhibit, Boston, MA, USA, 10–12 August 1998.
36. Huang, H.; Hoffmann, G.M.; Waslander, S.L.; Tomlin, C.J. Aerodynamics and control of autonomous quadrotor helicopters in aggressive maneuvering. In Proceedings of the IEEE International Conference on Robotics and Automation, Xi'an, China, 30 May–5 June 2021; pp. 3277–3282.
37. Nelson, R.C. *Flight Stability and Automatic Control*, 2nd ed.; McGraw-Hill Education: New York, NY, USA, 1997; pp. 96–130.

 **drones**

Article

Optimal Navigation of an Unmanned Surface Vehicle and an Autonomous Underwater Vehicle Collaborating for Reliable Acoustic Communication with Collision Avoidance

Andrey V. Savkin [1,*], Satish Chandra Verma [1] and Stuart Anstee [2]

1   School of Electrical Engineering and Telecommunications, University of New South Wales, Sydney 2052, Australia; z5220045@zmail.unsw.edu.au
2   Defence Science and Technology Group, Department of Defence, Canberra 2610, Australia; stuart.anstee@dst.defence.gov.au
*   Correspondence: a.savkin@unsw.edu.au

Abstract: This paper focuses on safe navigation of an unmanned surface vehicle in proximity to a submerged autonomous underwater vehicle so as to maximise short-range, through-water data transmission while minimising the probability that the two vehicles will accidentally collide. A sliding mode navigation law is developed, and a rigorous proof of optimality of the proposed navigation law is presented. The developed navigation algorithm is relatively computationally simple and easily implementable in real time. Illustrative examples with extensive computer simulations demonstrate the effectiveness of the proposed method.

Keywords: autonomous navigation; autonomous underwater vehicles; unmanned surface vehicles; AUVs; USVs; marine vehicles; cooperative control; sliding mode control; collision avoidance; acoustic communication; underwater communication; collaborating vehicles; optimal control

Citation: Savkin, A.V.; Verma, S.C.; Anstee, S. Optimal Navigation of an Unmanned Surface Vehicle and an Autonomous Underwater Vehicle Collaborating for Reliable Acoustic Communication with Collision Avoidance. *Drones* **2022**, *6*, 27. https://doi.org/10.3390/drones6010027

Academic Editor: Diego González-Aguilera

Received: 9 December 2021
Accepted: 14 January 2022
Published: 17 January 2022

Publisher's Note: MDPI stays neutral with regard to jurisdictional claims in published maps and institutional affiliations.

1. Introduction

Recent technological developments have made autonomous and unmanned maritime vehicles realistic alternatives to traditional vessel-based survey and monitoring systems [1]. Potential benefits of using unmanned maritime vehicles include reduced operational costs, improved safety and reliability, longer monitoring durations and mission repeatability. Two important vehicle classes are unmanned surface vehicles (USVs) and autonomous underwater vehicles (AUVs). Teams consisting of collaborating vehicles of both types can be used in applications that include safety and rescue missions, environmental disaster assessment, underwater geology, marine archeology, military survey, and detection and monitoring of marine fauna [2]. The latter area includes population surveys, migration tracking and vocal animal monitoring [3]. Some applications of AUVs and USVs for ocean wildlife monitoring and mapping are given in [4,5].

In this paper, we address the problem of a USV and an AUV navigating collaboratively to exchange data at close range. Under the considered scenario, the AUV collects some information in the submerged mode, e.g., from a seabed survey in a disaster assessment application, or intelligence information in military applications, with remote supervision from a USV. It must periodically rendezvous with the USV to offload a summary that the USV can transmit for operator inspection. The AUV could come to the surface for data transmission via one or more terrestrial communication systems (e.g., Wi-Fi), but in doing so it would lose most of its control authority while being exposed to rapidly changing sea surface conditions and the risk of collision with vessels [6]. We leave this difficult problem for future investigation and consider a different scenario; transmission of data while the AUV remains submerged, noting however that underwater data exchange using acoustic modems is notoriously slow, unreliable and range-dependent [7]. We are therefore motivated to find a solution that minimises the separation between the AUV and the USV to maximise the data transmission rate, while remaining sufficiently far apart to avoid the risk of collision with the supervising USV if the AUV is unexpectedly forced to come to the surface.

We state a constrained optimal control problem in which the AUV and USV navigate collaboratively to maximise the amount of data transmitted while keeping the probability of collision between them sufficiently small. We prove that the optimal solution to this problem is delivered by a navigation algorithm belonging to the class of sliding mode control laws [8,9].

Various problems of controlling multi-vehicle systems to achieve a certain objective are referred to as cooperative control; see, e.g., [10,11]. Thus, the problem studied in this paper falls into the domain of collaborative control.

The remainder of the paper is organised as follows. In Section 2, we give a brief survey of relevant publications in the field. In Section 3, we present the system model and state the problem under investigation. In Section 4, we present the proposed navigation law together with its theoretical analysis. Computer simulations illustrating the developed navigation method are conducted in Section 5 to show the performance of the proposed algorithm. Finally, a brief conclusion and possible directions for future research are given in Section 6.

2. Related Work

In recent years, research publications on using collaborating autonomous unmanned vehicles in monitoring applications have concentrated on unmanned aerial vehicles (UAVs); in particular, a review of recent results on deployment and navigation of teams of collaborating UAVs for surveillance can be found in the survey paper [12].

Research concerning autonomous marine vehicles working in collaboration has been more limited, possibly due to the relatively high cost associated with acquiring and operating such systems. However, the literature associated with this field is increasing.

Recent reviews of path planning for AUVs can be found in [13–15]. Teams of collaborating AUVs are studied in [16], where issues of navigation, localisation and underwater communication are analysed and various types of missions for AUV teams are discussed. Another approach for cooperative navigation of teams of AUVs was developed in [17].

Navigation of USVs is studied in [18]. The paper [19] addresses navigation of collaborating USVs for intruder interception on a marine region boundary. The publication [11] addresses a problem of path planning for a team of collaborating vehicles in marine safety and rescue missions.

Collision avoidance problems for various types of unmanned vehicles have attracted a lot of attention in recent decades; see, e.g., the survey papers [20,21]. Most of this research has concentrated on two-dimensional problems, usually for ground mobile robots [22]; however, many proposed collision avoidance algorithms may be extended to obstacle avoidance for USVs; see, e.g., [23,24]. Three-dimensional collision avoidance problems are also attracting attention; see, e.g., [25–28]. Most of the research on 3D collision avoidance concentrates on UAV navigation; however, many developed methods can be extended to collision-free navigation of AUVs [29,30]. Furthermore, some machine-learning-based approaches to AUV path planning with obstacle avoidance were recently proposed in [31,32].

At present, there are relatively few examples of systems involving collaborating AUVs and USVs [33,34], although Ocean Infinity used USV-AUV teams extensively for search operations connected with the disappearance of Malaysian Airlines flight MH370 [35]. This is a quite novel research area. The paper [6] addresses a navigation problem for an AUV-USV system for ocean sampling, environmental monitoring and providing real-time pollution measurement data. The paper [36] develops a method for a team of collaborating robots to sample primary production in the ocean. Furthermore, [37] addresses a control problem for a USV that deploys a remotely operated vehicle (ROV); however, unlike the current paper, [37] does not study a scenario with acoustic underwater communication, because the USV and the ROV are connected by an underwater cable. Finally, a problem involving collaboration between a UAV and an AUV was studied in [38].

3. Problem Statement

We consider an autonomous underwater vehicle (AUV), submerged in a 3D halfspace. Let $p(t) = [x(t), y(t), z(t)]$ be the AUV's coordinates at time t, where $x(t)$ and $y(t)$ are the coordinates in the horizontal plane parallel to the ground and $z(t)$ is the depth. We consider the following well-known model for the motion of the AUV:

$$\begin{cases} \dot{x}(t) = v(t)\cos(\theta(t)), \\ \dot{y}(t) = v(t)\sin(\theta(t)), \\ \dot{\theta}(t) = \omega(t), \\ \dot{z}(t) = u(t), \\ -V^{max} \leq v(t) \leq V^{max}, \\ -W^{max} \leq \omega(t) \leq W^{max}, \\ -U^{max} \leq u(t) \leq U^{max}. \end{cases} \tag{1}$$

Here $\theta(t)$ is the heading of the AUV; $v(t)$, $\omega(t)$ and $u(t)$ are its linear horizontal, angular and vertical speeds, respectively; and V^{max}, W^{max} and U^{max} are given positive constants indicating the maximum linear, angular and vertical speeds, respectively. The linear horizontal speed $v(t)$, angular speed $\omega(t)$ and the vertical speed $u(t)$ are the control inputs of the system (1). Furthermore, the AUV depth $z(t)$ should satisfy the constraints:

$$Z^{min} \leq z(t) \leq Z^{max} \tag{2}$$

where $0 < Z^{min} < Z^{max}$ are given constants. Furthermore, there is a moving unmanned surface vehicle (USV), with motion described by the following equations:

$$\begin{cases} \dot{x}_s(t) = v_s(t)\cos(\theta_s(t)), \\ \dot{y}_s(t) = v_s(t)\sin(\theta_s(t)), \\ \dot{\theta}_s(t) = \omega_s(t), \\ -V_s^{max} \leq v_s(t) \leq V_s^{max}, \\ -W_s^{max} \leq \omega(t) \leq W_s^{max}. \end{cases} \tag{3}$$

Here $\theta_s(t)$ is the heading of the USV; $v_s(t)$ and $\omega_s(t)$ are its linear horizontal and angular speeds, respectively; V_s^{max} and W_s^{max} are given positive constants indicating the maximum linear and angular speeds of the USV.

Typically, AUVs are battery-powered devices that move slowly—most cruise at between 1.5 and 2 ms^{-1} and have limited acceleration. In contrast, USVs that are used for supervision typically move and accelerate more quickly and have much better situational awareness. Therefore, we assume that it is appropriate to minimise the requirement for the AUV to manoeuvre; the USV is responsible for controlling its position relative to the AUV at all times.

The AUV collects some information while submerged, which it must periodically transmit to the USV. The USV (or surface vessel under autopilot control) informs the AUV that it is ready to exchange acoustic messages. The submerged AUV comes sufficiently close to the USV for acoustic communication and maintains a steady course, speed and depth.

In this scenario, we assume that at least one and potentially two sets of acoustic modems are transmitting data between the USV and the AUV: medium-range acoustic modems that transmit control instructions and telemetry and optionally short-range acoustic modems operating in a higher frequency band that transmit data at a higher rate. We assume that telemetry and control messages can be short and infrequent, but data transmission is an extended operation that is sensitive to the separation and relative position of the two vehicles; thus, it is advantageous that they maintain a constant separation and mutual orientation while data transmission is underway.

Because the telemetry channel has very low bandwidth, we assume that the AUV periodically transmits its state information to the USV, but it relies on the USV to manoeuvre safely during data transmission and does not monitor state transmissions from the USV. Thus, the USV can maintain an estimate of the coordinates, speed and heading of the AUV, but the AUV does not know, or does not react to, the state of the USV.

In this paper, we seek to develop an optimal strategy for reliably transmitting the data from the submerged AUV to the USV while avoiding collisions between the vehicles.

We consider a time interval $[t_0, t_0 + T]$. Let $N > 0$ be a given integer. We divide $[t_0, t_0 + T]$ into N equal intervals of length $\delta, \delta := \frac{T}{N}$. At any time instant $t_0, t_0 + \delta = \delta, t_0 + 2\delta, \dots, t_0 + T$, the AUV with probability p transmits some messages to the USV using an acoustic communication system. Here p is a given constant such that $0 < p \leq 1$. The probability of successful reception of the message by the USV is described by a given decreasing function $P(d(t_0 + k\delta))$ such that $0 < P(d(t_0 + k\delta)) < 1$, where $d(t)$ is the distance between the AUV and the USV at time t. Here we consider the problem of missed reception as the probability of successful reception is $P(d(t))$, where $d(t)$ is the distance between the autonomous vehicles. Since $P(d(t)) < 1$ for all $d(t) > 0$, with some non-zero probability the signal will be missed. Moreover, the USV needs to avoid collision with the AUV. Let $D_2 > D_1 > 0$ be given constants. We assume that to avoid collisions, the distance $d(t)$ between the AUV and the USV at time t should always satisfy the constraint:

$$d(t) \geq D_1. \tag{4}$$

Furthermore, we assume that any distance $d(t) \geq D_2$ is totally safe; however, it is desired that the USV would be at distances between D_1 and D_2 from the AUV in as short time as possible to minimise the risk of errors in mutual position estimates causing a collision between the USV and the AUV. More precisely, we introduce some function $C(d)$ defined for $d \geq D_1$ and decreasing such that $C(d) = 0$ for all $d \geq D_2$. We need to satisfy the following requirement:

$$\int_{t_0}^{t_0+T} C(d(t))dt \leq \epsilon, \tag{5}$$

where $\epsilon > 0$ is a given small probability, and the integral in (5) describes the probability of collision between the vehicles at a distance between D_1 and D_2 over the time interval $[t_0, t_0 + T]$.

Furthermore, the motion of the AUV and the USV should satisfy the following safety requirement that guarantees smooth enough changes of the distance between the vehicles:

$$-h \leq \dot{d}(t) \leq h, \tag{6}$$

where $h > 0$ is a given constant.

Now, let $N(t_0, t_0 + T)$ be the number of messages from the AUV successfully received by the USV over the time interval $[t_0, t_0 + T]$, and let $\mathcal{E}(\cdot)$ denote the mathematical expectation of a random variable.

Problem Statement: To construct control inputs $v(\cdot), \omega(\cdot), u(\cdot), v_s(\cdot), \omega_s(\cdot)$ for the AUV (1) and the USV (3) so that the optimal control problem

$$\mathcal{E}(N(t_0, t_0 + T)) \rightarrow \max \tag{7}$$

subject to the constraints (2), (4)–(6).

4. Navigation Law

We propose the following navigation law: at time $t = 0$, the USV selects some heading θ_0 and follows in this direction using the following sliding mode controller:

$$\omega_s(t) = -W_s^{max} sgn[\theta_s(t) - \theta_0] \tag{8}$$

where $sgn[\cdot]$ is the sign function defined as follows:

$$sgn[x] := \begin{cases} 1 & if \quad x > 0, \\ 0 & if \quad x = 0, \\ -1 & if \quad x < 0. \end{cases} \tag{9}$$

Furthermore, the USV transmits this direction θ_0 to the AUV, which uses an analogous controller:

$$\omega(t) = -W^{max} sgn[\theta(t) - \theta_0]. \tag{10}$$

It is obvious that after some time $t_0 \geq 0$, both vehicles will have headings $\theta_s(t) = \theta(t) = \theta_0$ for all $t \geq t_0$ with any control inputs $v(t), u(t), v_s(t)$. Notice that the controllers (8) and (10) belong to the class of sliding mode controllers [8] or switched controllers [39].

Let n_e and N_e be given positive integers. We consider the following class of piecewise constant functions $e(t)$ that can change values at discrete time instants $t_0, t_0 + \delta_e, t_0 + 2\delta_e, \dots$ where

$$\delta_e := \frac{T}{N_e}. \tag{11}$$

Moreover, $e(t)$ takes values only in the discrete set of $2n_e + 1$ numbers:

$$\left\{ \frac{j_e h}{n_e} \quad \forall j_e = -n_e, -n_e + 1, \dots, n_e \right\}. \tag{12}$$

Now, we propose the following optimisation scheme. We take all possible piecewise constant functions $e(t)$ with values from the set (12) changing values at discrete time instants $t_0, t_0 + \delta_e, t_0 + 2\delta_e, \dots$ and introduce the functions

$$D(t) := d(t_0) + \int_{t_0}^{t} e(\tau)d\tau \tag{13}$$

for all $t \in [t_0, t_0 + T]$.

Furthermore, we consider only pairs $(e(t), D(t))$ such that $d(t) := D(t)$ satisfies the constraints (4) and (5). Let $\mathcal{S}$ denote the set of all such pairs. Now, we select $(e^0(t), D^0(t))$ such that $D^0(t)$ delivers the maximum of the cost function

$$\mathcal{F}(d(\cdot)) := \sum_{k=0}^{N} P(D(t_0 + k\delta)) \tag{14}$$

over all $(e(t), D(t)) \in \mathcal{S}$.

In other words, we select the function $e(t)$ using the brute force method, i.e., complete search in the set of all possible piecewise constant functions $e(t)$ taking values from the set (12). Note that at any discrete time instant, there are $(2n_e + 1)$ options of values of $e(t)$. Then, for N_e instants, there are $(2n_e + 1)^{N_e}$ possible functions $e(t)$. However, in practice, as we only employ functions $e(t)$ that meet the constraints (4) and (5), we get a much smaller number of elements in the set $\mathcal{S}$, which greatly reduces the computational complexity of this optimisation scheme.

Now we consider the following navigation law over $[t_0, t_0 + T]$:

NL1: The AUV chooses an arbitrary control input $0 < v(t) \leq \hat{V}$, where $0 < \hat{V} \leq V^{max}$ some given constant, and the control input $\omega(t)$ is defined by (10). Furthermore, the AUV chooses the control input $u(t)$ defined by $u(t) = 0$.

NL2: The USV applies the control inputs $v_s(t)$ defined by

$$v_s(t) = v(t) - \frac{e^0(t)}{\cos(\alpha(t))}, \tag{15}$$

where $\alpha(t)$ is the angle between the heading θ_0 and the line connecting the current positions of the AUV and the USV. Furthermore, the control input $\omega_s(t)$ is defined by (8).

Our main theoretical result requires some assumptions.

Assumption 1. *The following inequalities hold:*

$$Z^{min} \le z(t_0) \le Z^{max}, \tag{16}$$

$$d(t_0) \ge D_1. \tag{17}$$

(2) and (4) hold at time $t = t_0$.

Assumption 2. *The following inequality holds:*

$$\hat{V} + \frac{D_1 h}{\sqrt{D_1^2 - z(t_0)^2}} \le V_s^{max}. \tag{18}$$

Theorem 1. *Consider the optimisation problem (7) for the AUV (1) and the USV (3) subject to the constraints (2), (4)–(6). As $n_e \to \infty$ and $N_e \to \infty$, the value of the cost function (7) delivered by the navigation law* **NL1, NL2** *converges to the global supremum in (7).*

Proof. Under the navigation law NL1, NL2, the AUV is moving in a plane that is parallel to the surface, so $z(t)$ is constant; therefore, Assumption 1 implies that the constraint (2) always holds. Furthermore, under the navigation law NL1, NL2, the AUV and the USV are moving along two parallel straight lines with the direction θ_0. Since the AUV and the USV are moving in the direction θ_0, it follows from (8) and (10) that $\omega_s(t) = \omega(t) = 0$ for all $t \in [t_0, t_0 + T]$. This implies that the derivative $\dot{d}(t)$ of the distance $d(t)$ between the AUV and the USV satisfies

$$\dot{d}(t) = -v(t)cos(\alpha(t)) + v_s(t)cos(\alpha(t)); \tag{19}$$

see, e.g., [22]. Moreover, (19) implies that if the USV's control input $v_s(t)$ is defined by (15) with some function $e^0(t)$, then

$$\dot{d}(t) = e^0(t), D(t) = d(t) \tag{20}$$

where $D(t)$ is defined by (13) with $e(t) = e^0(t)$. Furthermore, in this case, the cost function (7) is equal to the function (14). It is also obvious that the cost function (7) and the constraints (4)–(6) depend only on $d(t)$. Therefore, the navigation law NL1, NL2 delivers the maximum in the optimisation problem (7) subject to the constraints (2), (4)–(6) in the class of control inputs such that $\dot{d}(t)$ is piecewise constant with constant values on each of N_e intervals from the class (12). Any $\dot{d}(t)$ can be approximated with an arbitrary small precision by piecewise constant functions. Therefore, we can build a sequence of $e^0(t)$ from the above class for which the cost function (14) converges to the global supremum in (7) as $n_e \to \infty$ and $N_e \to \infty$. Finally, Assumption 2 obviously guarantees that the control input $v(t)$ of the AUV defined by (15) satisfies the constraint $-V^{max} \le v(t) \le V^{max}$. $\square$

Remark 1. *It should be pointed out that the proposed algorithm produces an asymptotically accurate approximation of the global optimum in the sense that as the parameters n_e and N_e are large enough, the obtained trajectories converge to the optimum.*

Remark 2. *A physical interpretation on the inequalities (16) from Assumption 1 is quite clear. They guarantee at the initial time that the AUV satisfies the depth constraint (2) and the safety constraint (4). Furthermore, a physical interpretation of the inequality (18) from Assumption 2 is* *that the maximum linear speed of the USV is larger than the current linear speed of the AUV with some margin that allows the USV to perform its manoeuvre as the USV is responsible for controlling its position relative to the AUV. Notice also that Assumption 2 is an assumption of a technical nature guaranteeing that the control input $v_s(t)$ of the USV defined by the navigation law NL1, NL2 satisfies the constraint $-V_s^{max} \leq v_s(t) \leq V_s^{max}$. The requirement (18) is quite conservative, and in practice, the constraint $-V_s^{max} \leq v_s(t) \leq V_s^{max}$ often holds even when Assumption 2 is not satisfied.*

Remark 3. *Notice that in practice, due to lack of communication and/or loss of data transmitted from the AUV to the USV, the USV would model the position of the AUV with a robust Kalman filter [40], algorithms of state estimation over communication channels [41] or similar techniques as uncertainty would increase whenever there was a significant period when the USV did not receive a state transmission from the AUV. However, we do not study this issue in this paper.*

5. Illustrative Examples and Computer Simulations

In this section, we present results of computer simulation that illustrate the performance of the navigation NL1, NL2. Simulations are performed in MATLAB software. We take times $t_0 = 20$ s and $T = 50$ s. Then we apply the navigation law NL1, NL2 ten times over the time intervals [20 s, 70 s], [70 s, 120 s], ..., [470 s, 520 s]. The values of the parameters used in simulations are $N_e = 5$, $n_e = 5$, $N = 50$, $D_1 = 10$ m, $D_2 = 110$ m, $v_s(t) = 1.7$ ms^{-1}, $h = 0.3$ ms^{-1}, $\theta_0 = 60°$, $t_0 = 20$ s, $V^{max} = 2.24$ ms^{-1}, $W^{max} = 0.12$ rads^{-1}, $Z^{min} = 2$ m, $Z^{max} = 150$ m, $U^{max} = 2$ ms^{-1}, $V_s^{max} = 2.74$ ms^{-1}, $W_s^{max} = 0.12$ rads^{-1}, $T = 50$ s, $\delta = 1$ and $\delta_e = 10$. Initially, the AUV and USV are at different coordinates and have different headings. The position of the AUV is $[0, 0, -5]$ with initial heading $25°$, while the USV is at $[27, 46.75, 0]$ with initial heading $0°$. The decreasing function used for the Equation (5) is defined as $C(d) = 1/d$ and the cost function for the Equation (14) is described as $P(d) = 1/(1.2 + d)$. The simulation runs for 520 s. Initially, in the first 20 s, both vehicles steer the same course θ_0. Thereafter, in the next 500 s, the AUV sends data to the USV. This interval is split into 10 equal intervals of 50 s each. The navigation algorithm is tested in three different scenarios as follows. Except where indicated, the parameters and initial conditions remain the same.

Note that the probability function $P(d)$ that we have assumed does not accurately reflect the performance of any particular acoustic or other underwater communications device, but the control law is not critically dependent on its particular form, so long as it is monotonically decreasing. Better models of particular modems could be included through the use of an underwater communication simulation system such as AquaSim NG (see, for example, [42]). The same is true for the function $C(d)$, as the proposed control law is not critically dependent on its particular form, so long as it is monotonically decreasing.

5.1. Scenario I

The AUV is travelling at a constant speed $v_s = 1.7$ ms^{-1}, and the USV transmits a new direction angle $\theta_0 = 60°$ to the AUV at time t = 0. Once the AUV obtains information, both vehicles travel in the same direction. Within 20 s, both of them achieve the same heading angle; see Figures 1 and 2. The chattering visible in the figures is due to the presence of a sign function in the controller. At the end of 20 s, Assumption III.1 is verified; subsequently, the value of the $e(t)$ function is generated for the next 50 s; see Figure 3. Once calculated, the navigation law NL2 applies to the USV, controlling the speed of the USV; see Figure 4. It also causes the value of $d(t)$ to fluctuate as shown in Figure 5. In a nutshell, at the end of each interval, $d(t)$ is calculated to compute $e(t)$ for the next interval. The whole process is repeated until the end of the simulation. In each interval, the piecewise constant ϵ has to be selected carefully before calculating $e(t)$; see Figure 6. In general, the distance is increased to avoid collision and decreased to improve data transmission. As observed from Figure 5, over ten intervals, the separation between the AUV and the USV decreases, and the probability of collision increases slightly. The total amount of data sent from AUV to

USV until time t, where $t \in [t_0, t_0 + 10T]$, is illustrated in Figure 7. As seen from the figure, when NL1 and NL2 are active, the cumulative number of data packets received by USV at time t is greater than when only NL1 is active. The trajectory of both vehicles looks straight after the first 20 s. The AUV follows the USV as shown in Figure 8.

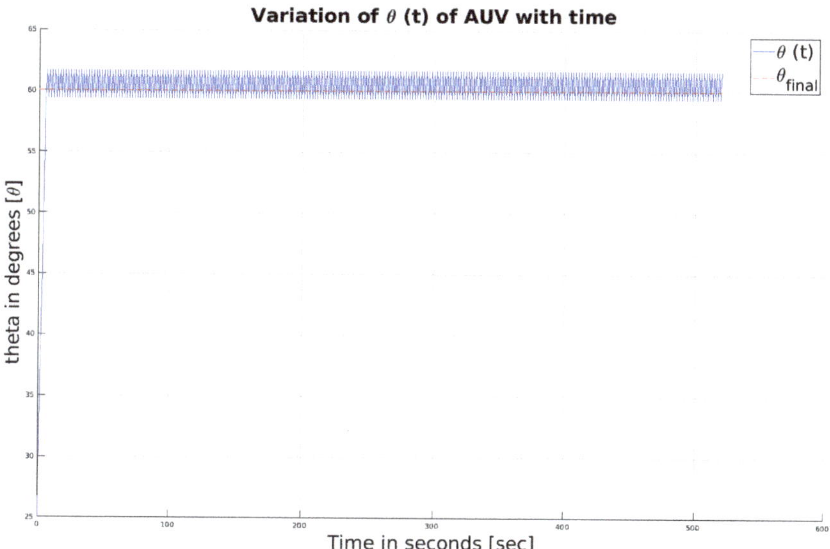

Figure 1. Scenario I: Variation of AUV's heading angle with time.

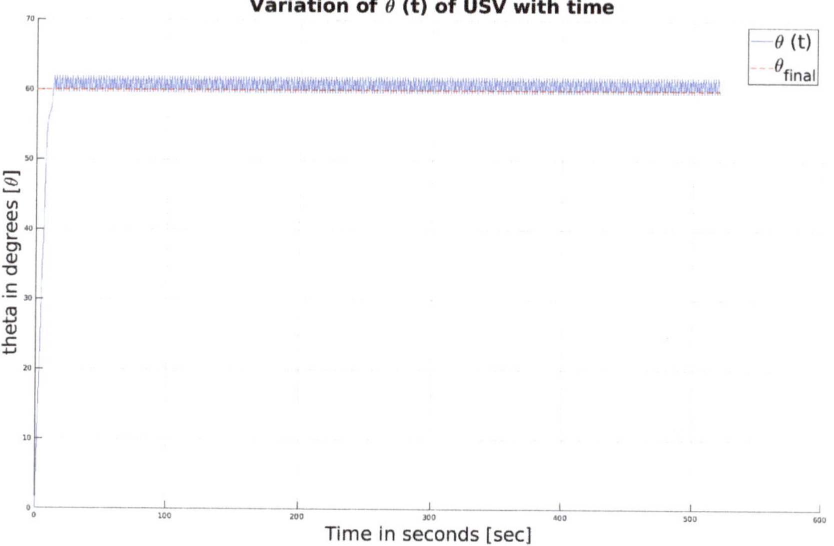

Figure 2. Scenario I: Variation of heading angle of USV with time.

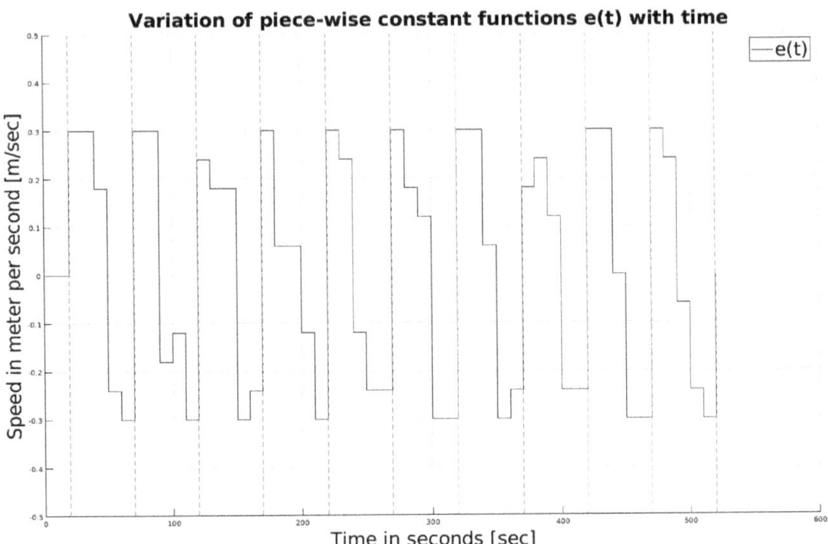

Figure 3. Scenario I: Variation of function e(t) with time.

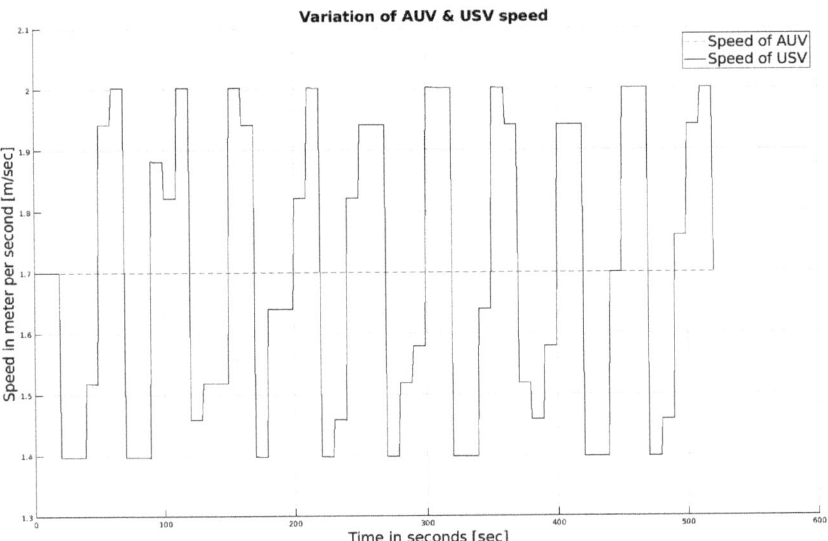

Figure 4. Scenario I: Variation of speed of AUV and USV with time.

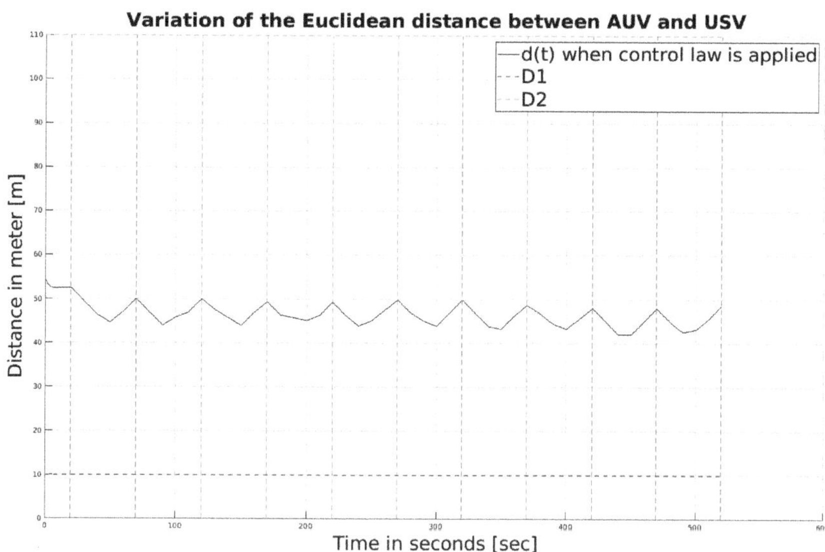

Figure 5. Scenario I: Variation of d(t) with time.

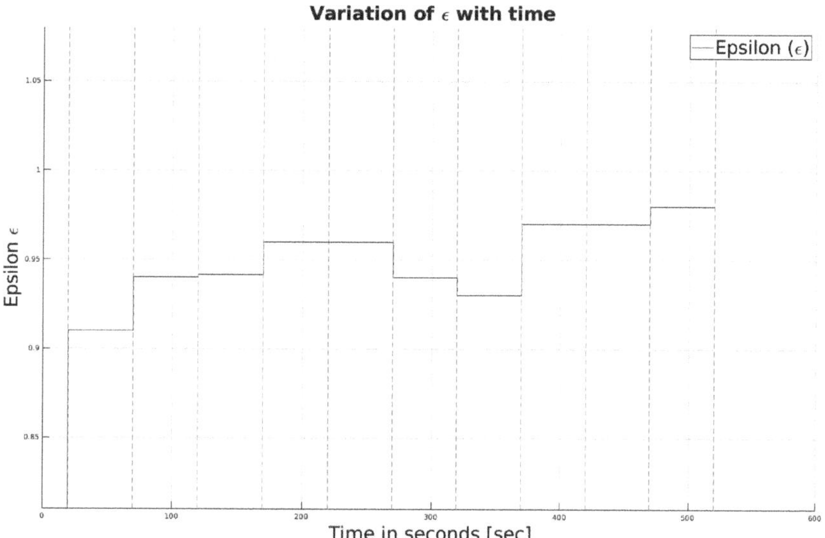

Figure 6. Scenario I: Variation of ϵ with time.

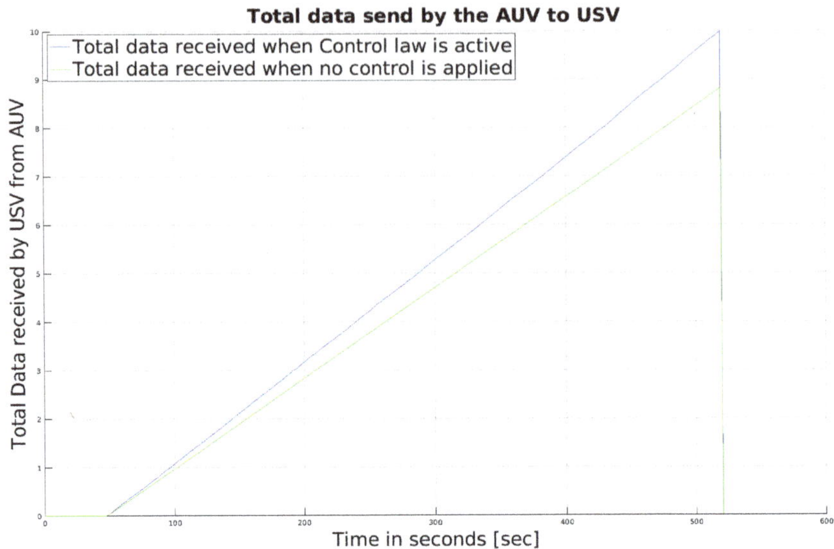

Figure 7. Scenario I: Cumulative data packets received by USV.

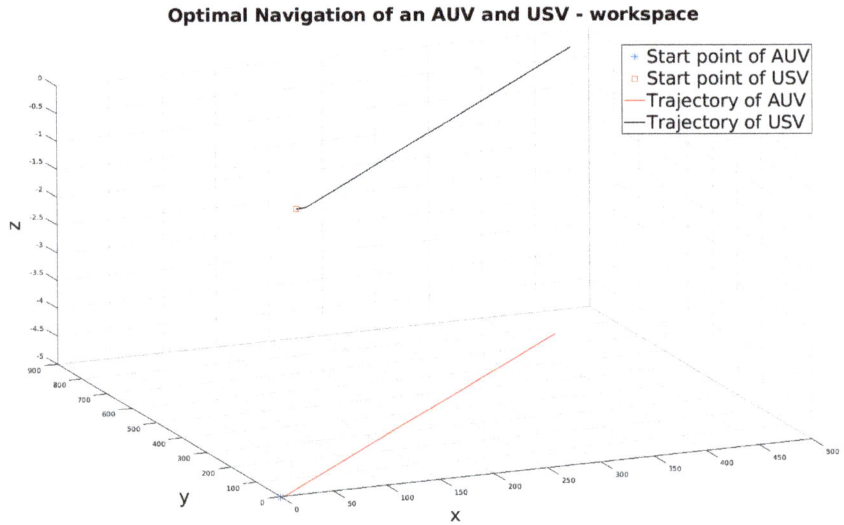

Figure 8. Scenario I: Trajectory of AUV and USV in 3D space.

5.2. Scenario II

White noise is added to the motion models of the USV and the AUV. A normally distributed random signal $W(t) = [w_1(t)\; w_2(t)\; w_3(t)\; w_4(t)]^T$ and $N(t) = [n_1(t)\; n_2(t)\; n_3(t)\; n_4(t)]^T$ with zero mean is added to the motion model of the AUV and the USV. The variance of $W(t)$ is $[1.5\;\; 1.5\;\; 0.001\;\; 0.009]^T$ and $N(t)$ is $[2.5\;\; 2.5\;\; 0.001\;\; 0.1]^T$, respectively. As a result, $d(t)$, the relative speed and bearing angles of the AUV and the USV, has noise components; see Figures 9–12. Overall, the system is stable in the presence of uncertainty in the model. In comparison with the previous scenario, trajectories of variables, such as $d(t)$, constant ϵ, speed of USV and $e(t)$ are different. The values of $d(t)$ are distinct at each time interval. The $e(t)$ function depends on $d(t)$ while the other parameters are constant. Therefore, the noisy model leads to different $d(t)$ from Scenario I after the initial 20 s. Dissimilar $e(t)$ functions are computed for subsequent intervals. Hence, by the end, completely different trajectories of $d(t)$, $e(t)$, ϵ and the speed of the USV are observed in Figures 9, 10, 13 and 14. The paths of the AUV and USV during the simulation time are shown in Figure 15. The uneven motion is due to the noise. The amount of data transmitted from AUV to USV at time t is similar to Scenario I as seen from Figure 16.

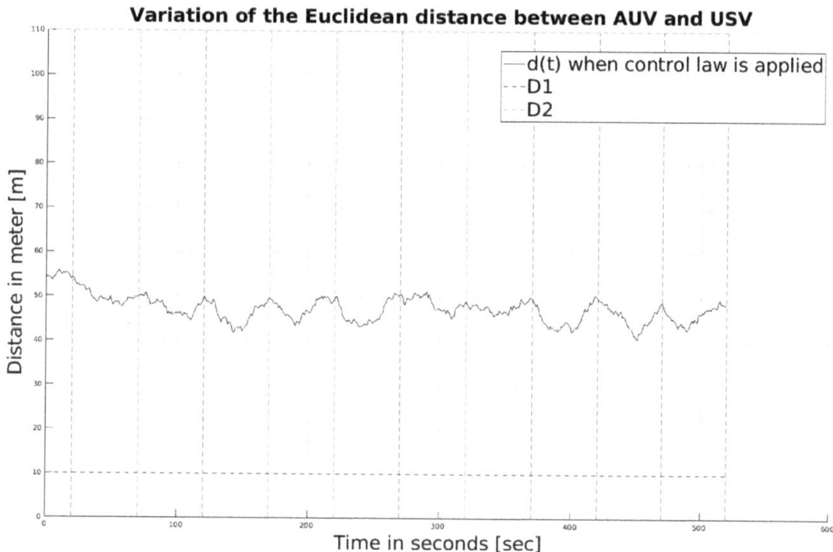

Figure 9. Scenario II: Variation of d(t) with time.

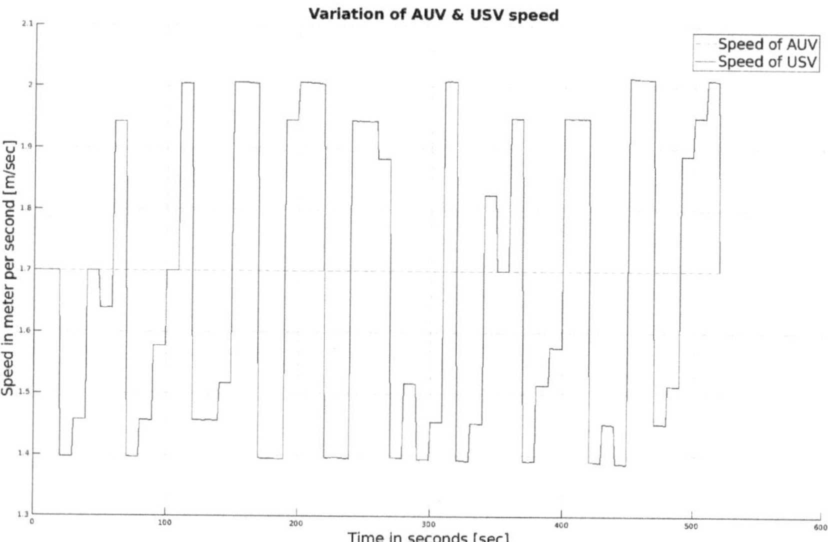

Figure 10. Scenario II: Variation of speed of AUV and USV with time.

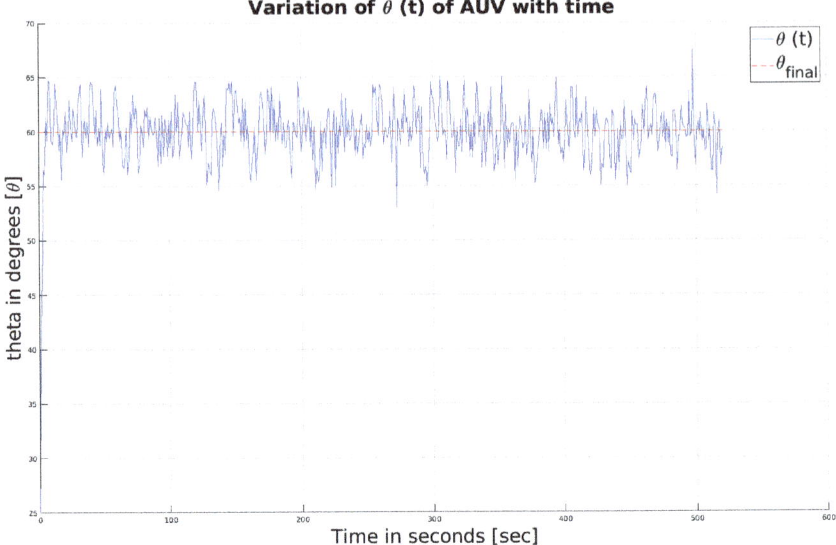

Figure 11. Scenario II: Variation of AUV's heading angle with time.

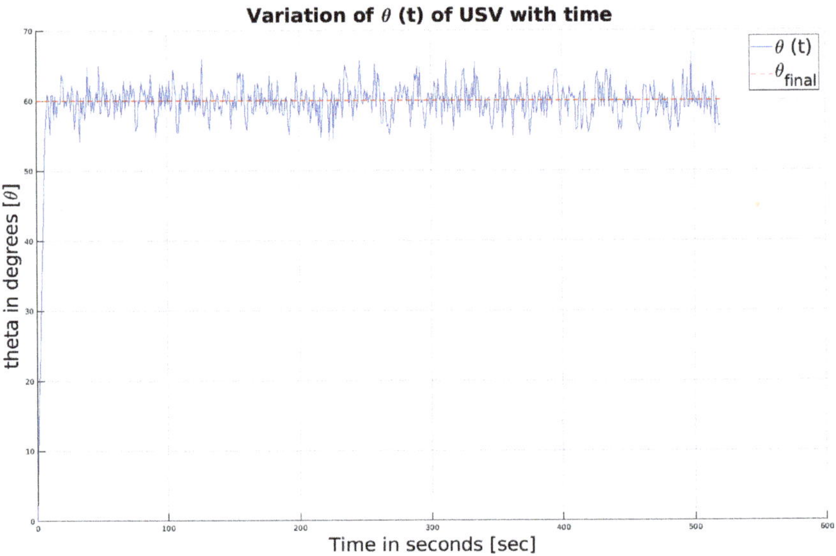

Figure 12. Scenario II: Variation of USV's heading angle with time.

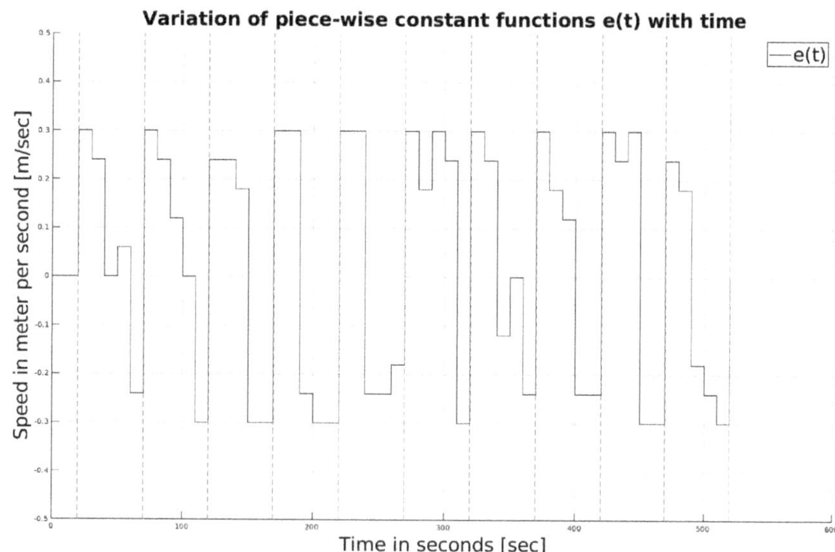

Figure 13. Scenario II: Variation of function e(t) with time.

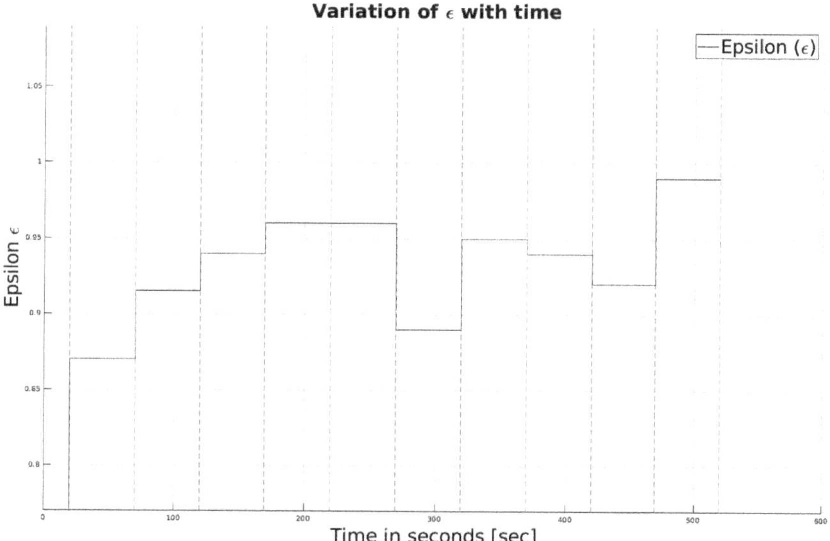

Figure 14. Scenario II: Variation of ϵ with time.

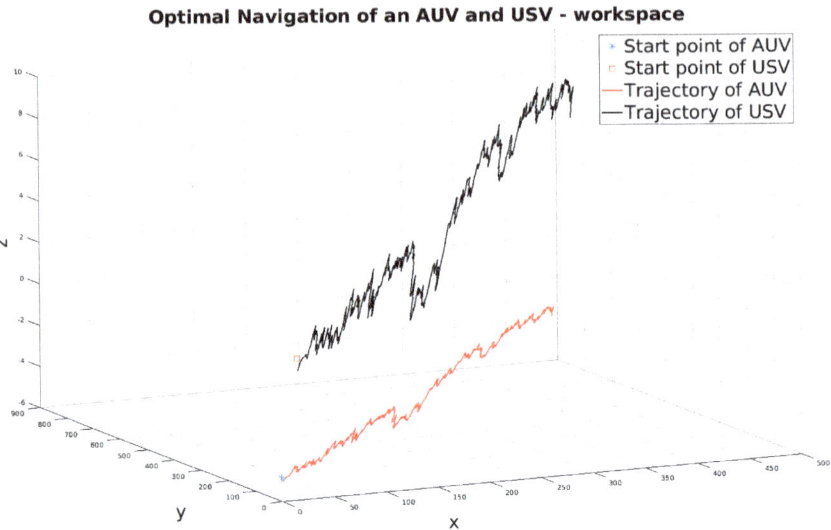

Figure 15. Scenario II: Trajectory of AUV and USV in 3D space.

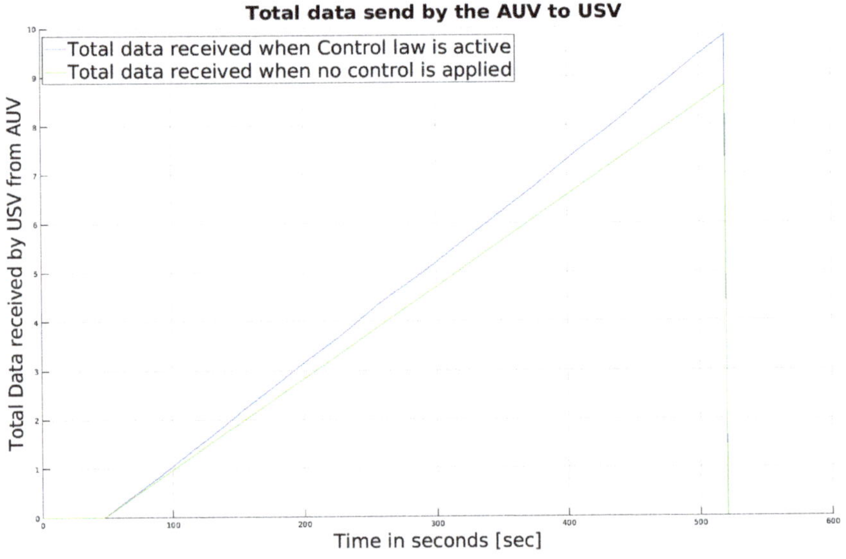

Figure 16. Scenario II: Cumulative data packets received by USV.

5.3. Scenario III

This scenario tries to mimic real-life conditions and test the performance of the algorithm. White noise is added to the speed of the USV and the motion models of both vehicles. Like the previous case, noise signals $W(t)$ and $N(t)$ are added to the motion models. Moreover, $M(t)$ is added to the speed of the USV. The variance of $W(t)$ is $[1 \; 1 \; 0.001 \; 0.01]^T$, $N(t)$ is $[1.5 \; 1.5 \; 0.002 \; 0.005]^T$ and $M(t)$ is 0.0001, respectively. A sinusoidal periodic function is also added to $v_s(t)$ to replicate the effect of sea surface waves on the USV. The amplitude of the wave is 0.2 ms^{-1} and its frequency is 0.07 rads^{-1}. Graphs of $v(t)$ and $v_s(t)$ are shown in Figure 17. As observed, the speed of the USV varies sinusoidally. And as expected, the distance $d(t)$ in Figure 18, constant ϵ in Figure 19 and the function $e(t)$ in

Figure 20 have different trajectories compared to the previous two scenarios. However, notice that in all the above scenarios, overall separation has decreased over the ten intervals. Therefore, the amount of data transfer is overall similar to previous scenarios; see Figure 21. The variation of the bearing angles of the AUV and the USV is similar to the previous case. See Figures 22 and 23. The path followed by both vehicles during the complete simulation shown in Figure 24 is similar to Figure 15.

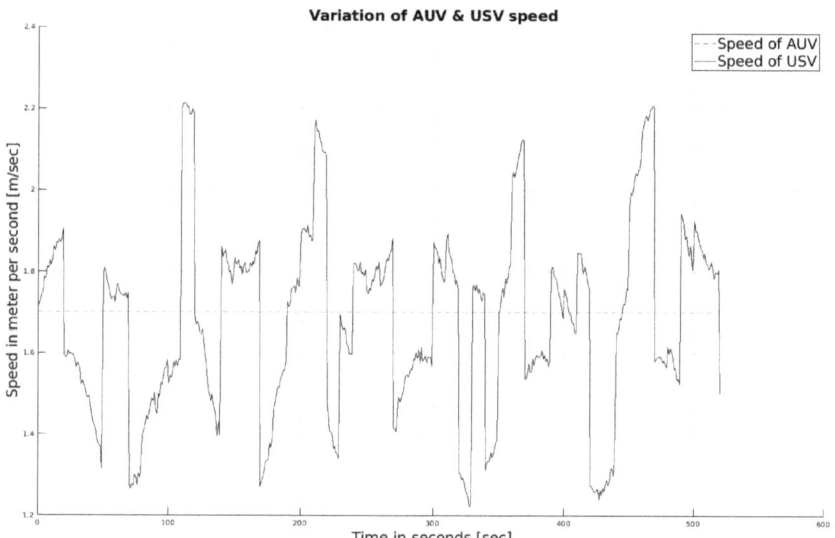

Figure 17. Scenario III: Variation of speed of AUV and USV with time.

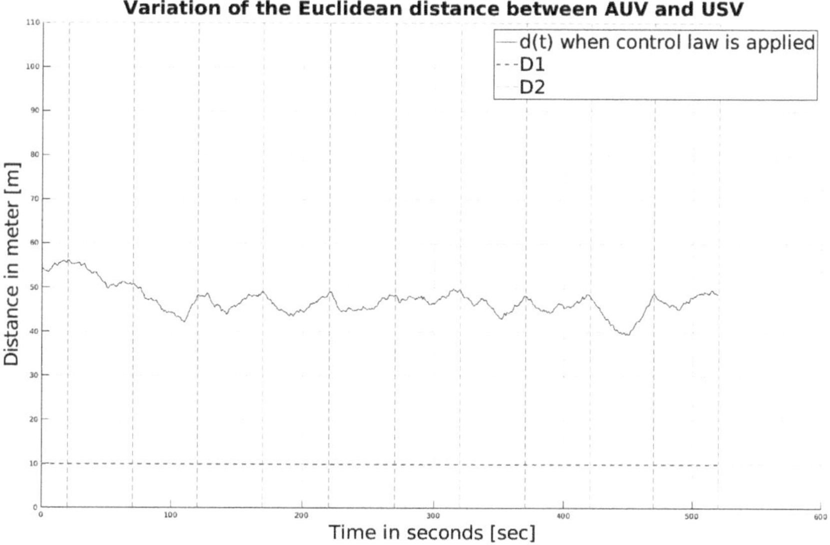

Figure 18. Scenario III: Variation of d(t) with time.

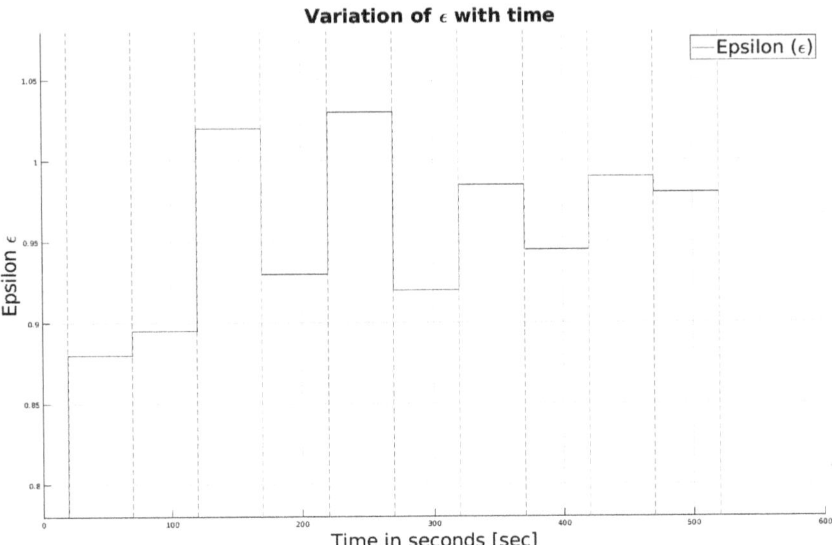

Figure 19. Scenario III: Variation of ϵ with time.

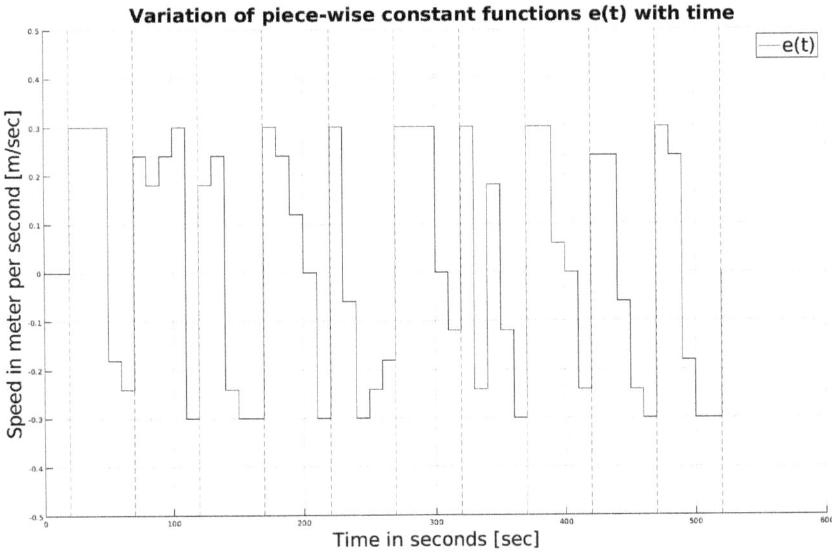

Figure 20. Scenario III: Variation of function e(t) with time.

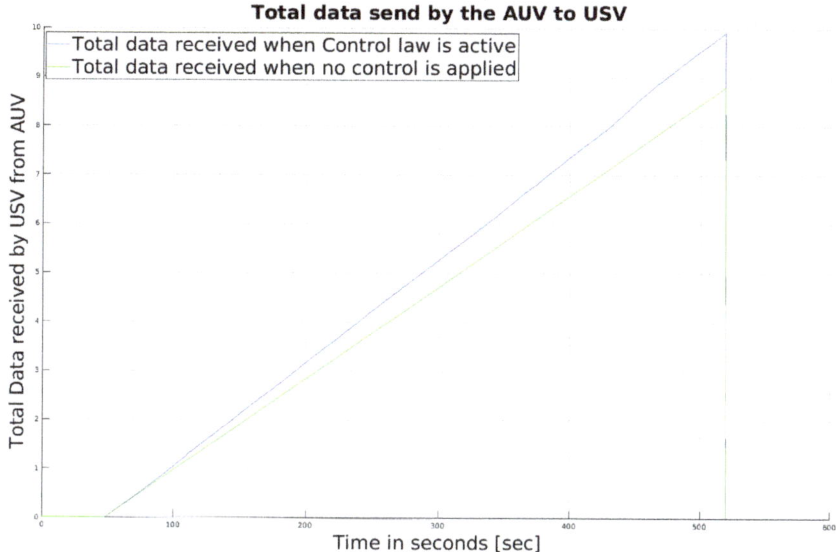

Figure 21. Scenario III: Cumulative data packets received by USV.

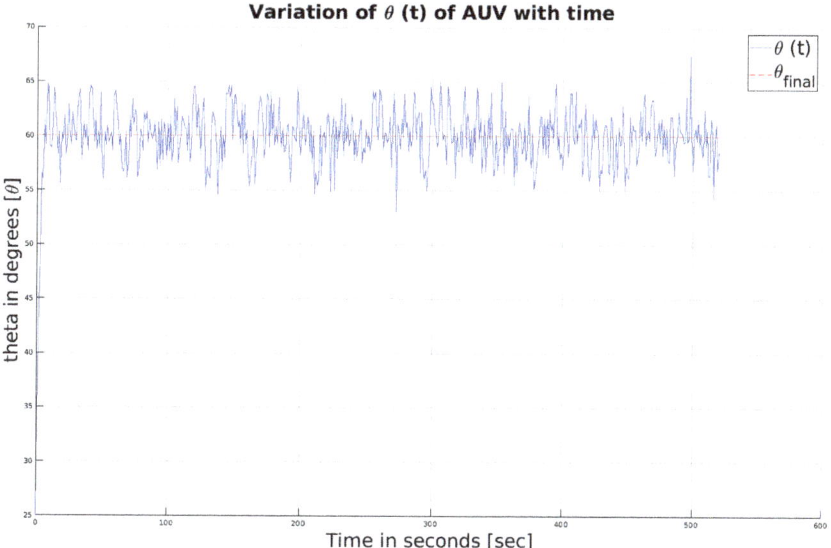

Figure 22. Scenario III: Variation of AUV's heading angle with time.

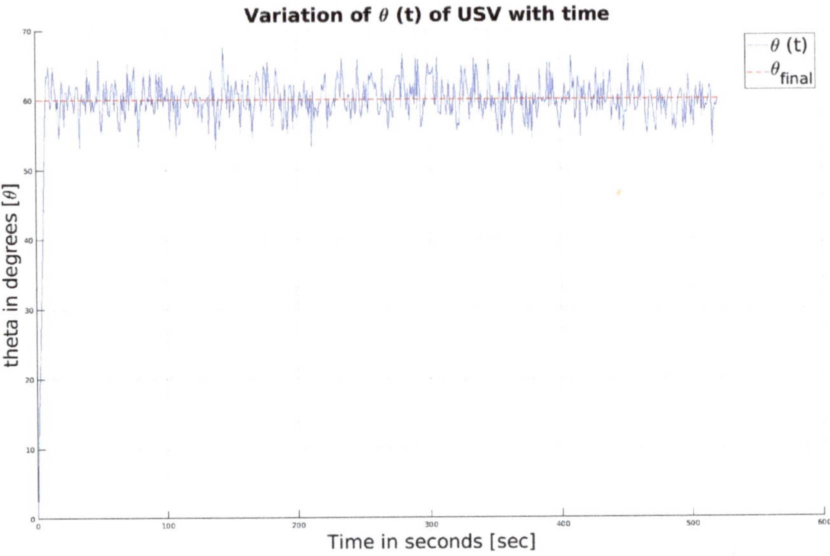

Figure 23. Scenario III: Variation of USV's heading angle with time.

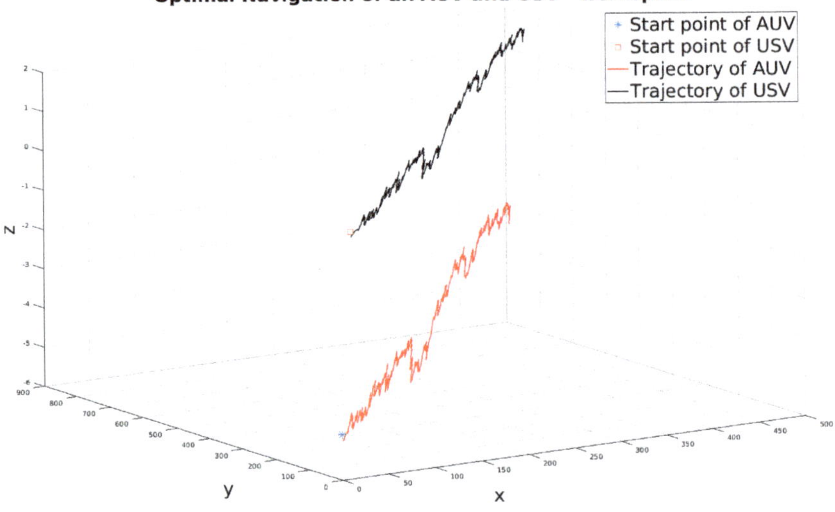

Figure 24. Scenario III: Trajectory of AUV and USV in 3D space.

6. Conclusions and Future Work

A novel navigation problem for two collaborating marine unmanned vehicles was introduced. In this problem, a submerged autonomous underwater vehicle acoustically transmits collected data to an unmanned surface vehicle. The aim is to navigate two autonomous vehicles so that the amount of data successfully transmitted between them using underwater acoustic communications is maximised while collisions between the vehicles are avoided. We proposed an easily implementable, real-time navigation algorithm belonging to the class of sliding mode control laws. Moreover, the developed autonomous navigation algorithm is asymptotically optimal in the sense that as the interval subdivision parameters of the algorithm tend to infinity, the value of the cost function describing the amount of successfully transmitted data converges to the global maximum. Illustrative examples and computer simulations showed the efficiency of the proposed approach.

There is considerable scope to improve the results from this preliminary investigation, particularly in regards to improved models for the system dynamics and the underwater communication systems. The models (1) and (3) studied in this paper are well-known kinematic models. An important direction for future research is to study more advanced models, taking into account the vehicles' dynamics, hydrodynamic effects and underwater acoustic communications characteristics. Then the navigation laws developed in this paper will be supplemented with robust controllers such as H^∞ controllers (see, e.g., [43] and references therein) that will stabilise vehicle motion and reduce fluctuations in vehicle headings. Moreover, the next stage of research will involve implementation and testing the developed control algorithms with real marine vehicles in a real environment. We hope to pursue these directions in future work.

In this paper, we have considered the case of a USV operating in proximity to a submerged AUV. A more challenging scenario is the case of a USV operating in proximity to a surfaced AUV. Under this scenario, the AUV has restricted or zero capability to manouevre. It is also subject to wave motions and surface currents that may cause it to sporadically submerge, at which times radio-frequency communications such as Wi-Fi will be lost, while acoustic connectivity will be particularly unreliable. This is an important and interesting direction for future research.

Finally, a challenging direction of future research is to extend the studied problem to scenarios with a USV collaborating with a few AUVs and even further to teams consisting of several USVs and AUVs. In such scenarios, we will need to develop some sophisticated decentralised scheduling algorithms that decide how each AUV selects a suitable USV and a time period for acoustic communication. Some existing algorithms for large teams of cooperating UAVs (see [12] and references therein) might be modified for this class of problems.

Author Contributions: Conceptualisation, S.A. and A.V.S.; methodology, A.V.S. and S.A.; software, S.C.V.; validation, S.C.V. and A.V.S.; formal analysis, A.V.S.; writing—original draft preparation, A.V.S. and S.C.V.; writing—review and editing, S.A. and A.V.S.; visualisation, S.C.V.; supervision, A.V.S.; project administration, A.V.S.; funding acquisition, A.V.S. All authors have read and agreed to the published version of the manuscript.

Funding: This work was supported by the Australian Research Council. In addition, this work received funding from the Australian Government via grant AUSMURIB000001 associated with ONR MURI grant N00014-19-1-2571.

Institutional Review Board Statement: Not applicable.

Informed Consent Statement: Not applicable.

Conflicts of Interest: The authors declare no conflict of interest.

References

1. Martin, B.; Tarraf, D.C.; Whitmore, T.C.; DeWeese, J.; Kenney, C.; Schmid, J.; DeLuca, P. *Advancing Autonomous Systems: An Analysis of Current and Future Technology for Unmanned Maritime Vehicles*; Technical report; Rand Corporation: Santa Monica, CA, USA, 2019.
2. Manley, J.E. Unmanned Maritime Vehicles, 20 years of commercial and technical evolution. In Proceedings of the OCEANS 2016 MTS/IEEE Monterey, Monterey, CA, USA, 19–23 September 2016; pp. 1–6.
3. Verfuss, U.K.; Aniceto, A.S.; Harris, D.V.; Gillespie, D.; Fielding, S.; Jiménez, G.; Johnston, P.; Sinclair, R.R.; Sivertsen, A.; Solbø, S.A.; et al. A review of unmanned vehicles for the detection and monitoring of marine fauna. *Mar. Pollut. Bull.* **2019**, *140*, 17–29. [CrossRef] [PubMed]
4. Raber, G.T.; Schill, S.R. Reef Rover: a low-cost small autonomous unmanned surface vehicle (USV) for mapping and monitoring coral reefs. *Drones* **2019**, *3*, 38. [CrossRef]
5. Butcher, P.A.; Colefax, A.P.; Gorkin, R.A.; Kajiura, S.M.; López, N.A.; Mourier, J.; Purcell, C.R.; Skomal, G.B.; Tucker, J.P.; Walsh, A.J.; et al. The drone revolution of shark science: a review. *Drones* **2021**, *5*, 8. [CrossRef]
6. Vasilijević, A.; Nađ, Đ.; Mandić, F.; Mišković, N.; Vukić, Z. Coordinated navigation of surface and underwater marine robotic vehicles for ocean sampling and environmental monitoring. *IEEE/ASME Trans. Mechatronics* **2017**, *22*, 1174–1184. [CrossRef]

7. Song, A.; Stojanovic, M.; Chitre, M. Editorial underwater acoustic communications: Where we stand and what is next? *IEEE J. Ocean. Eng.* **2019**, *44*, 1–6. [CrossRef]
8. Utkin, V.I. *Sliding Modes in Control and Optimization*; Springer: Berlin/Heidelberg, Germany, 2013.
9. Utkin, V.; Guldner, J.; Shi, J. *Sliding Mode Control in Electro-Mechanical Systems*; CRC Press: Boca Raton, FL, USA, 2017.
10. Khan, A.; Rinner, B.; Cavallaro, A. Cooperative robots to observe moving targets. *IEEE Trans. Cybern.* **2016**, *48*, 187–198. [CrossRef] [PubMed]
11. Yoon, S.; Do, H.; Kim, J. Collaborative mission and route planning of multi-vehicle systems for autonomous search in marine environment. *Int. J. Control Autom. Syst.* **2020**, *18*, 546–555. [CrossRef]
12. Li, X.; Savkin, A.V. Networked Unmanned Aerial Vehicles for Surveillance and Monitoring: A Survey. *Future Internet* **2021**, *13*, 174. [CrossRef]
13. Zeng, Z.; Lian, L.; Sammut, K.; He, F.; Tang, Y.; Lammas, A. A survey on path planning for persistent autonomy of autonomous underwater vehicles. *Ocean. Eng.* **2015**, *110*, 303–313. [CrossRef]
14. Li, D.; Wang, P.; Du, L. Path planning technologies for autonomous underwater vehicles-a review. *IEEE Access* **2018**, *7*, 9745–9768. [CrossRef]
15. Panda, M.; Das, B.; Subudhi, B.; Pati, B.B. A comprehensive review of path planning algorithms for autonomous underwater vehicles. *Int. J. Autom. Comput.* **2020**, *17*, 321–352. [CrossRef]
16. González-García, J.; Gómez-Espinosa, A.; Cuan-Urquizo, E.; García-Valdovinos, L.G.; Salgado-Jiménez, T.; Cabello, J.A.E. Autonomous underwater vehicles: localization, navigation, and communication for collaborative missions. *Appl. Sci.* **2020**, *10*, 1256. [CrossRef]
17. Yan, Z.; Wang, L.; Wang, T.; Yang, Z.; Chen, T.; Xu, J. Polar cooperative navigation algorithm for multi-unmanned underwater vehicles considering communication delays. *Sensors* **2018**, *18*, 1044. [CrossRef]
18. Xin, J.; Li, S.; Sheng, J.; Zhang, Y.; Cui, Y. Application of improved particle swarm optimization for navigation of unmanned surface vehicles. *Sensors* **2019**, *19*, 3096. [CrossRef]
19. Marzoughi, A.; Savkin, A.V. Autonomous navigation of a team of unmanned surface vehicles for intercepting intruders on a region boundary. *Sensors* **2021**, *21*, 297. [CrossRef] [PubMed]
20. Hoy, M.; Matveev, A.S.; Savkin, A.V. Algorithms for collision-free navigation of mobile robots in complex cluttered environments: a survey. *Robotica* **2015**, *33*, 463–497. [CrossRef]
21. Patle, B.; Pandey, A.; Parhi, D.; Jagadeesh, A. A review: On path planning strategies for navigation of mobile robot. *Def. Technol.* **2019**, *15*, 582–606. [CrossRef]
22. Matveev, A.S.; Savkin, A.V.; Hoy, M.; Wang, C. *Safe Robot Navigation Among Moving and Steady Obstacles*; Elsevier: Amsterdam, The Netherlands, 2015.
23. Polvara, R.; Sharma, S.; Wan, J.; Manning, A.; Sutton, R. Obstacle avoidance approaches for autonomous navigation of unmanned surface vehicles. *J. Navig.* **2018**, *71*, 241–256. [CrossRef]
24. Wu, X.; Chen, H.; Chen, C.; Zhong, M.; Xie, S.; Guo, Y.; Fujita, H. The autonomous navigation and obstacle avoidance for USVs with ANOA deep reinforcement learning method. *Knowl.-Based Syst.* **2020**, *196*, 105201. [CrossRef]
25. Li, H.; Savkin, A.V. Wireless sensor network based navigation of micro flying robots in the industrial internet of things. *IEEE Trans. Ind. Inform.* **2018**, *14*, 3524–3533. [CrossRef]
26. Wang, C.; Savkin, A.V.; Garratt, M. A strategy for safe 3D navigation of non-holonomic robots among moving obstacles. *Robotica* **2018**, *36*, 275–297. [CrossRef]
27. Elmokadem, T.; Savkin, A.V. A Hybrid Approach for Autonomous Collision-Free UAV Navigation in 3D Partially Unknown Dynamic Environments. *Drones* **2021**, *5*, 57. [CrossRef]
28. Elmokadem, T.; Savkin, A.V. A method for autonomous collision-free navigation of a quadrotor UAV in unknown tunnel-like environments. *Robotica* **2021**, 1–27. [CrossRef]
29. Zhang, W.; Wei, S.; Teng, Y.; Zhang, J.; Wang, X.; Yan, Z. Dynamic obstacle avoidance for unmanned underwater vehicles based on an improved velocity obstacle method. *Sensors* **2017**, *17*, 2742. [CrossRef] [PubMed]
30. Wang, S.M.; Fang, M.C.; Hwang, C.N. Vertical obstacle avoidance and navigation of autonomous underwater vehicles with H controller and the artificial potential field method. *J. Navig.* **2019**, *72*, 207–228. [CrossRef]
31. Yuan, J.; Wang, H.; Zhang, H.; Lin, C.; Yu, D.; Li, C. AUV Obstacle Avoidance Planning Based on Deep Reinforcement Learning. *J. Mar. Sci. Eng.* **2021**, *9*, 1166. [CrossRef]
32. Bhopale, P.; Kazi, F.; Singh, N. Reinforcement learning based obstacle avoidance for autonomous underwater vehicle. *J. Mar. Sci. Appl.* **2019**, *18*, 228–238. [CrossRef]
33. Thompson, F.; Guihen, D. Review of mission planning for autonomous marine vehicle fleets. *J. Field Robot.* **2019**, *36*, 333–354. [CrossRef]
34. Costanzi, R.; Fenucci, D.; Manzari, V.; Micheli, M.; Morlando, L.; Terracciano, D.; Caiti, A.; Stifani, M.; Tesei, A. Interoperability among unmanned maritime vehicles: Review and first in-field experimentation. *Front. Robot. AI* **2020**, *7*, 91. [CrossRef]
35. Huixi, X.; Jiang, C. Heterogeneous Oceanographic Exploration System Based on USV and AUV: A Survey of Developments and Challenges. *J. Univ. Chin. Acad. Sci.* **2021**, *38*, 145–159.
36. Ludvigsen, M. Collaborating robots sample the primary production in the ocean. *Sci. Robot.* **2021**, *6*, eabf4317. [CrossRef] [PubMed]

37. Cho, H.; Jeong, S.K.; Ji, D.H.; Tran, N.H.; Choi, H.S. Study on control system of integrated unmanned surface vehicle and underwater vehicle. *Sensors* **2020**, *20*, 2633. [CrossRef] [PubMed]
38. Shirakura, N.; Kiyokawa, T.; Kumamoto, H.; Takamatsu, J.; Ogasawara, T. Semi-automatic Collection of Marine Debris by Collaborating UAV and UUV. In Proceedings of the 2020 Fourth IEEE International Conference on Robotic Computing (IRC), Taichung, Taiwan, 9–11 November 2020; pp. 412–413.
39. Savkin, A.V.; Evans, R.J. *Hybrid Dynamical Systems: Controller and Sensor Switching Problems*; Birkhauser: Boston, MA, USA, 2002.
40. Petersen, I.R.; Savkin, A.V. *Robust Kalman Filtering for Signals and Systems with Large Uncertainties*; Birkhauser: Boston, MA, USA, 1999.
41. Matveev, A.S.; Savkin, A.V. *Estimation and Control over Communication Networks*; Birkhauser: Boston, MA, USA, 2009.
42. Centelles, D.; Soriano-Asensi, A.; Martí, J.V.; Marín, R.; Sanz, P.J. Underwater wireless communications for cooperative robotics with uwsim-net. *Appl. Sci.* **2019**, *9*, 3526. [CrossRef]
43. Petersen, I.R.; Ugrinovskii, V.A.; Savkin, A.V. *Robust Control Design Using H^∞ Methods*; Springer: Berlin/Heidelberg, Germany, 2000.

MDPI

Accuracy Assessment of a UAV Direct Georeferencing Method and Impact of the Configuration of Ground Control Points

Xiaoyu Liu [1], Xugang Lian [1,*], Wenfu Yang [2], Fan Wang [1], Yu Han [1] and Yafei Zhang [1]

1 School of Mining Engineering, Taiyuan University of Technology, Taiyuan 030024, China; Liuxiaoyu0586@link.tyut.edu.cn (X.L.); Wangfan1066@link.tyut.edu.cn (F.W.); Hanyu0747@link.tyut.edu.cn (Y.H.); Zhangyafei0754@link.tyut.edu.cn (Y.Z.)
2 Shanxi Provincial Key Lab of Resources, Environment and Disaster Monitoring, Shanxi Coal Geology Geophysical Surveying Exploration Institute, Jinzhong 030600, China; wcyywf@163.com
* Correspondence: lianxugang@tyut.edu.cn

Abstract: Unmanned aerial vehicles (UAVs) can obtain high-resolution topography data flexibly and efficiently at low cost. However, the georeferencing process involves the use of ground control points (GCPs), which limits time and cost effectiveness. Direct georeferencing, using onboard positioning sensors, can significantly improve work efficiency. The purpose of this study was to evaluate the accuracy of the Global Navigation Satellite System (GNSS)-assisted UAV direct georeferencing method and the influence of the number and distribution of GCPs. A FEIMA D2000 UAV was used to collect data, and several photogrammetric projects were established. Among them, the number and distribution of GCPs used in the bundle adjustment (BA) process were varied. Two parameters were considered when evaluating the different projects: the ground-measured checkpoints (CPs) root mean square error ($RMSE$) and the Multiscale Model to Model Cloud Comparison (M3C2) distance. The results show that the vertical and horizontal $RMSE$ of the direct georeferencing were 0.087 and 0.041 m, respectively. As the number of GCPs increased, the $RMSE$ gradually decreased until a specific GCP density was reached. GCPs should be uniformly distributed in the study area and contain at least one GCP near the center of the domain. Additionally, as the distance to the nearest GCP increased, the local accuracy of the DSM decreased. In general, UAV direct georeferencing has an acceptable positional accuracy level.

Keywords: unmanned aerial vehicle (UAV); structure from motion (SfM); direct georeferencing; ground control point (GCP); accuracy assessment; point cloud

Citation: Liu, X.; Lian, X.; Yang, W.; Wang, F.; Han, Y.; Zhang, Y. Accuracy Assessment of a UAV Direct Georeferencing Method and Impact of the Configuration of Ground Control Points. *Drones* **2022**, *6*, 30. https://doi.org/10.3390/drones6020030

Academic Editors: Arianna Pesci, Giordano Teza and Massimo Fabris

Received: 2 January 2022
Accepted: 17 January 2022
Published: 20 January 2022

Publisher's Note: MDPI stays neutral with regard to jurisdictional claims in published maps and institutional affiliations.

1. Introduction

As a new aerial survey platform, unmanned aerial vehicles (UAVs) have attracted increasing attention worldwide. Compared with methods based on satellite or airborne sensors, UAVs provide user-defined spatial and temporal resolution data at a relatively low cost, as well as flexible options for sensor use and data collection [1]. Images captured by UAVs and processed by structure-from-motion (SfM) photogrammetry are a combination of mature photogrammetry principles and modern computer vision technology [2] (hereafter UAV SfM). UAV SfM is widely used in Earth and Environmental Sciences to generate high-resolution topography (HRT) data [3,4], including precision agriculture [5,6], landslide monitoring [7,8], coastal change [9,10], and glacier dynamics [11,12].

Georeferencing is the process of referencing the results of bundle adjustment (BA) and photogrammetric processes to a specific coordinate system [13]. During bundle adjustment (BA), ground control point (GCP) coordinates on the ground are provided and measured in the images (indirect georeferencing) or known external elements of the images are directly used (direct georeferencing) [14]. Direct georeferencing requires cm-level positioning accuracy of the UAV, obtained by difference data. GCP field deployment, surveying, and recognition in images may require a significant amount of time and cost, while direct georeferencing based on IMU and GNSS can quickly collect data and significantly improve work efficiency. However, these benefits are only valuable if the accuracy of the directly georeferenced topographic product meets the requirements of the application. First, the GNSS mobile receiver on the UAV achieves centimeter-level positioning accuracy by receiving difference data provided by a virtual reference station (VRS), and then adds the position data solved by the fusion of real-time kinematic (RTK) and post-processed kinematic (PPK) into the BA process. Direct georeferencing requires obtaining the coordinates of a GNSS receiver at the exact moment the image is acquired, although RTK can provide high-accuracy single-point positioning. However, there might be distortions in signals from satellite constellations and interruptions in RTK connections for a fast-flying UAV. Fusion PPK technology can solve epoch data in the lockout period through a reverse Kalman filter to improve the fixed rate and positioning accuracy [15,16].

Considering the rapid development of UAV technology, it is necessary to evaluate the accuracy of UAV direct georeferencing methods with GNSS RTK/PPK technology. Many studies have evaluated the technology based on measured checkpoints on the ground. Padró et al. evaluated the data of a farm and showed that the horizontal and vertical *RMSE* of direct georeferencing were no more than 0.256 and 0.238 m, respectively [17]. Nolan et al. obtained data with a GSD of 10–20 cm in an area over tens of square kilometers and verified that the accuracy and precision (repeatability) of direct georeferencing were better than ±30 and ±8 cm, respectively, at 95% *RMSE* [18]. In the other two studies, the vertical *RMSE* did not exceed 10 and 20 cm, respectively [19,20]. The accuracy of the vertical direction is not always able to meet the requirements, and the accuracy can only reach the meter level in some studies [21,22]. Mian obtained a vertical *RMSE* of 40 cm using an image of 0.7 cm GSD [23]. Hugenholtz compared direct and indirect georeferencing methods, and recommended using GCPs where high accuracy is required [24].

The final product quality mainly depends on camera specifications, GCP configuration (accuracy, density and distribution), flight parameters (image overlap, GSD, etc.), land cover and terrain complexity, processing software, and flight platform (fixed wing or rotor wing). Measuring GCP is a time-consuming task, so a balance between appropriate accuracy and efficiency is needed. The literature to date shows consistently that accuracy increased with an increase in the number of GCPs and rapidly reached an asymptotic trend [25–27]. Conclusions differ among studies regarding the amount of GCPs required to produce a favorable outcome. In [28], 0.5~1 GCP per ha was the optimal GCP density, and GCPs were placed inside the area with stratified distribution to obtain the minimum total error. In [29], the optimal density was 1.8 GCPs per ha, uniformly distributed across the whole surface. Scott et al. reported setting control points in the center and edge of the study area, which is of great significance to reduce the height vertical error in spatial concentration [30].

The purpose of this study was to evaluate the accuracy of UAV RTK/PPK direct georeferencing and to determine whether the method could be used as a solution for rapid mapping applications with data generated from images of natural environments, including buildings, low vegetation and so on. The aim was to understand the difference in survey effectiveness between using direct georeferencing and GCPs. In addition, the effect of GCP quantity and distribution on the quality of the results was also studied. The evaluation was performed by calculating the vertical and horizontal *RMSE* of checkpoints on a digital surface model (DSM) and digital orthomosaic (DOM), respectively, and the Multiscale Model to Model Cloud Comparison (M3C2) distance based on point clouds [31]. The M3C2 method is the unique way to compute signed (and robust) distances directly between two point clouds.

2. Materials and Methods

The workflow of this study is shown in Figure 1. The study comprises four main steps: route planning (field survey, pre-flight, and setting flight parameters); data acquisition (ground GCP and CP layout and survey, UAV image acquisition); data processing considering different quantities of GCPs (BA and image-intensive matching); and horizontal and vertical quality assessment (data and error analysis).

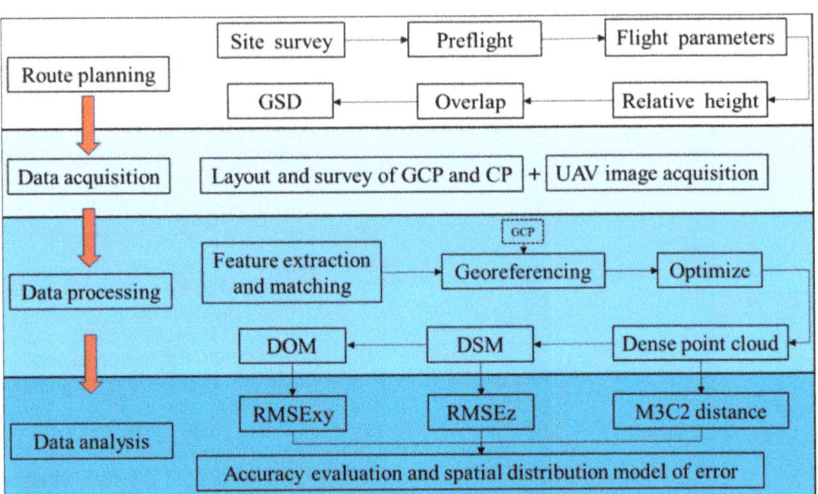

Figure 1. Workflow of methods, starting with the route planning of a UAV campaign until the determination of DSM and orthomosaic accuracy (*RMSE* and M3C2 distance).

2.1. The Study Area

The study area is located near Xishan, Taiyuan, Shanxi Province, People's Republic of China (Figure 2). The approximate coordinates in the geodetic reference system WGS84 are 112°27′5.64″ E and 37°52′0.98″ N. The region covers an area of approximately 0.5 km², with the highest point at 896 m and the lowest point at 840 m. The area features railways, factories, and low vegetation.

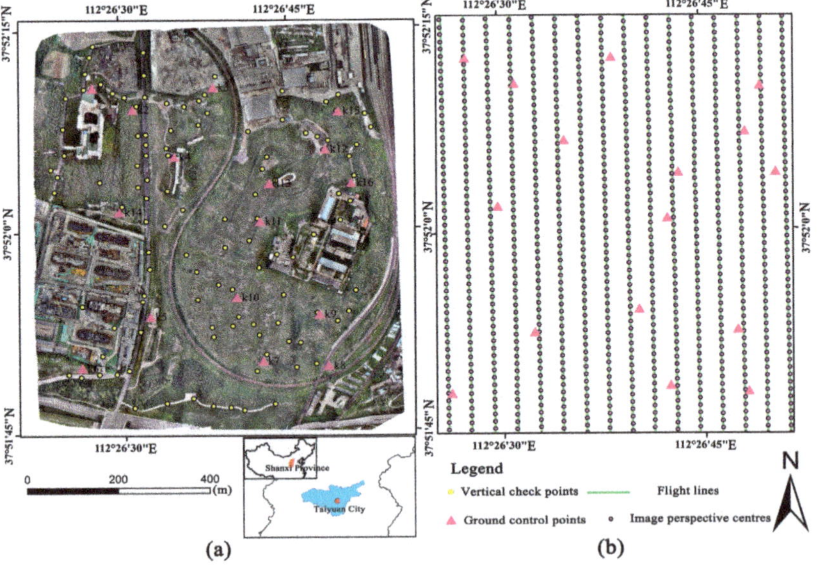

Figure 2. Study area with the GCP and checkpoint distributions. Projection coordinate system: WGS_1984_UTM_zone_49N; (**a**) DOM and (**b**) route map.

Accurate GCP coordinates are required for georeferencing UAV images. Sixteen ground survey markers were deployed, according to different photogrammetric projects—some of them were used as the input of the BA process, and the rest were used as horizontal checkpoints for cross verification of the horizontal accuracy. All control points were as evenly distributed as possible in the study area. In addition, coordinates of 120 points were obtained for vertical accuracy analysis of the generated DSM. Compared to CPs for horizontal accuracy assessment, vertical CPs do not require accurate identification in the image. Aerial markers consisted of 70×70 cm highly reflective red and yellow material fixed with nails to the ground, far away from high vegetation, buildings, and slopes. These markers were large enough to be identified in images and placed and measured prior to flight. A GNSS RTK receiver was used for field measurements. The receiver was connected to a virtual reference station and received differential signals through the network. Ten fixed solutions were recorded at each point, and the average was taken as the final result. Taking the average of the results of multiple measurements as the final result can exclude the influence of accidental factors and make the measurement results closer to the true value. At a control point marker, it takes about 3–5 min from deployment to the end of measurement. All coordinates were recorded in the WGS84 reference system. In the experiment, the means of the RTK coordinate residuals in the X, Y, and Z directions were 0.007, 0.006, and 0.012 m, respectively. The expected coordinate accuracy was higher than the spatial resolution of the UAV image (GSD about 1.7 cm). Figure 2 shows the position of GCPs and CPs relative to the study area.

2.2. Data Collection

In this study, a FEIMA D2000 multi-rotor UAV (Figure 3a) was used to collect data. This UAV included a fuselage, power motor, quick removal wing, differential antenna, magnetometer, and data transmission antenna. The 24.3-megapixel camera installed on the D2000 UAV provided a ground sampling distance (GSD) of 1.7 cm/pixel at an altitude of 110 m relative to the ground. The camera captured images at fixed intervals and stored them in JPG format. Detailed information about the UAV and camera are shown in Table 1.

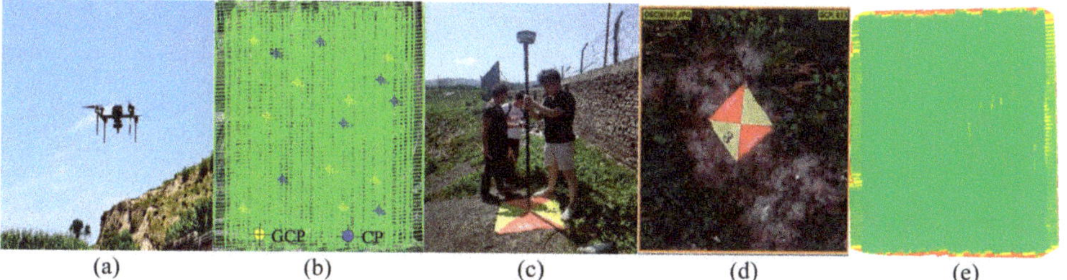

Figure 3. Experimental settings. (**a**) FEIMA D2000 UAV. (**b**) Distribution of GCPs and CPs in the bundle adjustment. (**c**) Measurement of field control points. (**d**) Display of control points in the image. (**e**) The number of overlapping images used in each pixel calculation. Red and yellow areas indicate low overlap, while green areas indicate more than 5 images per pixel.

Table 1. Technical specifications of the UAV platform.

UAV Body		D-CAM2000 Aerial Module	
System standard takeoff weight	2.8 kg	Camera	SONY a6000
Standard load	200 g	Effective pixels	24.3 million
Endurance	74 min	Sensor	23.5 × 15.6 mm (aps-c)
Remote control distance	20 km (max)	Focal length	25 mm

By pre-setting flight parameters in the UAV Manager software, the UAV flew autonomously from takeoff to landing. The ground control station consisted of computers, a ground-based data transmission radio, and antennas that communicated with the UAV to continuously monitor its flight status and allowed users to interrupt the flight if the UAV was in danger. The aerial survey was conducted on 10 August 2021 with clear weather and light wind.

2.3. Data Processing

In this study, Agisoft Photoscan Professional Version 1.7.0 [32] software was used to generate dense point clouds, a DSM, and a DOM based on the SfM algorithm. Before processing, position data were solved by the network RTK/PPK fusion differential job mode, which gave priority to the results of the PPK fixed solution, while for the non-fixed solution part of PPK, RTK fixed solution data were used for fusion, so as to ensure the quality of high-precision position data through complementary operation mode, and the RTK trajectory file was input during fusion difference resolution. Then, image EXIF was sequentially written—that is, the GPS positioning data were stored in the header file of the image, so that the GPS data could be directly read through the image in the software. The workflow in Photoscan is described as follows:

1. Image feature extraction and matching. The software automatically identifies many conspicuous points in each image, regardless of image scale or perspective, and similar feature points are recognized in multiple images. After locating the feature points in each image, similar feature points are recognized in multiple images. The quality of feature matching depends on the texture and overlap success of the image [33,34];
2. Iterative bundle adjustment. The purpose of BA is to determine internal and external orientation elements of the images by minimizing the reprojection errors between predicted and observed points, which can be converted into a nonlinear least-squares problem [35]. By applying the BA, the three-dimensional structure of the scene, the internal and external orientation elements of the camera are estimated at the same time;
3. Model optimization based on control points. GCPs provide additional external information about reconstructed scene geometry. The optimization process in Photoscan refines the camera position and reduces non-linear project deformations by incorporating GCPs [36];
4. Point cloud density matching. The MVS image matching algorithm operates on a single-pixel scale of the image to build dense clouds and increases the point density by several orders of magnitude;
5. Generate DSM and DOM. Using the dense point cloud as input, other results, such as DSM and DOM, can be produced. The outliers in the dense point cloud are removed before the dense point cloud is interpolated to generate DSM, and then the DOM is generated by digital differential correction based on DSM.

The images in all projects underwent the same photogrammetric processing, with differences in the number and distribution of GCPs used in the BA process. The number of GCPs in the experiment for evaluating the vertical *RMSE* ranged from 0 to 16, while the number for the horizontal *RMSE* ranged from 0 to 10. Seven other experiments evaluated the impact of the GCP distribution; the distribution schemes are shown in Figure 4.

2.4. Quality Assessment

Two methods were used to evaluate the quality of the UAV SfM results. The first method evaluates the accuracy of the DSM and DOM generated by different projects. The root mean square error (*RMSE*) was further calculated by comparing the CP coordinates estimated in the calculation results with the reference CP coordinates measured with the GNSS RTK receiver. Specifically, horizontal accuracy was verified on DOM, while vertical accuracy was verified by extracting the elevation of the corresponding DSM on a larger range of 120 vertical checkpoints.

$$RMSE_X = \sqrt{\frac{\sum_{i=1}^{n} \Delta x_i^2}{n}}, \tag{1}$$

$$RMSE_Y = \sqrt{\frac{\sum_{i=1}^{n} \Delta y_i^2}{n}}, \tag{2}$$

$$RMSE_Z = \sqrt{\frac{\sum_{i=1}^{n} \Delta z_i^2}{n}}, \tag{3}$$

where Δx, Δy, and Δz are the differences between RTK checkpoint coordinates and model-extracted coordinates, and n is the number of checkpoints. Calculation of the horizontal $RMSE$ ($RMSE_{XY}$) is as follows:

$$RMSE_{XY} = \sqrt{RMSE_X{}^2 + RMSE_Y{}^2}, \tag{4}$$

The second method uses the Multiscale Model to Model Cloud Comparison (M3C2) tool of the CloudCompare software version 2.10. [37], which calculates the distance of the reference cloud and comparison point cloud relative to the local surface normal direction through two parameters (user-defined normal proportion and projection proportion). The process runs directly on the point cloud and does not require grid partitioning, avoiding the uncertainties involved in the interpolation process.

To compare point clouds, the M3C2 distance between the reference point cloud generated by the 16 GCPs participating in BA and other project point clouds was calculated. Then, the mean and standard deviation of the M3C2 distance calculation were used to evaluate the accuracy and precision, respectively, of each point cloud. By plotting the error and its distribution curve, we determined the influence of the GCP distribution on the spatial distribution of the M3C2 distance difference and determined the possible pattern of the spatial distribution of error.

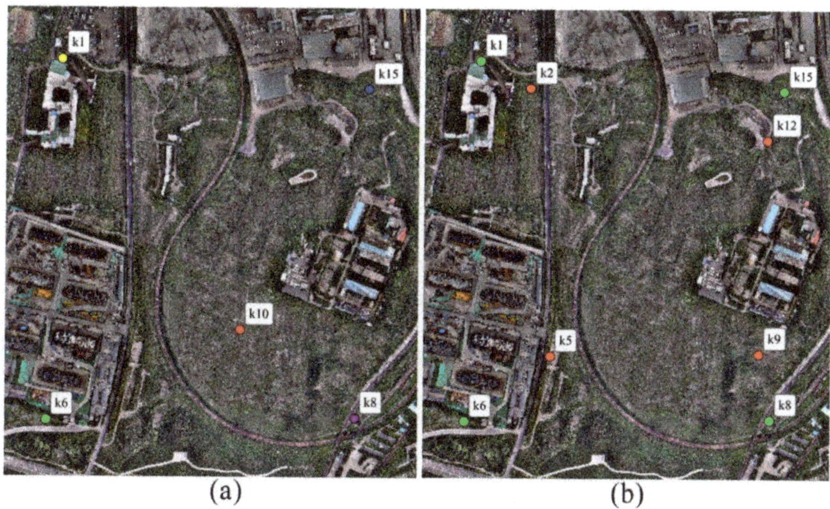

Figure 4. Schemes with different distributions of GCPs. (**a**) Five distributions of one GCP; (**b**) two distributions of four GCPs. The distributions are shown by different colours.

3. Results

3.1. Model Evaluation Based on RMSE

Based on the model extraction coordinates and measured checkpoint coordinates, the vertical and horizontal *RMSE* for each project were calculated using Equations (1)–(4). The GCP density was calculated from the number of GCPs used divided by the area investigated; the relationship between GCP density and *RMSE* was obtained, as shown in Figure 5. The results show that when the GCP density was highest, the vertical and horizontal *RMSE* were 0.032 and 0.015 m, respectively, representing approximately 1.88 and 0.88 GSD. When GCPs were not used, the vertical and horizontal *RMSE* increased to 0.087 and 0.041 m, respectively, representing a GSD of approximately 5.12 and 2.41. $RMSE_{XY}$ was less than $RMSE_Z$ in all projects. The mean ratio of $RMSE_Z$:$RMSE_{XY}$ is approximately 2.3. As the density of the GCP increased, the vertical and horizontal *RMSE* gradually decreased; this trend is fitted by the nonlinear curve in Figure 5. As shown in Figure 5a, when $d > 12 \, \text{GCP}/\text{km}^2$, the vertical *RMSE* did not significantly decrease (the change from the maximum was less than 20%). In Figure 5b, the same is true for $d > 10 \, \text{GCP}/\text{km}^2$ when considering the horizontal *RMSE* (the error does not show significant change).

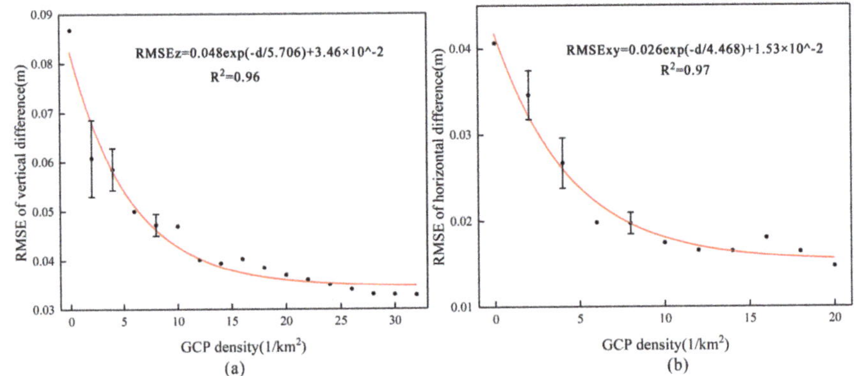

Figure 5. Effect of GCP density on the *RMSE* of (**a**) DSM vertical and (**b**) DOM horizontal. GCP density was calculated by dividing the quantity of GCPs used by the measurement area of 0.5 km². d represents the density of point clouds on the horizontal axis. Add error bars at ±1 standard error at GCP density of 2, 4 and 8, respectively.

There are some cases when the GCP density increased and the *RMSE* increased, which may be due to the error introduced by the ground measurement of the newly added GCPs. Our results for the horizontal error evaluation are influenced by manually setting the coordinates in the center of the GCPs. A similar uncertainty is expected when determining the coordinates of the GCP centers on the DOM.

Statistical analysis of the vertical differences of the three projects is shown in Figure 6. In Figure 6a–c, the histograms of the differences between RTK checkpoint elevations and DSM extraction elevations correspond to 0, 1, and 2 GCPs, respectively; all of the curves exhibit a Gaussian distribution. The mean value depicted in Figure 6a is −0.08 m, indicating a systematically biased distribution. The mean value in Figure 6b is −0.025 m, indicating that the addition of a single GCP is conducive to reducing the vertical error. When two GCPs are considered (Figure 6c), the mean value is 0.003 m, and the distribution is significantly improved. A single sample t-test was conducted, as shown in Table 2. The null hypothesis is that there is no significant difference between the mean and the test value 0. The significance of one GCP is less than 0.05, rejecting the original hypothesis, that is, there is a significant difference between the mean of one GCP and zero at a 95% confidence interval. The significance of two GCPs was 0.449, greater than 0.05, and the original hypothesis was accepted—that is, there was no significant difference between the mean of two GCPs and zero at a 95% confidence interval. Therefore, the DSM has systematic dome error in the vertical direction when zero or one GCP was used, and it is important to use at least two GCPs to improve DSM elevation accuracy. The standard deviation did not show a significant difference, with one GCP having the lowest standard deviation, at 0.033. Figure 6d shows the linear fit of the height difference between the DSM obtained by checkpoints and zero GCPs used; the coefficient of determination (R^2) is 0.99.

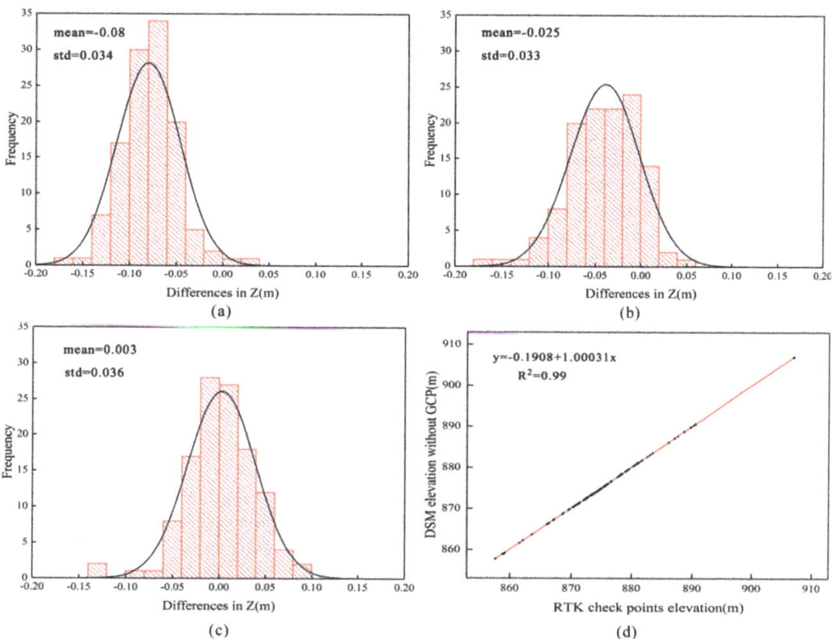

Figure 6. Statistical analysis of the UAV survey based on 120 checkpoints. (**a–c**) Histograms of elevation differences between the DSM and RTK checkpoints obtained by 0, 1, or 2 GCPs and their Gaussian fit. (**d**) Linear fitting function between RTK checkpoint elevations and DSM elevations obtained with 0 GCPs.

Table 2. Single sample *t*-test results.

Number of GCPs	t	Degree of Freedom	Significance	Mean Difference Value	95% Confidence Interval	
					Lower Limit	Upper Limit
1	−8.175	119	3.64×10^{-13}	−0.025	−0.031	−0.019
2	0.760	119	0.449	0.003	−0.004	0.009

3.2. Point Cloud Evaluation Based on M3C2 Distance

The M3C2 distance between the reference point cloud (16 GCPs of the BA process) and point cloud obtained from different projects was calculated. Mean and standard deviation were used as indicators to evaluate the accuracy and precision, respectively, of the point clouds of different projects (Table 3 and Figure 7). The mean distance between point clouds obtained by zero GCPs and the reference point cloud was 0.062 m. Adding a GCP at any position can reduce the mean distance, and the improvement is most obvious after adding K8 GCP in the lower right corner. Regardless of the distribution, when using two GCPs, the mean error can be reduced by about 50% compared with 0 GCPs. With an increase in the number of GCPs, the mean distance decreased, and the point cloud became more accurate. The mean distance decreased to a minimum of 0.01 m at eight and nine GCPs. The standard deviations of different projects were distributed between 0.02 and 0.03, showing no obvious difference. Several projects achieved a minimum standard deviation of 0.021 m.

Table 3. M3C2 distance values between reference point clouds and point clouds obtained by GCPs with different numbers and distributions. Refer to Figure 4 to identify these points.

Project	Mean (m)	std (m)	Project	Mean (m)	std (m)
16-0	0.062	0.025	16-3	0.027	0.022
16-1/k1	0.046	0.023	16-4/k2,k5,k9,k12	−0.015	0.028
16-1/k8	0.023	0.022	16-4/k1,k6,k8,k15	0.023	0.021
16-1/k15	0.053	0.025	16-5	0.021	0.021
16-1/k10	0.041	0.023	16-6	0.013	0.021
16-1/k6	0.031	0.022	16-7	0.016	0.021
16-2/k1, k10	0.032	0.023	16-8	0.01	0.023
16-2/k6, k15	0.024	0.028	16-9	0.01	0.021

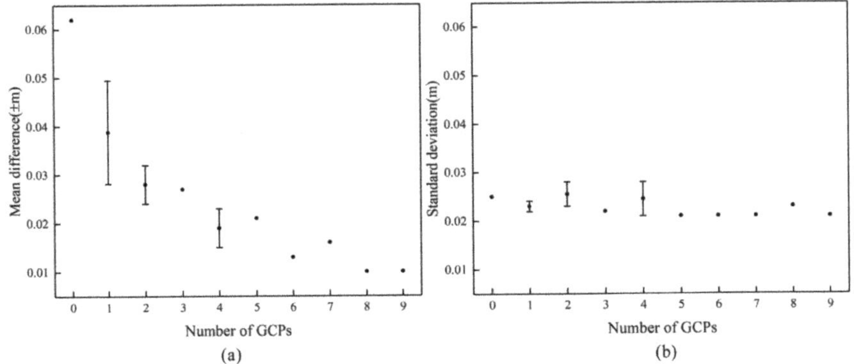

Figure 7. M3C2 distance between reference point clouds (16 GCPs) and point clouds obtained from different photogrammetric projects. (**a**) Average difference (accuracy). (**b**) Standard deviation (precision). Add error bars at +/−1 standard error at GCP number of 1, 2 and 4, respectively.

The error space distributions of the M3C2 distance between the reference point cloud and the point clouds of the 0-, 3-, 6-, and 9-GCP projects were calculated, and the results are displayed with the same legend and histogram distribution of the distance (Figure 8). A significant dome effect (red area) can be observed at zero GCPs in Figure 8a; the upper left corner of Figure 8b shows a clear error compared with the reference point cloud due to a lack of control; the error from Figure 8c to Figure 8d is not significantly reduced, which is only reflected in the mean distance (from 0.013 to 0.01 m).

Overall, the accuracy near the boundary of the study area is not as high as that in the center, which is related to the low overlap of images near the boundary (Figure 3e). There are obvious errors along the north–south road and the boundaries of buildings. The reasons for the large M3C2 distance error along the road are as follows: (1) tall plants on both sides of the road block the ground; (2) sample data from the road area are extracted, and the calculated point cloud density is approximately 12 per m^2, which is less than the average density of approximately 35 per m^2 in the overall study area. The uniform surface of the road (asphalt pavement) lacks adequate feature points, resulting in a sparse area within the point cloud. The M3C2 distance increased at the boundaries of buildings due to the sudden change in topographic characteristics.

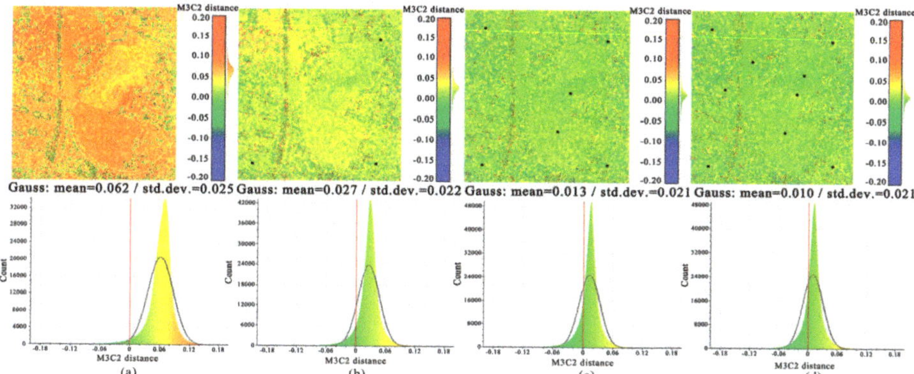

Figure 8. M3C2 distance between the point cloud obtained by using different numbers of GCPs as control points and the reference point cloud (obtained by using all 16 GCPS). (**a–d**) Zero-, 3-, 6-, and 9-GCP and Gaussian distribution. Solid black dots indicate the location of GCPs. The vertical red line in the figure are zero error lines.

3.3. Influence of GCPs Distribution

The experimental results of evaluating the influence of the GCP distribution are shown in Figure 9. The box diagram represents the difference between the elevation of checkpoints and the elevation of the DSM extraction point. In the layout of one GCP, four tests (excluding K15) showed low variability, in which K10 was near the center of the region and the error was the smallest. In the experiments with different distributions of four GCPs, compared with the results of using K1, 6, 8, and 15, the error of using K2, 5, 9, and 12 was significantly reduced. The difference is that the distance between K2, 5, 9, and 12 decreased.

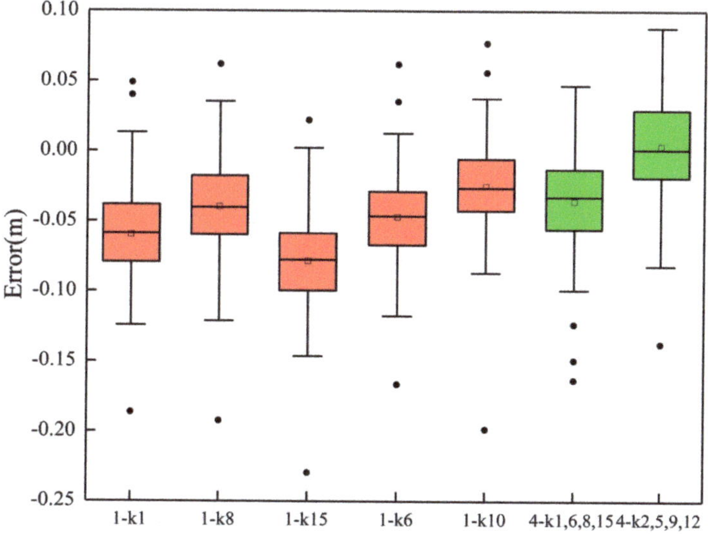

Figure 9. Boxplots represent vertical errors obtained by seven layout schemes divided by the number of GCPs (the horizontal line represents the median, the lower part of the hinge represents the 25th percentile, and the upper part represents the 75th percentile). The whiskers represent 1.5× IQR (interquartile range) in both directions. The black squares and circles represent the mean and outliers, respectively. Refer to the schematic diagram of different GCP distribution in Figure 4. 1 and k1 in 1-k1 represent the number and code of control points respectively, and others are similar.

In addition, ArcGIS was used for differential calculations. The difference was assessed by subtracting the DSM produced using a uniform distribution of five GCPs and the DSM produced using the optimal distribution of 16 GCPs; the differential DSM resolution was 0.2 m (Figure 10a). The results show that when the distance from the nearest GCP increases, the DSM local vertical difference tends to increase. There are many scattered points in Figure 10b, indicating that the local DSM accuracy is not solely determined by the distance to the nearest GCP; therefore, a nonlinear curve is used for fitting.

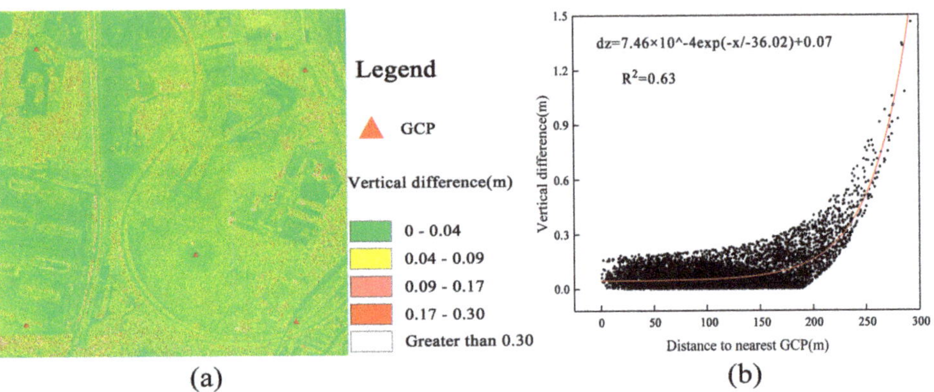

Figure 10. Relationship between local DSM accuracy and nearest GCP. (**a**) Differential DSM; (**b**) the scatterplot and nonlinear curve fit obtained by sampling from the differential DSM.

4. Discussion

Using direct georeferencing can significantly improve the efficiency of UAV measurement. This study evaluated the geospatial accuracy of photogrammetric products obtained by direct georeferencing, and the impact of the configuration of GCPs on the quality of results. The *RMSE*, the M3C2 distance of the point cloud, the GCP distribution, and measures to improve the accuracy of the results when direct georeferencing are discussed below.

1. Error analysis based on the *RMSE* shows that adding one GCP can help to reduce the deviation, but there may still be dome error, as shown in the study of Rosnell and Javernick [27,38]. When two GCPs were used, the mean vertical difference was reduced to 0.003 m, and the horizontal *RMSE* was 0.0198 m, approximately 1.1 GSD, when 3 GCPs were used. From the work in some natural environments to the investigation of infrastructure, such as modelling water runoff during rain, different projects have different requirements for the fineness of ground features, so the required accuracy depends on the purpose of generating DSM. Therefore, using only one GCP may not meet the high accuracy standards; two to three GCPs are recommended for a trade-off between accuracy and work efficiency. The influence of each error on *RMSE* is directly proportional to the size of the square error, and therefore *RMSE* is sensitive to large differences and does not reflect terrain changes. Because of scale differences, errors that do not occur in flat areas may also occur in sloped areas [15]. The natural environment presents a series of complexities, including changing vegetation cover, strong topographic relief, and changes in texture. Future studies will need to assess the impact of these complexities on the accuracy of the results. Calculating *RMSE* is a common error assessment method when the actual dataset of the ground surface is a set of distribution points rather than a continuous, real surface. Error evaluation benefits from a larger number and more evenly distributed checkpoints. Gomes et al. arranged 270 vertical checkpoints in an area of about 0.22 km², with a density of 1227 checkpoints per km² [39]. In the study of Tomaštík, the density of checkpoints at the three sites was approximately 11363,

3674 and 2749 per km^2 [8]. Thus, even if the number of GCP measurements on the ground is minimized, the role of checkpoints in the error assessment is critical. In the future, the plan is to deploy as many checkpoints as possible in the study area, build an accurate error surface, analyze the spatial distribution characteristics of errors, and better verify the results; these tasks will help to understand and reduce the potential error sources in the UAV SfM workflow [36];

2. The M3C2 algorithm eliminates the error introduced by the interpolation process. Lower error measurements with M3C2 are comparable to point-to-point or point-to-mesh; the method has been widely used in research based on point cloud change detection [40–43]. Due to the high point density of UAV matching point clouds, an intercomparison can actually be regarded as continuous [44]. At zero GCP, the M3C2 distance error shows randomness in the study area. The standard deviation range of the comparison between the reference point cloud and the point cloud of different projects is 0.021–0.028 m. The change is relatively small, and the point cloud only deviates in the vertical direction, which is similar to the study of Tomaštík and Štroner et al. [6,44]. Standard deviation is an indicator of precision. In some applications of UAV SfM, the accuracy of geolocation is not as important as the repeatability (precision) of data. For example, comparing multi temporal measurement data to study the change in terrain with time, more attention is paid to the relative change between data, and the quality of multi temporal data can be improved through cooperative registration;

3. GCP distribution experiments show that the uniform distribution of GCPs is crucial when using more than one GCP. Figure 10b uses nonlinear curve fitting. In Gindraux's study, linear fitting was used to determine that, on average, the vertical accuracy decreased by 0.09 m when the distance from the nearest GCP increased by 100 m [45]. In the experiment with four GCPs, the accuracy was improved after moving the GCP slightly towards the center of the study area compared with placing the GCP at the edge of the study area, which is similar to the study of Martínez. Martínez's study concluded that the best horizontal accuracies are achieved by placing GCPs around the edges of the study area, but it is also essential to place GCPs inside the area with a stratified distribution to optimize vertical accuracy [28];

4. The DSM vertical *RMSE* and DOM horizontal *RMSE* obtained by direct georeferencing with the GSD set to 1.7 cm/pixel were 0.087 and 0.041 m, respectively. Without GCPs, the accuracy of the results was highly dependent on the accuracy of the image position data. The following measures can be taken to improve the accuracy of results obtained without GCPs. UAV cross flights and imagery with large overlap can provide redundant data and improve the reliability of image matching, which requires high computing power [1,39]. The addition of oblique images helps to accurately estimate the internal and external orientation elements in the process of bundle adjustment, extract vertical features such as building sidewalls, and obtain the best vertical accuracy [13]. A more accurate GCP measurement method can be used, rather than simply increasing the number of GCPs. Another recommendation is to use a tripod-mounted prism instead of a pole-mounted prism, as well as the RTK static measurement method when time permits. If inaccurate coordinates are introduced when measuring GCPs, a more complex error surface will be introduced, as opposed to reducing the initial deformation [14].

The direct georeferencing technology integrated by UAV and GNSS RTK has advantages in monitoring locations with large ranges or difficult access. However, further development is needed. High-quality optical lenses and multi-frequency GPS can obtain higher quality images and positioning accuracy.

5. Conclusions

In this study, the geospatial accuracy of photogrammetric products obtained by the FEIMA D2000 direct georeferencing method was evaluated using ground-measured data.

In addition, we evaluated the effect of the quantity and spatial distribution of GCPs on the quality of the results. The research results are summarized as follows:

1. UAV SfM is a flexible and efficient method to obtain high-resolution topographic data. The direct georeferencing method based on the RTK/PPK fusion difference to obtain high-accuracy image positions has potential for improving the accuracy of the products, especially when GPS measurements are difficult, as well as reducing the dependence on the GCPs in the bundle adjustment, and decreasing the field work time and cost.
2. The research results show that the vertical *RMSE* of the DSM obtained by direct georeferencing was 0.087 m, approximately equal to 5.12 GSD. The horizontal *RMSE* of the DOM was 0.041 m, approximately equal to 2.41 GSD. Both values reached the centimeter positioning accuracy and achieve the application research of decimeter-error scale. The accuracy of UAV direct georeferencing could be guaranteed through careful flight planning, an appropriate survey, and accurate data post-processing. In the study of terrain change detection, we suggest evenly deploying two to three GCPs to achieve a good compromise between appropriate accuracy, repeatability and efficiency.
3. GCPs should be uniformly distributed in the study area and contain at least one GCP near the center of the domain to reduce the dome effect. With an increase in the number of GCPs in the bundle adjustment, both the horizontal error and vertical error decreased, and the horizontal error was always lower than the vertical error. When the density of the GCPs was greater than 12 GCP/km^2 and 10 GCP/km^2, respectively, the decrease in the vertical and horizontal errors was not obvious. The minimum vertical and horizontal *RMSE* were 0.032 (~1.88 GSD) and 0.015 m (~0.88 GSD), respectively.

Author Contributions: Conceptualization, X.L. (Xiaoyu Liu) and X.L. (Xugang Lian); methodology, X.L. (Xiaoyu Liu) and X.L. (Xugang Lian); software, F.W.; validation, F.W., Y.H. and Y.Z.; formal analysis, X.L. (Xugang Lian); investigation, X.L. (Xiaoyu Liu); resources, X.L. (Xugang Lian) and W.Y.; data curation, Y.Z.; writing—original draft preparation, X.L. (Xiaoyu Liu); writing—review and editing, X.L. (Xugang Lian); visualization, Y.H.; supervision, X.L. (Xugang Lian); project administration, W.Y.; funding acquisition, X.L. (Xugang Lian) and W.Y. All authors have read and agreed to the published version of the manuscript.

Funding: This research was funded by the National Natural Science Foundation of China, grant numbers 42101414 and 51704205; the Basic Applied Research Projects of Shanxi Province, grant number 201901D111466; the Key projects of Shanxi coal based low carbon joint fund of National Natural Science Foundation of China, grant number U1810203.

Institutional Review Board Statement: Not applicable.

Informed Consent Statement: Not applicable.

Acknowledgments: We would like to thank the reviewers for their helpful comments.

Conflicts of Interest: The authors declare no conflict of interest.

References

1. Zhang, H.; Aldana-Jague, E.; Clapuyt, F.; Wilken, F.; Vanacker, V.; Van Oost, K. Evaluating the potential of post-processing kinematic (PPK) georeferencing for UAV-based structure-from-motion (SfM) photogrammetry and surface change detection. *Earth Surf. Dyn.* **2019**, *7*, 807–827. [CrossRef]
2. Ullman, S. The interpretation of structure from motion. *Proc. R. Soc. Lond. Ser. B Biol. Sci.* **1979**, *203*, 405–426.
3. Eltner, A.; Kaiser, A.; Castillo, C.; Rock, G.; Neugirg, F.; Abellán, A. Image-based surface reconstruction in geomorphometry—merits, limits and developments. *Earth Surf. Dyn.* **2016**, *4*, 359–389. [CrossRef]
4. James, M.R.; Robson, S. Straightforward reconstruction of 3D surfaces and topography with a camera: Accuracy and geoscience application. *J. Geophys. Res. Earth Surf.* **2012**, *117*, F3. [CrossRef]
5. Kim, D.-W.; Yun, H.S.; Jeong, S.-J.; Kwon, Y.-S.; Kim, S.-G.; Lee, W.S.; Kim, H.-J. Modeling and testing of growth status for Chinese cabbage and white radish with UAV-based RGB imagery. *Remote Sens.* **2018**, *10*, 563. [CrossRef]
6. Tomaštík, J.; Mokroš, M.; Saloň, Š.; Chudý, F.; Tunák, D. Accuracy of photogrammetric UAV-based point clouds under conditions of partially-open forest canopy. *Forests* **2017**, *8*, 151. [CrossRef]

7. Lian, X.; Li, Z.; Yuan, H.; Hu, H.; Cai, Y.; Liu, X. Determination of the Stability of High-Steep Slopes by Global Navigation Satellite System (GNSS) Real-Time Monitoring in Long Wall Mining. *Appl. Sci.* **2020**, *10*, 1952. [CrossRef]

8. Godone, D.; Allasia, P.; Borrelli, L.; Gullà, G. UAV and Structure from Motion Approach to Monitor the Maierato Landslide Evolution. *Remote Sens.* **2020**, *12*, 1039. [CrossRef]

9. Long, N.; Millescamps, B.; Guillot, B.; Pouget, F.; Bertin, X. Monitoring the topography of a dynamic tidal inlet using UAV imagery. *Remote Sens.* **2016**, *8*, 387. [CrossRef]

10. Long, N.; Millescamps, B.; Pouget, F.; Dumon, A.; Lachaussée, N.; Bertin, X. Accuracy assessment of coastal topography derived from UAV images. *Int. Arch. Photogramm. Remote Sens. Spat. Inf. Sci.* **2016**, *41*, B1.

11. Immerzeel, W.W.; Kraaijenbrink, P.D.A.; Shea, J.; Shrestha, A.; Pellicciotti, F.; Bierkens, M.F.P.; De Jong, S.M. High-resolution monitoring of Himalayan glacier dynamics using unmanned aerial vehicles. *Remote Sens. Environ.* **2014**, *150*, 93–103. [CrossRef]

12. Mallalieu, J.; Carrivick, J.L.; Quincey, D.J.; Smith, M.W.; James, W.H. An integrated Structure-from-Motion and time-lapse technique for quantifying ice-margin dynamics. *J. Glaciol.* **2017**, *63*, 937–949. [CrossRef]

13. Zeybek, M. Accuracy assessment of direct georeferencing UAV images with onboard global navigation satellite system and comparison of CORS/RTK surveying methods. *Meas. Sci. Technol.* **2021**, *32*, 065402. [CrossRef]

14. Sanz-Ablanedo, E.; Chandler, J.H.; Rodríguez-Pérez, J.R.; Ordóñez, C. Accuracy of unmanned aerial vehicle (UAV) and SfM photogrammetry survey as a function of the number and location of ground control points used. *Remote Sens.* **2018**, *10*, 1606. [CrossRef]

15. Cucchiaro, S.; Fallu, D.J.; Zhang, H.; Walsh, K.; Van Oost, K.; Brown, A.G.; Tarolli, P. Multiplatform-SfM and TLS data fusion for monitoring agricultural terraces in complex topographic and landcover conditions. *Remote Sens.* **2020**, *12*, 1946. [CrossRef]

16. McMahon, C.; Mora, O.E.; Starek, M.J. Evaluating the Performance of sUAS Photogrammetry with PPK Positioning for Infrastructure Mapping. *Drones* **2021**, *5*, 50. [CrossRef]

17. Padró, J.C.; Muñoz, F.J.; Planas, J.; Pons, X. Comparison of four UAV georeferencing methods for environmental monitoring purposes focusing on the combined use with airborne and satellite remote sensing platforms. *Int. J. Appl. Earth Obs. Geoinf.* **2019**, *75*, 130–140. [CrossRef]

18. Nolan, M.; Larsen, C.; Sturm, M. Mapping snow depth from manned aircraft on landscape scales at centimeter resolution using structure-from-motion photogrammetry. *Cryosphere* **2015**, *9*, 1445–1463. [CrossRef]

19. Tomaštík, J.; Mokroš, M.; Surový, P.; Grznárová, A.; Merganič, J. UAV RTK/PPK method—An optimal solution for mapping inaccessible forested areas? *Remote Sens.* **2019**, *11*, 721. [CrossRef]

20. Erenoglu, R.C.; Erenoglu, O. A case study on the comparison of terrestrial methods and unmanned aerial vehicle technique in landslide surveys: Sarıcaeli landslide, Çanakkale, NW Turkey. *Int. J. Environ. Geoinf.* **2018**, *5*, 325–336. [CrossRef]

21. Turner, D.; Lucieer, A.; Watson, C. An automated technique for generating georectified mosaics from ultra-high resolution unmanned aerial vehicle (UAV) imagery, based on structure from motion (SfM) point clouds. *Remote Sens.* **2012**, *4*, 1392–1410. [CrossRef]

22. Shahbazi, M.; Sohn, G.; Théau, J.; Menard, P. Development and evaluation of a UAV-photogrammetry system for precise 3D environmental modeling. *Sensors* **2015**, *15*, 27493–27524. [CrossRef]

23. Mian, O.; Lutes, J.; Lipa, G.; Hutton, J.J.; Gavelle, E.; Borghini, S. Accuracy assessment of direct georeferencing for photogrammetric applications on small unmanned aerial platforms. *Int. Arch. Photogramm. Remote Sens. Spat. Inf. Sci.* **2016**, *40*, 77. [CrossRef]

24. Hugenholtz, C.; Brown, O.; Walker, J.; Barchyn, T.E.; Nesbit, P.; Kucharczyk, M.; Myshak, S. Spatial accuracy of UAV-derived orthoimagery and topography: Comparing photogrammetric models processed with direct geo-referencing and ground control points. *Geomatica* **2016**, *70*, 21–30. [CrossRef]

25. Agüera-Vega, F.; Carvajal-Ramírez, F.; Martínez-Carricondo, P. Assessment of photogrammetric mapping accuracy based on variation ground control points number using unmanned aerial vehicle. *Measurement* **2017**, *98*, 221–227. [CrossRef]

26. Tahar, K.N. An evaluation on different number of ground control points in unmanned aerial vehicle photogrammetric block. *Int. Arch. Photogramm. Remote Sens. Spat. Inf. Sci.* **2013**, *40*, 93–98. [CrossRef]

27. Rosnell, T.; Honkavaara, E. Point cloud generation from aerial image data acquired by a quadrocopter type micro unmanned aerial vehicle and a digital still camera. *Sensors* **2012**, *12*, 453–480. [CrossRef]

28. Martínez-Carricondo, P.; Agüera-Vega, F.; Carvajal-Ramírez, F.; Mesas-Carrascosa, F.J.; García-Ferrer, A.; Pérez-Porras, F.-J. Assessment of UAV-photogrammetric mapping accuracy based on variation of ground control points. *Int. J. Appl. Earth Obs. Geoinf.* **2018**, *72*, 1–10. [CrossRef]

29. Reshetyuk, Y.; Mårtensson, S.G. Generation of highly accurate digital elevation models with unmanned aerial vehicles. *Photogramm. Rec.* **2016**, *31*, 143–165. [CrossRef]

30. Stott, E.; Williams, R.D.; Hoey, T.B. Ground control point distribution for accurate kilometre-scale topographic mapping using an RTK-GNSS unmanned aerial vehicle and SfM photogrammetry. *Drones* **2020**, *4*, 55. [CrossRef]

31. Lague, D.; Brodu, N.; Leroux, J. Accurate 3D comparison of complex topography with terrestrial laser scanner: Application to the Rangitikei canyon (NZ). *ISPRS J. Photogram. Remote Sens.* **2013**, *82*, 10–26. [CrossRef]

32. Agisoft LCC. Agisoft PhotoScan. Available online: http://www.agisoft.com (accessed on 20 February 2017).

33. Yu, Z.; Zhou, H.; Li, C. Fast non-rigid image feature matching for agricultural UAV via probabilistic inference with regularization techniques. *Comput. Electron. Agric.* **2017**, *143*, 79–89. [CrossRef]

34. Jiang, S.; Jiang, W. On-board GNSS/IMU assisted feature extraction and matching for oblique UAV images. *Remote Sens.* **2017**, *9*, 813. [CrossRef]
35. Snavely, N.; Seitz, S.M.; Szeliski, R. Modeling the world from internet photo collections. *Int. J. Comput. Vis.* **2008**, *80*, 189–210. [CrossRef]
36. James, M.R.; Robson, S.; d'Oleire-Oltmanns, S.; Niethammer, U. Optimising UAV topographic surveys processed with structure-from-motion: Ground control quality, quantity and bundle adjustment. *Geomorphology* **2017**, *280*, 51–66. [CrossRef]
37. CloudCompare v2.10.2. Available online: https://www.danielgm.net/cc/ (accessed on 18 July 2020).
38. Javernick, L.; Brasington, J.; Caruso, B. Modeling the topography of shallow braided rivers using Structure-from-Motion photogrammetry. *Geomorphology* **2014**, *213*, 166–182. [CrossRef]
39. Pessoa, G.G.; Carrilho, A.C.; Miyoshi, G.T.; Amorim, A.; Galo, M. Assessment of UAV-based digital surface model and the effects of quantity and distribution of ground control points. *Int. J. Remote Sens.* **2021**, *42*, 65–83. [CrossRef]
40. Stumpf, A.; Malet, J.P.; Allemand, P.; Deseilligny, M.P.; Skupinski, G. Ground-based multi-view photogrammetry for the monitoring of landslide deformation and erosion. *Geomorphology* **2015**, *231*, 130–145. [CrossRef]
41. Cucchiaro, S.; Cavalli, M.; Vericat, D.; Crema, S.; Llena, M.; Beinat, A.; Marchi, L.; Cazorzi, F. Monitoring topographic changes through 4D-structure-from-motion photogrammetry: Application to a debris-flow channel. *Environ. Earth Sci.* **2018**, *77*, 632. [CrossRef]
42. Milan, D.J.; Heritage, G.L.; Large, A.R.G.; Fuller, I.C. Filtering spatial error from DEMs: Implications for morphological change estimation. *Geomorphology* **2011**, *125*, 160–171. [CrossRef]
43. Cook, K.L. An evaluation of the effectiveness of low-cost UAVs and structure from motion for geomorphic change detection. *Geomorphology* **2017**, *278*, 195–208. [CrossRef]
44. Štroner, M.; Urban, R.; Reindl, T.; Seidl, J.; Brouček, J. Evaluation of the georeferencing accuracy of a photogrammetric model using a quadrocopter with onboard GNSS RTK. *Sensors* **2020**, *20*, 2318. [CrossRef] [PubMed]
45. Gindraux, S.; Boesch, R.; Farinotti, D. Accuracy assessment of digital surface models from unmanned aerial vehicles' imagery on glaciers. *Remote Sens.* **2017**, *9*, 186. [CrossRef]

Article

Low-Altitude Sensing of Urban Atmospheric Turbulence with UAV

Alexander Shelekhov [1,*], Alexey Afanasiev [2], Evgeniya Shelekhova [1], Alexey Kobzev [1], Alexey Tel'minov [1], Alexander Molchunov [1] and Olga Poplevina [1]

1   Institute of Monitoring of Climatic and Ecological Systems SB RAS, 10/3, Academichesky Ave, 634055 Tomsk, Russia; sea1125@mail.ru (E.S.); alexey-kobzev@mail.ru (A.K.); talexey@imces.ru (A.T.); anm@imces.ru (A.M.); olgalevina1711@mail.ru (O.P.)
2   V.E. Zuev Institute of Atmospheric Optics SB RAS, 1, Academician Zuev square Tomsk, 634055 Tomsk, Russia; afanasiev@iao.ru
*   Correspondence: ash@imces.ru; Tel.: +7-952-883-9923

Abstract: The capabilities of a quadcopter in the hover mode for low-altitude sensing of atmospheric turbulence with high spatial resolution in urban areas characterized by complex orography are investigated. The studies were carried out in different seasons (winter, spring, summer, and fall), and the quadcopter hovered in the immediate vicinity of ultrasonic weather stations. The DJI Phantom 4 Pro quadcopter and AMK-03 ultrasonic weather stations installed in different places of the studied territory were used in the experiment. The smoothing procedure was used to study the behavior of the longitudinal and lateral spectra of turbulence in the inertial and energy production ranges. The longitudinal and lateral turbulence scales were estimated by the least-square fit method with the von Karman model as a regression curve. It is shown that the turbulence spectra obtained with DJI Phantom 4 Pro and AMK-03 generally coincide, with minor differences observed in the high-frequency region of the spectrum. In the inertial range, the behavior of the turbulence spectra shows that they obey the Kolmogorov–Obukhov "5/3" law. In the energy production range, the longitudinal and lateral turbulence scales and their ratio measured by DJI Phantom 4 Pro and AMK-03 agree to a good accuracy. Discrepancies in the data obtained with the quadcopter and the ultrasonic weather stations at the territory with complex orography are explained by the partial correlation of the wind velocity series at different measurement points and the influence of the inhomogeneous surface.

Keywords: quadcopter; ultrasonic weather station; turbulence; longitudinal and lateral spectra; scales; urban environment

Citation: Shelekhov, A.; Afanasiev, A.; Shelekhova, E.; Kobzev, A.; Tel'minov, A.; Molchunov, A.; Poplevina, O. Low-Altitude Sensing of Urban Atmospheric Turbulence with UAV. *Drones* **2022**, *6*, 61. https://doi.org/10.3390/drones6030061

Academic Editors: Arianna Pesci, Giordano Teza and Massimo Fabris

Received: 14 February 2022
Accepted: 18 February 2022
Published: 27 February 2022

Publisher's Note: MDPI stays neutral with regard to jurisdictional claims in published maps and institutional affiliations.

1. Introduction

Recently, unmanned aerial vehicles (UAVs) have become widespread in our life, and they are now an important component of the airspace. Analysis and forecast of modern trends in their use show that, in the next decade, there will be an explosive growth in the number of commercial and military UAVs. This growth will require efficient systems capable of controlling UAV traffic, in particular, under bad weather conditions [1–6]. Atmospheric turbulence is the main factor that most strongly affects the efficiency of future UAV traffic management systems. Many UAVs are small in size and light in weight. As a result, their trajectory can deviate significantly in a turbulent atmosphere, and loss of control is highly probable in this case.

Most UAVs fly at altitudes of up to 500 m. The atmospheric boundary layer at these altitudes is considerably affected by local orography. For example, for flights in an urban environment, i.e., an environment with complex orography, atmospheric turbulence is characterized by strong spatial inhomogeneity due to the presence of buildings, park zones, highways, etc.

The main technologies to obtain information about turbulence profiles of the urban atmosphere are lidar, sodar, and radar sensing methods. For these sensing technologies, the spatial resolution is determined by the size of the sensing volume. It ranges from a few tens to hundreds of meters [7–15]. Spatial variations of turbulent air flows can reach several meters, which is much smaller than the spatial resolution of lidars, sodars, and radars. This discrepancy between the spatial resolution of the applied technologies and the scale of turbulent flows in the atmosphere can lead to their significant averaging and, consequently, to significant errors in atmospheric turbulence measurements. Thus, the future UAV traffic management systems must use data on the turbulent atmosphere obtained with high spatial resolution.

Acoustic anemometry methods can be used to obtain atmospheric data with high spatial resolution [11,16,17]. Complete information about the state of the atmosphere at different heights can be obtained with acoustic anemometry methods if acoustic devices are set on weather towers or a tethered balloon, which is not always possible in an urban environment.

One of the main trends in low-altitude sensing, i.e., to heights of about 500 m, is the development of methods for diagnostics of the turbulent atmosphere with UAVs. The results of diagnostics of the speed of air mass flows with UAVs are reported in [18–35]. In [36–47], the fundamental possibility of measuring the turbulence spectra with fixed-wing UAVs of various sizes and weights was shown. In the process of sensing, UAVs of this type move in space for a long time. The typical flight pattern in a turbulence measurement experiment consists of straight sections about one kilometer long [47]. As a result, the spatial resolution in measurement of the turbulence spectrum is approximately comparable with the length of a straight section. Thus, for the non-even and non-homogeneous underlying surface and in the non-stationary atmosphere, the use of a fixed-wing UAV can lead, as in the case of lidars, sodars, and radars, to significant averaging and, as a consequence, to significant errors in measurements of atmospheric turbulence.

In contrast to a fixed-wing UAV, a quadcopter can hover at a needed point in space for a long time and obtain atmospheric data with the high spatial resolution in an area characterized by complex orography. The results of studying the turbulence spectrum in the inertial and energy production ranges with the DJI Mavic Mini quadcopter in the altitude hold mode are reported in [48]. The measurements were carried out over territory characterized by a flat and uniform surface with a slight slope; it bordered a cottage village on one side and a forest on the other side. The results obtained are in a good agreement with the theory of homogeneous and isotropic turbulence and the data measured by the methods of acoustic anemometry.

Knowledge of the state of atmospheric turbulence allows us to study its effect on the efficiency of UAV management systems. It is well-known that turbulence reduces the possibility of efficient use of wind energy and causes accelerated wear. Techniques of low-altitude sensing of the atmospheric state, including turbulence, allow us to estimate the climate change caused by urban growth and are needed to address current and future urbanization challenges.

In addition to turbulent fluctuations of the wind velocity field in the atmosphere, there are random temperature oscillations. These oscillations lead to random fluctuations of the refractive index, which should be taken into account when studying the propagation of optical radiation in the atmosphere. Turbulent fluctuations of the wind velocity field and temperature in the atmosphere are correlated. For example, the relation between the turbulence scales that characterize fluctuations of wind velocity and temperature is given in [49]. The results of investigation of the structure characteristic of turbulent fluctuations of the refractive index with UAV and acoustic anemometry methods are reported in [50,51].

When atmospheric turbulence is studied over a territory with complex orography, diagnostic methods providing data with high spatial resolution are preferable. As already mentioned, a quadcopter—in contrast to fixed-wing UAV [36–47] and to lidars, sodars, and radars—allows us to obtain atmospheric data with high spatial resolution. Quadcopter capabilities under ideal conditions were examined in [48], in particular, to compare theory with experiment. In this paper, we study the capabilities of a quadcopter for monitoring the state of atmospheric turbulence in an urban area over a territory with complex orography. One of the main fundamental parameters that describe the state of atmospheric turbulence quite completely and accurately and that is investigated in this paper is the turbulence spectrum [52–56]. The theoretical part of the paper describes the coordinate systems, introduces the concept of a spectral tensor of turbulence, and presents the Taylor hypothesis, which allows us to obtain a relationship between the spatial and temporal spectra of turbulence and the basic equations of the von Karman model. In this part, we derive the equations for the wind velocity components measured with the quadcopter in the hover mode, as well as the equations for longitudinal and lateral velocity fluctuations.

The second part of the paper provides general information about the experiment, presents the results of measurements of the quadcopter speed and the longitudinal and lateral components of the wind velocity measured with the DJI Phantom 4 Pro quadcopter and AMK-03 ultrasonic weather stations [16,17]. In addition, longitudinal and lateral turbulence spectra measured with the DJI Phantom 4 Pro and AMK-03 are presented, and their behavior in the inertial and energy production ranges is studied. In conclusion, the main results are summarized.

2. Theory of a Quadcopter in a Turbulent Atmosphere

In the sensing of the mean wind velocity, its projections onto the axes of the coordinate system used in meteorology are of particular interest. In contrast to diagnostics of the mean wind, the fluctuation wind velocity is studied in the coordinate system, one of whose axes is directed along the mean wind. In this section, we examine theoretically the ideal hovering of a quadcopter in a turbulent atmosphere as applied to the problem of sensing of wind velocity fluctuations in the case of horizontal air mass transfer in the coordinate system related to the mean horizontal wind.

2.1. Coordinate Systems

Figure 1 shows schematically the arrangement of the equipment and the direction of the average wind speed during the experiments. It can be seen that the AMK-03 weather station and the quadcopter are oriented differently relative to each other and relative to the direction of the average wind. The ultrasonic weather station is oriented in space along the cardinal points, and the quadcopter can be oriented arbitrarily. Thus, in the case of AMK-03, the measured wind speed data correspond to the {E, N} coordinate system used in meteorology in which one axis is directed to the east (E) and the other axis is directed to the north (N) [53,54].

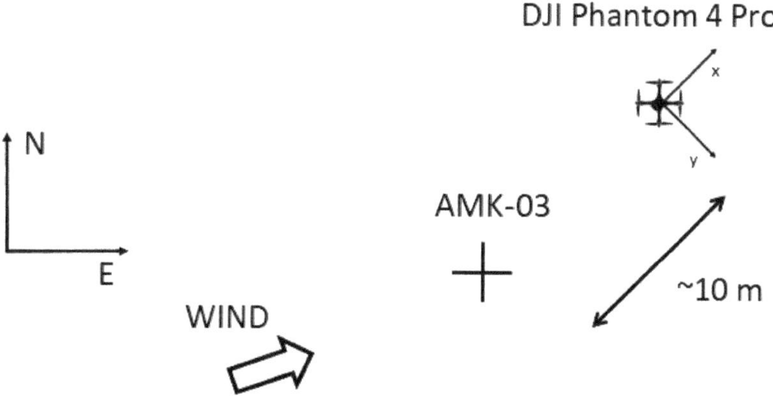

Figure 1. Block diagram of the experiments.

The quadcopter dynamics may be described in a coordinate system other than the $\{E, N\}$ meteorological system. The coordinate system in which the quadcopter dynamics is described is denoted as $\{x, y\}$. The axes of this coordinate system are shown as x and y in Figure 1 [21,31,57,58].

In the theory of turbulence [52–54], fluctuations of the wind velocity are described in the coordinate system related to the mean wind. In the coordinates system related to the mean horizontal wind, one axis is directed along the mean wind, and the two other coordinate axes are directed normally to the mean wind. As a result of this choice of the coordinate system, one longitudinal and two lateral fluctuations of the wind velocity are clearly distinguished.

2.2. Taylor Hypothesis

It is well known that, in the atmosphere, the horizontal air mass transfer often prevails over the vertical motion. The average vertical component of the wind speed is small and can be neglected. Thus, we can take into account only the horizontal component. In this case, one of the axes of the coordinate system can be directed along the mean horizontal wind. Two other axes are directed normally to the mean horizontal wind, and one of them lies in a horizontal plane, while another is directed vertically upward. As a result, in this coordinate system related to the mean horizontal wind, the velocity field of the turbulent air flow at the point $\mathbf{r} = \{\xi, 0, 0\}$ has the form [52–54]:

$$u(\mathbf{r}, t) = \langle u \rangle + u'(\mathbf{r}, t), \tag{1}$$

$$v(\mathbf{r}, t) = v'(\mathbf{r}, t), \tag{2}$$

$$w(\mathbf{r}, t) = w'(\mathbf{r}, t), \tag{3}$$

where $\langle u \rangle$ is the mean wind speed, $u'(\mathbf{r}, t)$, $v'(\mathbf{r}, t)$, and $w'(\mathbf{r}, t)$ are fluctuations of the wind speed, $\langle \ldots \rangle$ is the operator of statistical averaging. It can be seen from Equations (1)–(3) that longitudinal fluctuations of the wind speed $u'(\mathbf{r}, t)$ are directed along the direction of the mean wind speed and that two lateral fluctuations of the wind speed $v'(\mathbf{r}, t)$ and $w'(\mathbf{r}, t)$ are directed normally to the mean wind speed. The component $v'(\mathbf{r}, t)$ lies in the horizontal plane, and the component $w'(\mathbf{r}, t)$ is directed vertically upward and describes vertical fluctuations.

In the mathematical description of fluctuations of the velocity field, the concepts of the second-rank correlation tensor and the turbulence spectrum tensor are introduced [52–54]. For isotropic turbulence, the diagonal elements of the correlation tensor can be written in the form

$$B_u(\xi; t) = \langle u'(\xi, 0, 0; t)u'(0, 0, 0; 0) \rangle, \tag{4}$$

$$B_v(\xi; t) = \langle v'(\xi, 0, 0; t)v'(0, 0, 0; 0) \rangle, \tag{5}$$

$$B_w(\xi; t) = \langle w'(\xi, 0, 0; t)w'(0, 0, 0; 0) \rangle. \tag{6}$$

The diagonal components of the one-dimensional spatial spectral tensor of turbulence take the form

$$\phi_u(\kappa) = \int_{-\infty}^{\infty} B_u(\xi; t)e^{2\pi i \kappa \xi} d\xi, \tag{7}$$

$$\phi_v(\kappa) = \int_{-\infty}^{\infty} B_v(\xi; t)e^{2\pi i \kappa \xi} d\xi, \tag{8}$$

$$\phi_w(\kappa) = \int_{-\infty}^{\infty} B_w(\xi; t)e^{2\pi i \kappa \xi} d\xi, \tag{9}$$

and the diagonal components of the temporal spectral tensor of turbulence are determined by the following equations:

$$\Phi_u(f) = \int_{-\infty}^{\infty} B_u(\xi;t)e^{2\pi i f t}dt, \tag{10}$$

$$\Phi_v(f) = \int_{-\infty}^{\infty} B_v(\xi;t)e^{2\pi i f t}dt, \tag{11}$$

$$\Phi_w(f) = \int_{-\infty}^{\infty} B_w(\xi;t)e^{2\pi i f t}dt, \tag{12}$$

The diagonal components of the temporal spectral tensor of turbulence $\Phi_u(f)$ and $\Phi_v(f)$ are the longitudinal and lateral spectra of turbulence, and those of $\Phi_w(f)$ form the vertical spectrum of turbulence.

In the coordinate system related to the mean wind, we have the opportunity to use Taylor's hypothesis of "frozen" turbulent fluctuations [52–54]. The essence of this hypothesis is that the entire spatial turbulent pattern moves in time with the mean wind speed $\langle u \rangle$. The application of Taylor's hypothesis leads to the relation between the spatiotemporal and purely spatial characteristics of fluctuations of the wind velocity field in the form

$$u'(\xi,0,0;t) = u'(\xi - \langle u \rangle t, 0, 0), \tag{13}$$

$$v'(\xi,0,0;t) = v'(\xi - \langle u \rangle t, 0, 0), \tag{14}$$

$$w'(\xi,0,0;t) = w'(\xi - \langle u \rangle t, 0, 0). \tag{15}$$

The application of Taylor's hypothesis (13)–(15) to Equations (7)–(9) with allowance for Equations (10)–(12) leads to the well-known relation between the spatial and temporal spectra

$$\phi_u(\kappa) = \langle u \rangle \Phi_u(f), \tag{16}$$

$$\phi_v(\kappa) = \langle u \rangle \Phi_v(f), \tag{17}$$

$$\phi_w(\kappa) = \langle u \rangle \Phi_w(f). \tag{18}$$

The relation between the spatial and temporal frequencies is given as $\kappa = f / \langle u \rangle$. In experiments, we measure temporal spectra, while the theory deals with spatial spectra. Thus, Equations (16)–(18) allow us to compare the behavior of experimentally measured temporal spectra with theoretical results.

2.3. Model of Atmospheric Turbulence

One of the most commonly used turbulence spectra models is the von Karman model [52–56,58], which allows us to study the behavior of the spectrum in the energy production and inertial ranges. In addition to the von Karman model, a suitable approximation in problems of UAV dynamics in a turbulent atmosphere is the Dryden turbulence model [52,56,58]. Other models, such as the unified turbulence model, can also be used to describe atmospheric turbulence [56].

In this study, we use the von Karman model to analyze turbulence spectra. With allowance for Equations (16)–(18), the equations for the longitudinal, lateral, and vertical temporal spectra of turbulence for the von Karman model have the form

$$\frac{\Phi_u(f)}{\sigma_u^2} = \frac{2L_u}{\pi} \frac{1}{\left[1 + (1.339 L_u \cdot 2\pi f / \langle u \rangle)^2\right]^{5/6}}, \tag{19}$$

$$\frac{\Phi_v(f)}{\sigma_v^2} = \frac{2L_v}{\pi} \frac{1 + 8/3(2.678L_v \cdot 2\pi f / \langle w \rangle)^2}{\left[1 + (2.678L_v \cdot 2\pi f / \langle u \rangle)^2\right]^{11/6}}, \tag{20}$$

$$\frac{\Phi_w(f)}{\sigma_w^2} = \frac{2L_w}{\pi} \frac{1 + 8/3(2.678L_w \cdot 2\pi f / \langle w \rangle)^2}{\left[1 + (2.678L_w \cdot 2\pi f / \langle u \rangle)^2\right]^{11/6}}, \tag{21}$$

where L_u is the longitudinal turbulence scale, L_v is the lateral turbulence scale, and L_w is the vertical turbulence scale and σ_u^2, σ_v^2, and σ_v^2 are turbulence intensities.

2.4. Wind Velocity Components

The dynamic equations for the quadcopter's center of gravity can be written in the inertial coordinates associated with the Earth as [21,31,57]

$$\ddot{x} = \left(s_\phi s_\psi + c_\phi s_\theta c_\psi\right)\frac{T}{m} + \frac{F_x}{m} \tag{22}$$

$$\ddot{y} = \left(-s_\phi c_\psi + c_\phi s_\theta s_\psi\right)\frac{T}{m} + \frac{F_y}{m} \tag{23}$$

$$\ddot{z} = c_\phi c_\theta \frac{T}{m} - g + \frac{F_z}{m} \tag{24}$$

where $s_{(\bullet)} = \sin(\bullet)$; $c_{(\bullet)} = \cos(\bullet)$; ϕ is the roll angle; θ is the pitch angle; ψ is the yaw angle; T is the aerodynamic force generated by propellers; F_x, F_y, and F_z are the drag force components along the x, y, and z axes; m is the quadcopter mass; and g is the acceleration due to gravity.

The components of the drag force along the x, y, and z axes, which arise during the quadcopter flight, have the form [18,21,31,57]

$$F_j = -c_j\left(v_j - w_j\right) \tag{25}$$

in the linear case, and

$$F_j = -\frac{1}{2}\rho C_j A_j \mathrm{sgn}\left(v_j - w_j\right) \times \left(v_j - w_j\right)^2 \tag{26}$$

in the square-law case. In Equations (25) and (26), c_j and C_j are the drag coefficient along the x, y, and z axes; j is the subscript for enumeration of the orthogonal components of vectors, i.e., $j \in \{x, y, z\}$; v_j are the quadcopter speed components; and w_j are components of the turbulent flow velocity in the atmosphere in the coordinate system $\{x, y\}$; ρ is the air density; A_j are the projections of the quadcopter area on the corresponding axes; and $\mathrm{sgn}(\bullet)$ is the sign function.

Let us consider the case of ideal hover, which can be achieved by compensating all the forces acting on the quadcopter and at $v_j = 0$. Equations (22)–(24) can be transformed to the case of ideal hover through their linearization. The roll, pitch, and yaw angles in a turbulent atmosphere are sums of the average and fluctuation components: $\phi = \langle \phi \rangle + \phi'$, $\theta = \langle \theta \rangle + \theta'$ and $\psi = \langle \psi \rangle + \psi'$. In the small-angle approximation, $\phi, \theta \ll \pi$ and at $\psi' \ll \pi$, as well as if the conditions $\ddot{x} = \ddot{y} = \ddot{z} = 0$ and $v_j = 0$ are fulfilled, the equations for estimation of the horizontal velocity components of the wind velocity field w_x and w_y take the form

$$w_x = -\frac{mg}{c_x}\left(\langle \phi \rangle s_{\langle \psi \rangle} + \langle \theta \rangle c_{\langle \psi \rangle}\right) - \frac{mg}{c_x}\left(\phi' s_{\langle \psi \rangle} + \theta' c_{\langle \psi \rangle}\right), \tag{27}$$

$$w_y = -\frac{mg}{c_y}\left(-\langle \phi \rangle c_{\langle \psi \rangle} + \langle \theta \rangle s_{\langle \psi \rangle}\right) - \frac{mg}{c_y}\left(-\phi' c_{\langle \psi \rangle} + \theta' s_{\langle \psi \rangle}\right) \tag{28}$$

in the linear case, and

$$w_x = -\mathrm{sgn}\left(\langle\varphi\rangle s_{\langle\psi\rangle} + \langle\theta\rangle c_{\langle\psi\rangle}\right)\sqrt{\frac{2mg}{\rho C_x A_x}\left|\left(\langle\varphi\rangle s_{\langle\psi\rangle} + \langle\theta\rangle c_{\langle\psi\rangle}\right)\right|}\left\{1 + \frac{\varphi' s_{\langle\psi\rangle} + \theta' c_{\langle\psi\rangle}}{2\left(\varphi s_{\langle\psi\rangle} + \theta c_{\langle\psi\rangle}\right)}\right\},\tag{29}$$

$$w_y = -\mathrm{sgn}\left(-\langle\varphi\rangle c_{\langle\psi\rangle} + \langle\theta\rangle s_{\langle\psi\rangle}\right)\sqrt{\frac{2mg}{\rho C_x A_x}\left|\left(-\langle\varphi\rangle c_{\langle\psi\rangle} + \langle\theta\rangle s_{\langle\psi\rangle}\right)\right|}\left\{1 + \frac{-\varphi' c_{\langle\psi\rangle} + \theta' s_{\langle\psi\rangle}}{2\left(-\varphi c_{\langle\psi\rangle} + \theta s_{\langle\psi\rangle}\right)}\right\}\tag{30}$$

in the square-law case.

It follows from Equations (27)–(30) that, regardless of the model of the drag force, the estimates of the horizontal components of the turbulent flow velocity are the sum of the regular and fluctuation parts. The regular part of the estimates is determined by the average values of the roll, pitch, and yaw angles, whereas the fluctuation part is proportional to the fluctuations of the roll φ' and pitch θ' angles.

2.5. Longitudinal and Lateral Velocity Fluctuations

In the case of predominance of the horizontal air mass transfer over the vertical motion, the longitudinal and lateral turbulent fluctuations of the wind velocity take the form

$$\mathrm{u}' = n_E w_E' + n_N w_N',\tag{31}$$

$$\mathrm{v}' = -n_N w_E' + n_E w_N'\tag{32}$$

$$\mathbf{n} = \{n_E, n_N, 0\} = \left\{\frac{\langle w_E\rangle}{\langle \mathrm{u}\rangle}, \frac{\langle w_N\rangle}{\langle \mathrm{u}\rangle}, 0\right\}\tag{33}$$

for the ultrasonic weather station, and

$$\mathrm{u}' = n_x w_x' + n_y w_y'\tag{34}$$

$$\mathrm{v}' = -n_x w_y' + n_y w_x'\tag{35}$$

$$\mathbf{n} = \{n_x, n_y, 0\} = \left\{\frac{\langle w_x\rangle}{\langle \mathrm{u}\rangle}, \frac{\langle w_y\rangle}{\langle \mathrm{u}\rangle}, 0\right\}\tag{36}$$

for the quadcopter. Here, w_E' and w_N' are fluctuations of wind velocity components along the E and N axes, i.e., data of the ultrasonic weather station; w_x' and w_y' are fluctuations of the wind velocity components obtained from the results of quadcopter telemetry; $\langle w_E\rangle$ and $\langle w_N\rangle$ are the average components of the horizontal velocity along the E and N axes; $\langle w_x\rangle$ and $\langle w_y\rangle$ are estimates of the velocity components along the x and y axes. Thus, Equations (31)–(36) allow us to compare the longitudinal and lateral turbulence spectra measured by the quadcopter and the ultrasonic weather station.

3. Experiment

Usually, turbulence spectra are studied experimentally over territories having a flat and uniform underlying surface and under weather conditions corresponding to the stationary state of the atmosphere. At such a territory and under such weather conditions, obtained experimental data agree well with theoretical results. The capabilities of a hovering quadcopter as applied to the study of homogeneous and isotropic turbulence were examined in [48]. It was shown that the obtained results are in a good agreement with the theory of homogeneous and isotropic turbulence and the data measured by acoustic anemometry methods. However, from the practical point of view, it is interesting to analyze the capabilities of a hovering quadcopter when studying atmospheric turbulence over an urban territory with complex orography in different seasons: winter, spring, summer, and fall. From the viewpoint of the theory of turbulence, the behavior of the turbulence spectrum for this territory and under bad weather conditions is poorly studied. Therefore, the quadcopter data were compared with the data of ultrasonic weather stations.

3.1. General Information about the Experiment

Experimental studies were carried out at the territory of the Institute for Monitoring of Climatic and Ecological Systems of the Siberian Branch of the Russian Academy of Sciences (IMCES SB RAS), which is located in Academgorodok, one of the districts of the city of Tomsk (Russian Federation). This area is a territory with complex orography: it is a forested area with the buildings of the Academgorodok institutes and highways. Figure 2 shows a Google map of the experimental area. The arrows show the location of the used AMK-03 ultrasonic weather stations, and the measurement dates are indicated.

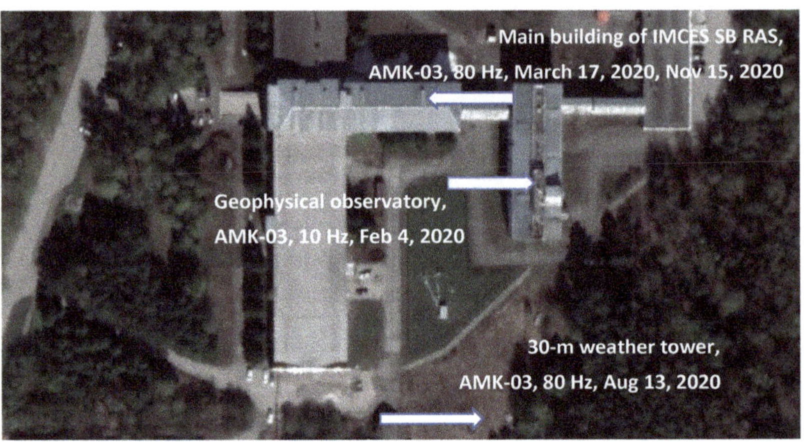

Figure 2. Arrangement of ultrasonic weather stations.

The studies were carried out in different seasons: winter (4 February 2020), spring (17 March 2020), summer (13 August 2020), and fall (15 November 2020). The quadcopter hovered at different places of the territory with complex orography: over the building of the Geophysical Observatory and the IMCES SB RAS main building, as well as in the immediate vicinity of a small grove where a 30 m weather tower is installed (see Figures 2 and 3). In the experiment on February 20, the launch point of the DJI Phantom 4 Pro quadcopter was chosen in close proximity to the foundation of the Geophysical Observatory building, and on August 13 the quadcopter started from the foundation of the 30 m weather tower. During the measurements on March 17 and November 15, the quadcopter took off in the immediate vicinity of the foundation of the IMCES SB RAS main building.

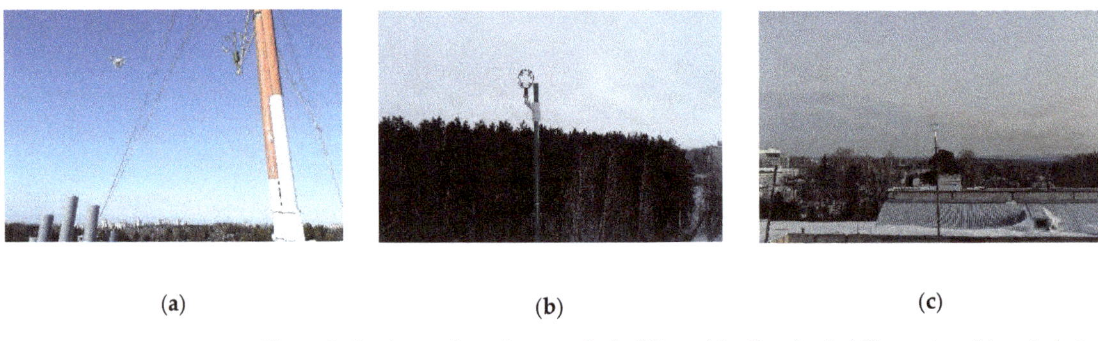

(a)   (b)   (c)

Figure 3. Quadcopter hovering over the building of the Geophysical Observatory (**a**), and photos of the ultrasonic weather stations installed on the 30 m weather tower (**b**) and on the IMCES SB RAS main building (**c**).

The AMK-03 ultrasonic weather station serves to measure and record the wind speed and direction using acoustic anemometry methods, as well as recording temperature, relative air humidity, and atmospheric pressure [16,17]. We used AMK-03 data of two types, which recorded the wind speed and direction with a frequency of 10 and 80 Hz. The locations of the weather stations of different types are shown by the arrows in Figure 2. The data on the state of the quadcopter in the flight logs were recorded at a frequency of 10 Hz in CSV format.

Figure 4 shows the quadcopter's flight paths during the experiments. Table 1 presents the dates and times of the start and end of the experiments, as well as the DJI Phantom 4 Pro quadcopter's flight heights. After takeoff, the quadcopters flew up to the AMK-03 ultrasonic weather stations located on the roofs of the buildings and on the weather tower. After the end of the experiments, the quadcopters returned to the starting point.

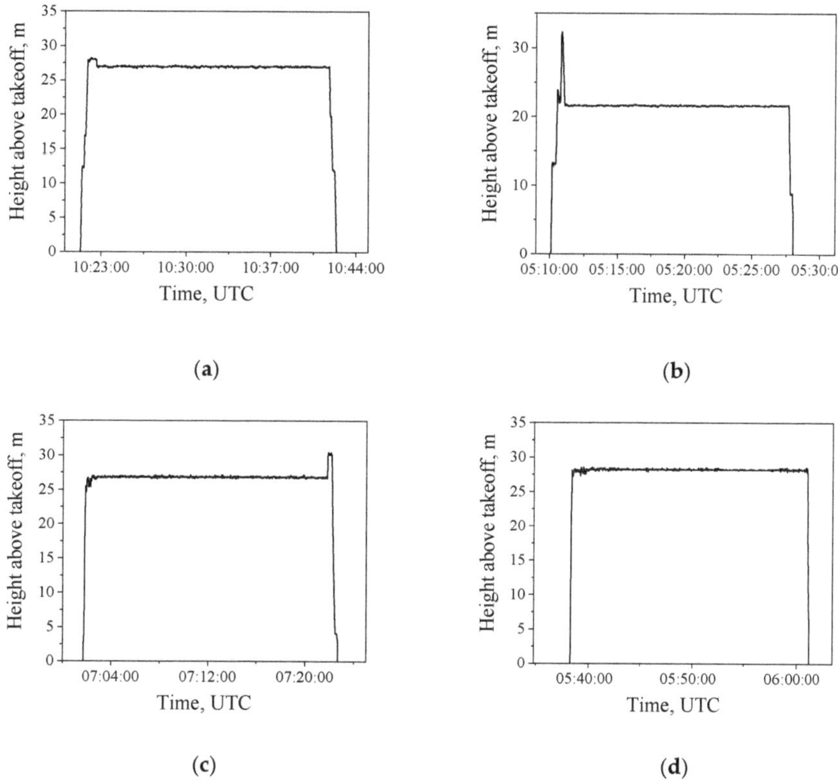

Figure 4. Quadcopter flight trajectory during the experiments on (**a**) 4 February, (**b**) 17 March, (**c**) 13 August, and (**d**) 15 November 2020.

Table 1. Date and time of the start and end of the study and the hover height.

Date	Start, UTC	End, UTC	Hover Height, m
4 February 2020	10:21	10:42	27
17 March 2020	05:10	05:28	22
13 August 2020	07:01	07:23	27
15 November 2020	05:38	06:01	28

According to the data of the Tomsk International Airport spaced by ~10 km from IMCES SB RAS, the following weather conditions were observed during the experiments.

On February 4, weather conditions were satisfactory in terms of the quadcopter flight: south–south-west wind, speed of 3 m/s, air temperature of −4 °C, air humidity of 86%, light snow, horizontal visibility range of 8 km. On March 17, good weather conditions were recorded: variable wind, speed of 1 m/s, air temperature of −1 °C, air humidity of 55%, no precipitation, horizontal visibility range of 10 km or more. On August 13, the weather in the airport was excellent: southeast wind, speed of 4 m/s, air temperature of 27 °C, air humidity of 42%, no precipitation, horizontal visibility range of 10 km or more. On November 15, the weather in the airport was satisfactory in terms of the quadcopter flight: south–south-west wind, speed of 2 m/s, air temperature of −9 °C, air humidity of 90%, no precipitation, horizontal visibility range of 10 km or more.

Thus, the experiments were carried out in different seasons, under different weather conditions, and the hover took place at different places of the IMCES SB RAS territory, which is characterized by complex orography.

3.2. Quadcopter Velocity

Figure 5 shows the dynamics of the v_x, v_y, and v_z components of the quadcopter velocity during hovering. It can be seen that, generally, the quadcopter velocity components are equal to zero during the measurements. In short periods of time, the forces acting on the quadcopter exceed the capabilities of the control system and high-precision positioning is disrupted. After regaining control, the quadcopter begins to move to its original position and, upon reaching this position, it stops.

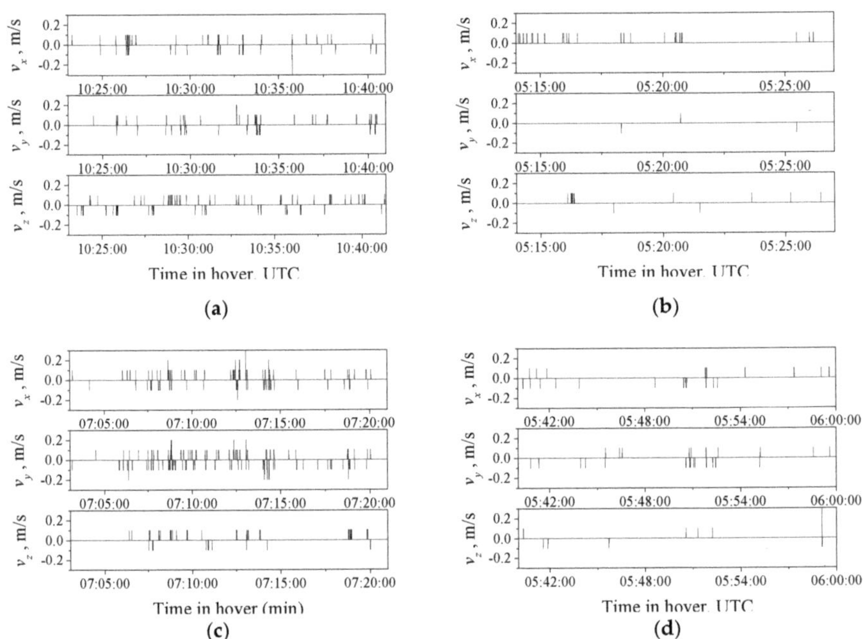

Figure 5. Quadcopter velocity components along the x, y, and z axes during hovering. (**a**) 4 February, (**b**) 17 March, (**c**) 13 August, and (**d**) 15 November 2020.

Thus, the periods in which the precision positioning of the quadcopter in space is disrupted can be neglected due to their insignificance, and we can believe that ideal hovering was observed during the experiment.

3.3. Longitudinal and Lateral Wind Velocity Components

Let us consider the behavior of the estimates of the longitudinal and lateral components of the wind velocity from the quadcopter data in the altitude hold mode in the turbulent atmosphere and compare it with the results obtained from the data of the AMK-03 ultrasonic weather station.

Figure 6 shows the temporal dynamics of the longitudinal and lateral wind velocity components measured with AMK-03 (red curves) and DJI Phantom 4 Pro (black curves) on (**a**) 4 February, (**b**) 17 March, (**c**) 13 August, and (**d**) 15 November 2020. It follows from Figure 6 that the time series of u and v measured in different ways generally coincide, and discrepancies are observed only in the high-frequency range of fluctuations.

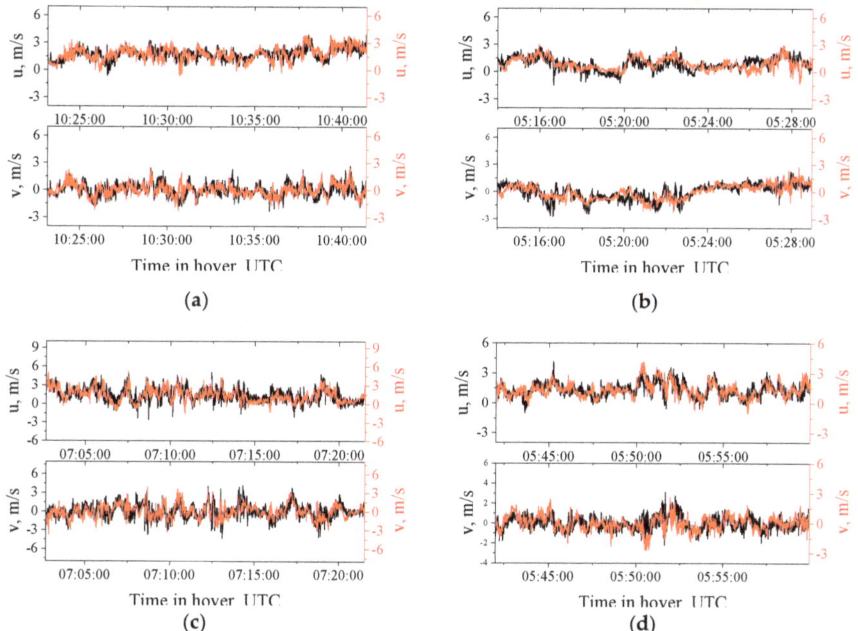

Figure 6. Temporal dynamics of the longitudinal and lateral components of wind velocity: quadcopter data (black curve) and AMK-03 data (red curve) on (**a**) 4 February, (**b**) 17 March, (**c**) 13 August, and (**d**) 15 November 2020.

The difference between the two different ways of wind velocity measurement can be characterized by the variance defined as $\sigma_u = \sqrt{\langle (u^{Dron} - u^{AMK-03})^2 \rangle}$ and $\sigma_v = \sqrt{\langle (v^{Dron} - v^{AMK-03})^2 \rangle}$. Here, u^{Dron}, u^{AMK-03}, and v^{Dron}, v^{AMK-03} are the longitudinal and lateral wind velocity components measured by the quadcopter and the weather station. Table 2 presents the variances σ_u and σ_v calculated for original and smoothed time series. It can be seen from Table 2 that, for the original time series, the variances range within 0.60–0.93 for the longitudinal component and 0.59–1.00 for the lateral component. When the time series are smoothed by moving average for the period of 60 s, which corresponds to suppression of the high-frequency part of the spectrum, σ_u and σ_v decrease down to 0.21–0.36 and 0.18–0.31 for, respectively, the longitudinal and lateral components.

Table 2. Variances σ_u and σ_v.

Date	Longitudinal Component		Lateral Component	
	No Smoothing	Smoothing	No Smoothing	Smoothing
4 February 2020	0.63	0.24	0.66	0.18
17 March 2020	0.64	0.36	0.59	0.22
13 August 2020	0.93	0.32	1.00	0.25
15 November 2020	0.60	0.21	0.72	0.31
Average	0.7	0.28	0.74	0.24

Other parameters characterizing the difference between the two measurement methods are the correlation coefficients. Table 3 presents the calculated coefficients of correlation between the wind velocities measured by the quadcopter and the weather station for the original and smoothed time series. It can be seen from Table 3 that, for the original time series, the correlation coefficients range within 0.54–0.63 for the longitudinal component and 0.37–0.73 for the lateral component. When the time series are smoothed by moving average for the period of 60 s, which corresponds to suppression of the high-frequency part of the spectrum, the correlation coefficients increase up to 0.72–0.90 and 0.61–0.95 for, respectively, the longitudinal and lateral components.

Table 3. Correlation coefficients.

Date	Longitudinal Component		Lateral Component	
	No Smoothing	Smoothing	No Smoothing	Smoothing
4 February 2020	0.55	0.87	0.48	0.83
17 March 2020	0.54	0.72	0.73	0.95
13 August 2020	0.59	0.90	0.52	0.86
15 November 2020	0.63	0.88	0.37	0.61
Average	0.58	0.84	0.53	0.81

It follows from Tables 2 and 3 that, upon smoothing, the variance decreases 2.5–3.1 times, while the correlation increases 1.5 times. This leads to a decrease of the average variances from $\sigma_u = 0.70$ and $\sigma_v = 0.74$ down to $\sigma_u = 0.28$ and $\sigma_v = 0.24$, whereas the average correlation coefficients increase from 0.58 and 0.53 up to 0.84 and 0.81 for the longitudinal and lateral components of the wind velocity, respectively. Thus, the smoothing suppresses the high-frequency component of the signal, which is accompanied by a significant decrease in the variances and an increase in the correlation. Taking this into account, we can conclude that the quadcopter data well describe the behavior of the large-scale turbulent vortices. Small-scale vortices, due to their low energy and quadcopter inertia, do not always give the correct response to the resulting signal.

Table 4 presents the average values of the longitudinal, lateral, and vertical components of the wind velocity measured with AMK-03 in the experiments. It can be seen that the average wind speed $\langle u \rangle$ differs from the corresponding values observed at the airport. This difference is explained by two circumstances. First, the airport is located approximately 10 km from the IMCES SB RAS buildings. Second, the airport territory, where the measurements were carried out, has a flat underlying surface, whereas the IMCES SB RAS territory has complex orography.

It can also be seen from Table 4 that, in the experiment, the horizontal transfer of air masses predominated over the vertical motion, i.e., $\langle w \rangle \approx 0$. The fulfillment of the condition $\langle w \rangle \approx 0$ in the experiment means that the assumption of the predominance of the horizontal air mass transfer over the vertical motion when calculating longitudinal and lateral turbulent fluctuations of the wind velocity using Equations (31)–(36) is justified.

It should be noted that the maximal values of the longitudinal and lateral components of the wind velocity exceed the average wind, which is indicative of strong turbulence during measurements.

Table 4. Average values of the longitudinal, lateral, and vertical wind velocity components and their maximal values.

Date	$\langle u \rangle$, m/s	$\langle v \rangle$, m/s	max (u), m/s	max (v), m/s	$\langle w \rangle$, m/s
4 February 2020	1.78	0	4.9	2.6	0.3
17 March 2020	0.89	0	3.03	2.76	0.04
13 August 2020	1.36	0	5.20	3.91	−0.1
15 November 2020	1.35	0	4.13	2.58	0.15

3.4. Spectra of Turbulence

Turbulence spectra were calculated by the well-known methods with standard FFT software. Figures 7 and 8 show the results of measurements of the longitudinal and lateral relative turbulence spectra $\Phi_u(f)$ and $\Phi_v(f)$. The turbulence spectra obtained from the data of the AMK-03 ultrasonic weather station and the DJI Phantom 4 Pro quadcopter are shown by the black curves, σ^2 is the normalization coefficient.

It can be seen from Figures 7 and 8 that the values of spectra $\Phi_u(f)$ and $\Phi_v(f)$ vary significantly at minor variations of the frequency f. These variations are random oscillations about the main regularities of the turbulence spectra. To reveal these regularities in the turbulence spectra, a smoothing procedure was used.

The result of applying the smoothing procedure is shown in Figures 7 and 8 by continuous colored curves: red and blue curves for the turbulence spectra of the longitudinal velocity component $\Phi_u(f)$ and pink and purple curves for the turbulence spectra of the lateral velocity component $\Phi_v(f)$. Figure 7a,c,e,g and Figure 8a,c,e,g depict the results of the application of the smoothing procedure of the relative turbulence spectra obtained from the data of the AMK-03 ultrasonic weather station, while Figure 7b,d,f,h and Figure 8b,d,f,h show those for the DJI Phantom 4 Pro quadcopter data. The figures correspond to the following dates: (a, b) 4 February, (c, d) 17 March, (e, f) 13 August, and (g, h) 15 November 2020.

Figure 9 compares the smoothed turbulence spectra obtained from the AMK-03 and DJI Phantom 4 Pro data. Similarly to the case in Figures 7 and 8, red and blue curves represent the turbulence spectra of the longitudinal velocity component $\Phi_u(f)$ and pink and purple curves represent the turbulence spectra of the lateral velocity component $\Phi_v(f)$. It can be seen from Figure 9 that the turbulence spectra obtained from the AMK-03 and DJI Phantom 4 Pro data generally coincide, and slight differences are observed in the high-frequency range of the spectrum.

It is well known [52,54] that the turbulence spectrum has three main spectral ranges: the energy production range, the inertial range, and the dissipation range. In the energy production range, which contains the main part of the turbulent energy, the energy is generated by buoyancy and shear. In the inertial range, the energy is neither generated nor dissipated but transferred from large scales to smaller ones. In the dissipation range, the kinetic energy is converted into the internal energy. Next, we consider the behavior of the turbulence spectra in the inertial and energy production ranges.

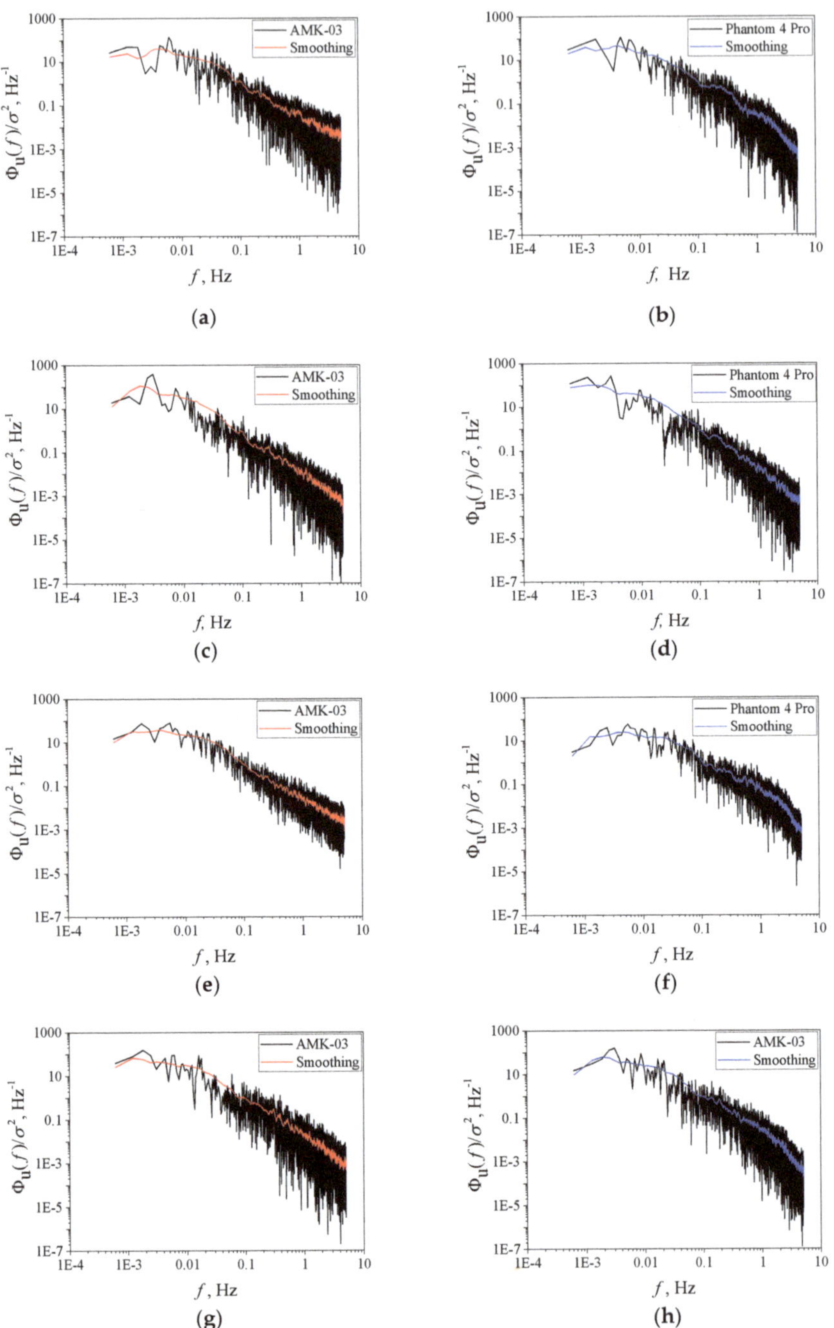

Figure 7. Longitudinal spectra of turbulence $\Phi_u(f)$: measured turbulence spectra (black curves), smoothed turbulence spectra obtained from AMK-03 (red curves), and DJI Phantom 4 Pro data (blue curves); σ^2 is the normalization coefficient; (**a,b**) 4 February, (**c,d**) 17 March, (**e,f**) 13 August, and (**g,h**) 15 November 2020.

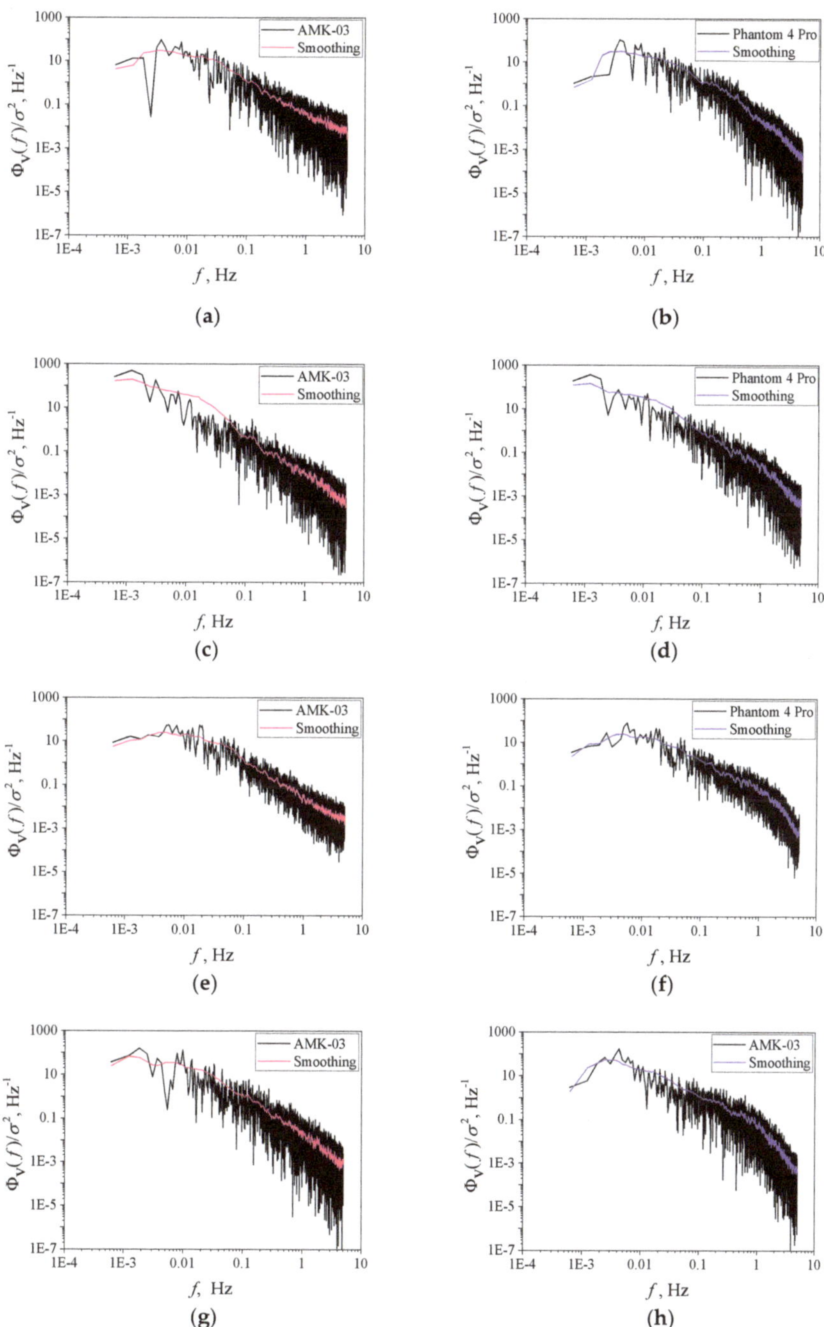

Figure 8. Lateral spectra of turbulence $\Phi_v(f)$: measured turbulence spectra (black curves), smoothed turbulence spectra obtained from AMK-03 (pink curves), and DJI Phantom 4 Pro data (purple curves); σ^2 is the normalization coefficient; (**a**,**b**) 4 February, (**c**,**d**) 17 March, (**e**,**f**) 13 August, and (**g**,**h**) 15 November 2020.

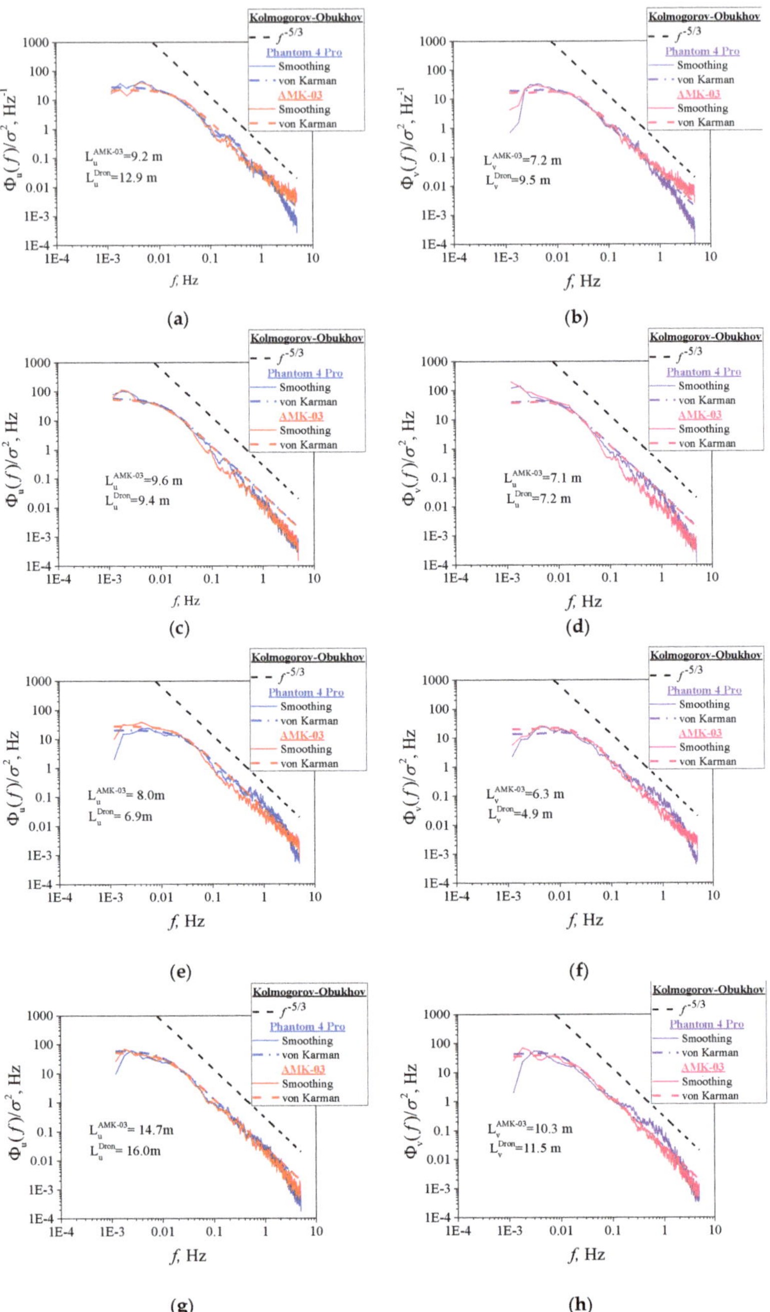

Figure 9. Longitudinal $\Phi_u(f)$ and lateral $\Phi_v(f)$ turbulence spectra upon application of the smoothing procedure: (red and blue curves) smoothed longitudinal turbulence spectra obtained from the AMK-03 and DJI Phantom 4 Pro data, respectively, (pink and purple curves) smoothed turbulence spectra obtained from the AMK-03 and DJI Phantom 4 Pro data; σ^2 is the normalization coefficient; (**a**,**b**) 4 February, (**c**,**d**) 17 March, (**e**,**f**) 13 August, and (**g**,**h**) 15 November 2020.

3.5. Inertial and Energy Production Ranges

In the inertial range, the turbulence spectrum obeys the Kolmogorov–Obukhov "5/3" law [52–56], which was found from dimensional considerations. The von Karman model is a generalization of the "5/3" law to the frequencies of the energy production range. Therefore, the Kolmogorov–Obukhov law can be found from model (19)–(21) at $L_u f / \langle u \rangle \gg 1$, $L_v f / \langle u \rangle \gg 1$ and $L_w f / \langle u \rangle \gg 1$. In the inertial range, the spectral curve has the form

$$\frac{\Phi_u(f)}{\sigma_u^2} \sim f^{-5/3}, \tag{37}$$

$$\frac{\Phi_v(f)}{\sigma_v^2} \sim f^{-5/3}, \tag{38}$$

$$\frac{\Phi_w(f)}{\sigma_w^2} \sim f^{-5/3}. \tag{39}$$

In Figure 9, the dashed curves show the turbulence spectra corresponding to the Kolmogorov–Obukhov "5/3" law, which, as already noted, holds true for a homogeneous surface. It was already mentioned that the IMCES SB RAS territory is not homogeneous and has complex orography. Despite the complex orography, the turbulence spectrum in the inertial range obeys the $f^{-5/3}$ law, as is clearly seen from Figure 9.

Studying the behavior of turbulence spectra in the inertial range by the least-squares fit method with the frequency dependence of the turbulence spectrum in the form $f^{-\gamma}$ taken as a regression curve, it follows from Equations (37) to (39) that $\gamma = 5/3 \approx 1.7$. Table 5 presents the exponents γ_u and γ_v calculated for the longitudinal and lateral turbulence spectra, respectively, and the frequency ranges Δf_u and Δf_v used in the least-squares fit. It can be seen from Table 5 that the exponents γ_u and γ_v for the longitudinal and lateral turbulence spectra agree with the theoretical results within the statistical error. In some cases, the frequency ranges corresponding to the inertial range differ for the different measurement methods. Insignificant discrepancies are explained by additional vortices arising as a result of the turbulent flow around the IMCES SB RAS buildings or near the small grove where the 30 m weather tower is installed.

Table 5. Exponents γ_u and γ_v and frequency ranges Δf_u and Δf_v.

	Longitudinal Spectra		Lateral Spectra	
	γ_u	Δf_u	γ_v	Δf_v
4 February 2020				
AMK-03	1.7	[0.0281, 0.639]	1.6	[0.0281, 0.639]
DJI Phantom 4 Pro	1.6	[0.0281, 0.639]	1.7	[0.0281, 0.639]
17 March 2020				
AMK-03	1.7	[0.0281, 0.639]	1.8	[0.0281, 0.944]
DJI Phantom 4 Pro	1.8	[0.0281, 0.639]	1.7	[0.0281, 0.639]
13 August 2020				
AMK-03	1.7	[0.0281, 0.639]	1.7	[0.0281, 0.639]
DJI Phantom 4 Pro	1.7	[0.0372, 0.251]	1.6	
15 November 2020				
AMK-03	1.7	[0.0281, 0.639]	1.8	[0.0617, 0.672]
DJI Phantom 4 Pro	1.8	[0.0281, 0.639]	1.7	[0.0220, 0.639]

The main characteristics of the energy production range include the longitudinal and lateral scales of turbulence L_u and L_v, respectively. The information about these scales is contained in relative turbulence spectra. In the case of an isotropic atmosphere, the equation for the scale ratio has the form

$$\frac{L_v}{L_u} = 0.5 \tag{40}$$

To estimate the scales L_u and L_v, the von Karman model is used in the approximation of experimental data. The values of the turbulence scales are determined by the equations relating the turbulence scales and the values of the maxima of the functions $f\Phi_u(f)$ and $f\Phi_v(f)$ [51]. In this study, we used the least-square fit method for approximation of the experimental data with the von Karman model as the regression curve. In contrast to [55], the maximum was not sought, and the scales L_u and L_v were determined directly as the parameters of the best fit when applying the least-square fit method. The calculation procedure used in this study is equivalent to the approach outlined in [55].

The values of the turbulence scales are given in Table 6. Keeping in mind that the IMCES SB RAS territory is not homogeneous and has complex orography, we can conclude that the scale ratio is also true to a good accuracy. In Figure 9, the colored dashed curves are for the best fit curves, red and blue curves are for the turbulence spectra of the longitudinal component $\Phi_u(f)$, and pink and purple curves are for the turbulence spectra of the lateral component $\Phi_v(f)$. Figure 9 demonstrates the good agreement for the best fit curves for the turbulence spectra obtained from the ANK-03 and DJI Phantom 4 Pro data.

Table 6. Integral scales of turbulence.

	L_u	L_v	L_v/L_u
	4 February 2020		
AMK-03	9.2	7.2	0.78
DJI Phantom 4 Pro	12.9	9.5	0.74
	17 March 2020		
AMK-03	9.6	7.1	0.74
DJI Phantom 4 Pro	9.4	7.2	0.77
	13 August 2020		
AMK-03	8.0	6.3	0.79
DJI Phantom 4 Pro	6.9	4.9	0.71
	15 November 2020		
AMK-03	14.7	10.3	0.70
DJI Phantom 4 Pro	16.0	11.5	0.72

4. Discussion

The problem of ideal hovering of a quadcopter in a turbulent atmosphere in the case of horizontal air mass transfer as applied to sensing of wind velocity fluctuations is considered in this paper based on dynamic equations. From the viewpoint of the theory of turbulence and correct calculation of the turbulence spectrum, the coordinate system, one of whose axes is directed along the mean wind, is of interest. In this paper, the equations for fluctuations of the longitudinal and lateral wind speed, which characterize random air mass motions in the coordinate system with the axis directed along the mean horizontal wind, are obtained. It is shown that fluctuations of the longitudinal and lateral wind velocity are proportional to fluctuations of the pitch and roll angles.

The studies of the use of DJI Phantom 4 Pro in the hover mode in combination with the AMK-03 ultrasonic weather stations show that the quadcopter allows us to obtain turbulence spectra with high spatial resolution in the atmosphere in areas with complex orography in hard-to-reach places, under various weather conditions, as well as in different seasons: winter, spring, summer, and fall.

The measured values of the longitudinal and lateral spectra vary significantly with insignificant variations of the frequency. Therefore, we used the smoothing procedure to study the main regularities in the behavior of turbulence in the inertial and energy production ranges. To estimate the longitudinal and lateral scales of turbulence, the least square fit method was used with the von Karman model as a regression curve.

The longitudinal and lateral turbulence spectra obtained with the DJI Phantom 4 Pro and AMK-03 are generally the same, with minor differences observed in the high-frequency range of the spectrum. Discrepancies in the high-frequency spectral range are also observed in the behavior of the time series of the longitudinal and lateral components of wind velocity measured by different methods. The behavior of the turbulence spectra in the inertial range shows that they obey the Kolmogorov–Obukhov "5/3" law. In the energy production range, the longitudinal and lateral turbulence scales and their ratio measured by both the DJI Phantom 4 Pro and AMK-03 coincide to a good accuracy.

Discrepancies in the behavior of the turbulence spectra obtained experimentally by different methods can be explained as follows. First, for safety reasons, the quadcopter was at a distance of ~10 m from AMK-03 during the experiment. Measurements of the integral turbulence scales show that they are of the same order of magnitude as the distance from AMK-03 to DJI Phantom 4 Pro. This relation between the scales and the distances means that the wind velocity fields at the AMK-03 and quadcopter locations are partially correlated. Therefore, it makes no sense to talk about complete equality of the measured data and, consequently, the turbulence spectra should differ.

Second, experimental studies were carried out at a territory with complex orography. The presence of the park zone, institute buildings, and highways means that the territory is inhomogeneous, which leads to significant deviations from homogeneity and isotropy. As a result, the behavior of the wind velocity field and its characteristics is different for the AMK-03 and quadcopter locations. This difference is most pronounced in the behavior of the ratio of turbulence scales. For the homogeneous and isotropic atmosphere, this ratio is $L_v/L_u = 0.5$, but in our experiment, as can be seen from Table 6, $L_v/L_u \sim 0.8$ for AMK-03 and $L_v/L_u \sim 0.7$ for DJI Phantom 4 Pro on August 13.

It was shown in [48] that the results of investigation of the turbulence spectra in the inertial and energy production ranges with the quadcopter in the altitude holding mode over a homogeneous territory are in a good agreement with the theory of homogeneous and isotropic turbulence and with the measurement data obtained using acoustic anemometry. This study allows us to conclude that, in these ranges, at the territory with complex orography, the behavior of the turbulence spectrum measured by the quadcopter agrees with objective data on the state of atmospheric turbulence. Thus, the results obtained allow us to assert that a rotary-wing UAV can serve as a tool having great potential for diagnostics of the atmospheric boundary layer. Due to the capability of providing data on the state of atmospheric turbulence with high spatial resolution, the quadcopter is a promising tool for solving problems of controlling the UAV movement under bad weather conditions, as well as problems of wind energy, climatic measurements in an urban environment, etc.

From the scientific and practical points of view, it is of great interest to monitor the state of atmospheric turbulence at various spatial points of the studied area. From this point of view, further work on the use of a quadcopter for low-altitude sensing is associated with the use of a quadcopter swarm to determine profiles of atmospheric turbulence in both the vertical and horizontal planes.

Author Contributions: Conceptualization, A.S.; methodology, A.S.; software, A.A. and E.S.; validation, A.K., A.T. and A.M.; investigation, A.S. and A.A.; data curation, A.K., A.T. and A.M.; writing—original draft preparation, O.P.; writing—review and editing, A.S.; supervision, A.S.; project administration, A.S. All authors have read and agreed to the published version of the manuscript.

Funding: The study was supported by the Russian Foundation for Basic Research (project no. 19-29-06066 mk).

Acknowledgments: The authors are grateful to A.B. Gonchar for the help in preparing this paper.

Conflicts of Interest: The authors declare no conflict of interest. The funders had no role in the design of the study; in the collection, analyses, or interpretation of data; in the writing of the manuscript; or in the decision to publish the results.

References

1. Cornman, L.B.; Chan, W.N. Summary of a workshop on integrating weather into unmanned aerial system traffic management. *Bull. Am. Meteorol. Soc.* **2017**, *98*, ES257–ES259. [CrossRef]
2. Shakhatreh, H.; Sawalmeh, A.; Al-Fuqaha, A.; Dou, Z.; Almaita, E.; Khalil, I.; Othman, N.; Khreishah, A.; Guizani, M. Unmanned aerial vehicles (UAVs): A survey on civil applications and key research challenges. *IEEE Access* **2019**, *7*, 48572–48634. [CrossRef]
3. Hatfield, M.; Cahill, C.; Webley, P.; Garron, J.; Beltran, R. Integration of unmanned aircraft systems into the national airspace system-efforts by the university of alaska to support the FAA/NASA UAS traffic management program. *Remote Sens.* **2020**, *12*, 3112. [CrossRef]
4. Kazarin, P.; Golubev, V.; MacKunis, W.; Moreno, C. Robust nonlinear tracking control for unmanned aircraft in the presence of wake vortex. *Electronics* **2021**, *10*, 1890. [CrossRef]
5. Borener, S.; Trajkov, S.; Balakrishna, P. Design and development of an integrated safety assessment model for NextGen. In Proceedings of the International Annual Conference of the American Society for Engineering Management, San Antonio, TX, USA, 10–13 June 2012.
6. Abichandani, P.; Lobo, D.; Ford, G.; Bucci, D.; Kam, M. Wind measurement and simulation techniques in multi-rotor small unmanned aerial vehicles. *IEEE Access* **2020**, *8*, 54910–54927. [CrossRef]
7. Stith, J.L.; Baumgardner, D.; Haggerty, J.; Hardesty, M.; Lee, W.; Lenschow, D.; Pilewskie, P.; Smith, P.L.; Steiner, M.; Vömel, H. 100 Years of progress in atmospheric observing systems. *Meteorol. Monogr.* **2018**, *59*, 2.1–2.55. [CrossRef]
8. Frehlich, R.; Meillier, Y.; Jensen, M.A.; Balsley, B.B. Turbulence measurements with the CIRES tethered lifting system during CASES-99: Calibration and spectral analysis of temperature and velocity. *J. Atmos. Sci.* **2003**, *60*, 2487–2495. [CrossRef]
9. Yang, S.; Petersen, G.N.; von Löwis, S.; Preißler, J.; Finger, D. Determination of eddy dissipation rate by Doppler lidar in Reykjavik, Iceland. *Meteorol. Appl.* **2020**, *27*, e1951. [CrossRef]
10. Hocking, W.K.; Röttger, J.; Palmer, R.D.; Sato, T.; Chilson, P.B. *Atmospheric Radar*; Cambridge University Press: Cambridge, UK, 2016.
11. Lundquist, J.K.; Wilczak, J.M.; Ashton, R.; Bianco, L.; Brewer, W.A.; Choukulkar, A.; Clifton, A.; Debnath, M.; Delgado, R.; Friedrich, K.; et al. Assessing state-of-the-art capabilities for probing the atmospheric boundary layer: The XPIA field campaign. *Bull. Am. Meteorol. Soc.* **2017**, *98*, 289–314. [CrossRef]
12. Kral, S.T.; Reuder, J.; Vihma, T.; Suomi, I.; O'Connor, E.; Kouznetsov, R.; Wrenger, B.; Rautenberg, A.; Urbancic, G.; Jonassen, M.O.; et al. Innovative strategies for observations in the arctic atmospheric boundary layer (ISOBAR)—The Hailuoto 2017 campaign. *Atmosphere* **2018**, *9*, 268. [CrossRef]
13. Leosphere, Windcube, Vaisala. Available online: https://www.vaisala.com/en/wind-lidars/wind-energy/windcube/ (accessed on 21 February 2022).
14. METEK Meteorologische Messtechnik GmbH. Available online: https://metek.de/product-group/doppler-sodar/ (accessed on 21 February 2022).
15. Scintec. Available online: https://www.scintec.com/ (accessed on 21 February 2022).
16. Azbukin, A.A.; Bogushevich, A.Y.; Korolkov, V.A.; Tikhomirov, A.A.; Shelevoi, V.D. A field version of the AMK-03 automated ultrasonic meteorological complex. *Russ. Meteorol. Hydrol.* **2009**, *34*, 133–136. [CrossRef]
17. Azbukin, A.A.; Bogushevich, A.Y.; Kobzev, A.A.; Korolkov, V.A.; Tikhomirov, A.A.; Shelevoy, V.D. AMK-03 Automatic weather stations, their modifications and applications. *Sens. Syst.* **2012**, *3*, 47–52.
18. Palomaki, R.T.; Rose, N.T.; van den Bossche, M.; Sherman, T.J.; De Wekker, S.F.J. Wind estimation in the lower atmosphere using multirotor aircraft. *J. Atmos. Ocean. Technol.* **2017**, *34*, 1183–1190. [CrossRef]
19. González-Rocha, J.; De Wekker, S.F.J.; Ross, S.D.; Woolsey, C.A. Wind profiling in the lower atmosphere from wind-induced perturbations to multirotor UAS. *Sensors* **2020**, *20*, 1341. [CrossRef]
20. González-Rocha, J.; Woolsey, C.A. Cornel Sultan measuring atmospheric winds from quadrotor motion. In Proceedings of the AIAA Atmospheric Flight Mechanics Conference, Grapevine, TX, USA, 9–13 January 2017.
21. González-Rocha, J.; Woolsey, C.A.; Cornel Sultan, C.; De Wekker, S.F.J. Sensing Wind from Quadrotor Motion. *J. Guid. Control Dyn.* **2019**, *42*, 1–18. [CrossRef]
22. Reuder, J.; Brisset, P.; Jonassen, M.; Muller, M.; Mayer, S. The small unmanned meteorological observer SUMO: A new tool for atmospheric boundary layer research. *Meteorol. Z.* **2009**, *18*, 141–147. [CrossRef]
23. Varentsov, M.; Stepanenko, V.; Repina, I.; Artamonov, A.; Bogomolov, V.; Kuksova, N.; Marchuk, E.; Pashkin, A.; Varentsov, A. Balloons and quadcopters: Intercomparison of two low-cost wind profiling methods. *Atmosphere* **2021**, *12*, 380. [CrossRef]
24. Rautenberg, A.; Graf, M.; Wildmann, N.; Platis, A.; Bange, J. Reviewing wind measurement approaches for fixed-wing unmanned aircraft. *Atmosphere* **2018**, *9*, 422. [CrossRef]
25. Shimura, T.; Inoue, M.; Tsujimoto, H.; Sasaki, K.; Iguchi, M. Estimation of wind vector profile using a hexarotor unmanned aerial vehicle and its application to meteorological observation up to 1000 m above surface. *J. Atmos. Ocean. Technol.* **2018**, *35*, 1621–1631. [CrossRef]

26. Neumann, P.P.; Bartholmai, M. Real-time wind estimation on a micro unmanned aerial vehicle using its internal measurement unit. *Sens. Actuators* **2015**, *235A*, 300–310. [CrossRef]
27. Allison, S.; Bai, H.; Jayaraman, B. Wind estimation using quadcopter motion: A machine learning approach. *Aerosp. Sci. Technol.* **2020**, *98*, 105699. [CrossRef]
28. Wang, L.; Misra, G.; Bai, X. A K Nearest neighborhood-based wind estimation for rotary-wing VTOL UAVs. *Drones* **2019**, *3*, 31. [CrossRef]
29. Chechin, D.G.; Artamonov, A.Y.; Bodunkov, N.E.; Zhivoglotov, D.N.; Zaytseva, D.V.; Kalyagin, M.Y.; Kouznetsov, D.D.; Kounashouk, A.A.; Shevchenko, A.M.; Shestakova, A.A. Experience of studying the turbulent structure of the atmospheric boundary layer using an unmanned aerial vehicle. *Izv. Atmos. Ocean. Phys.* **2021**, *57*, 526–532. [CrossRef]
30. Simma, M.; Mjøen, H.; Boström, T. Measuring wind speed using the internal stabilization system of a quadrotor drone. *Drones* **2020**, *4*, 23. [CrossRef]
31. Wang, J.Y.; Luo, B.; Zeng, M.; Meng, Q. A Wind estimation method with an unmanned rotorcraft for environmental monitoring tasks. *Sensors* **2018**, *18*, 4504. [CrossRef]
32. Brossard, M.; Condomines, J.-P.; Bonnabel, S. Tightly coupled navigation and wind estimation for Mini UAVs. AIAA 2018-1843. In Proceedings of the AIAA Guidance, Navigation, and Control Conference, Kissimmee, FL, USA, 8–12 January 2018.
33. Wang, B.H.; Wang, D.B.; Ali, Z.A.; Ting, B.T.; Wang, H. An overview of various kinds of wind effects on unmanned aerial vehicle. *Meas. Control* **2019**, *52*, 731–739. [CrossRef]
34. Al-Ghussain, L.; Bailey, S.C.C. An approach to minimize aircraft motion bias in multi-hole probe wind measurements made by small unmanned aerial systems. *Atmos. Meas. Tech.* **2021**, *14*, 173–184. [CrossRef]
35. Sekula, P.; Zimnoch, M.; Bartyzel, J.; Kud, M.; Necki, J.; Bokwa, A. Ultra-light airborne measurement system for investigation of urban boundary layer dynamics. *Sensors* **2021**, *21*, 2920. [CrossRef]
36. Reuder, J.; Jonassen, M.O. First results of turbulence measurements in a wind park with the small unmanned meteorological observer SUMO. *Energy Procedia* **2012**, *24*, 176–185. [CrossRef]
37. Reuder, J.; Jonassen, M.O. Proof of concept for wind turbine wake investigations with the RPAS SUMO. *Energy Procedia* **2016**, *94*, 452–461. [CrossRef]
38. Reineman, B.D. Development and testing of instrumentation for UAV-based flux measurements within terrestrial and marine atmospheric boundary layers. *J. Atmos. Ocean. Technol.* **2013**, *30*, 1295–1319. [CrossRef]
39. Reineman, B.D.; Lenain, L.; Melville, W.K. The use of ship-launched fixed-wing UAVs for measuring the marine atmospheric boundary layer and ocean surface processes. *J. Atmos. Ocean. Technol.* **2016**, *33*, 2029–2052. [CrossRef]
40. Balsley, B.B.; Lawrence, D.A.; Fritts, D.C.; Wang, L.; Wan, K.; Werne, J. Fine structure, instabilities, and turbulence in the lower atmosphere: High-resolution in situ slant-path measurements with the datahawk uav and comparisons with numerical modeling. *J. Atmos. Ocean. Technol.* **2018**, *35*, 619–642. [CrossRef]
41. Rautenberg, A.; Allgeier, J.; Jung, S.; Bange, J. Calibration procedure and accuracy of wind and turbulence measurements with five-hole probes on fixed-wing unmanned aircraft in the atmospheric boundary layer and wind turbine wakes. *Atmosphere* **2019**, *10*, 124. [CrossRef]
42. Fuertes, F.C.; Wilhelm, L.; Porté-Agel, F. Multirotor UAV-based platform for the measurement of atmospheric turbulence: Validation and signature detection of tip vortices of wind turbine blades. *J. Atmos. Ocean. Technol.* **2019**, *36*, 941–955. [CrossRef]
43. Kim, M.; Kwon, B.H. Estimation of sensible heat flux and atmospheric boundary layer height using an unmanned aerial vehicle. *Atmosphere* **2019**, *10*, 363. [CrossRef]
44. Luce, H.; Kantha, L.; Hashiguchi, H.; Lawrence, D. Estimation of turbulence parameters in the lower troposphere from ShUREX (2016–2017) UAV Data. *Atmosphere* **2019**, *10*, 384. [CrossRef]
45. Witte, B.M.; Singler, R.F.; Bailey, S.C.C. Development of an Unmanned Aerial Vehicle for the Measurement of Turbulence in the Atmospheric Boundary Layer. *Atmosphere* **2017**, *8*, 195. [CrossRef]
46. Wildmann, N.; Eckert, R.; Dörnbrack, A.; Gisinger, S.; Rapp, M.; Ohlmann, K.; van Niekerk, A. In situ measurements of wind and turbulence by a motor glider in the andes. *J. Atmos. Ocean. Technol.* **2021**, *38*, 921–935. [CrossRef]
47. Båserud, L.; Reuder, J.; Jonassen, M.O.; Kral, S.T.; Paskyabi, M.B.; Lothon, M. Proof of concept for turbulence measurements with the RPAS SUMO during the BLLAST campaign. *Atmos. Meas. Tech.* **2016**, *9*, 4901–4913. [CrossRef]
48. Shelekhov, A.; Afanasiev, A.; Shelekhova, E.; Kobzev, A.; Tel'minov, A.; Molchunov, A.; Poplevina, O. Using small unmanned aerial vehicles for turbulence measurements in the atmosphere. *Izv. Atmos. Ocean. Phys.* **2021**, *57*, 533–545. [CrossRef]
49. Yamada, T.; Mellor, G. A simulation of the wangara atmospheric boundary layer Data. *J. Atmos. Sci.* **1975**, *32*, 2309–2329. [CrossRef]
50. Sucher, E.; Sprung, D.; Kremer, M.; Kociok, T.; van Eijk, A.M.; Stein, K. Investigation of optical turbulence from an unmanned aerial system. In Proceedings of the Environmental Effects on Light Propagation and Adaptive Systems, Berlin, Germany, 12–13 September 2018; Volume 10787, p. 1078706.
51. Sprung, D.; Sucher, E.; Grossmann, P.; Kociok, T.; van Eijk, A.; Stein, K. Using ultrasonic anemometers for temperature measurements and implications on Cn2. In Proceedings of the Environmental Effects on Light Propagation and Adaptive Systems II, Online, 21–25 September 2020; Volume 11153, p. 111530B.
52. Monin, A.S.; Yaglom, A.M. Statistical Hydromechanics. Part 2. In *Turbulent Mechanics*; Nauka: Moscow, Russia, 1967.
53. Stull, R.B. *An Introduction to Boundary Layer Meteorology*; Kluwer Academic Publishers: Dordrecht, The Netherlands, 1989.

54. Kaimal, J.C.; Finnigan, J.J. *Atmospheric Boundary Layer Flows. Their Structure and Measurement*; Oxford University Press: Oxford, UK, 1994.
55. Teunissen, H.W. Characteristics of the mean wind and turbulence in the planetary boundary layer. *UTIAS Rev.* **1970**, *32*, 57.
56. Tieleman, H.W. Universality of velocity spectra. *J. Wind Eng. Ind. Aerodyn.* **1995**, *56*, 55–69. [CrossRef]
57. Mahony, R.; Kumar, V.; Corke, P. Multirotor aerial vehicles: Modeling, estimation, and control of quadrotor. *IEEE Robot. Autom. Mag.* **2012**, *19*, 20–32. [CrossRef]
58. Beard, R.; McLain, T. *Small Unmanned Aircraft. Theory and Practice*; Princeton University Press: Princeton, NJ, USA, 2010.

Multi-Camera Networks for Coverage Control of Drones

Sunan Huang *[iD], Rodney Swee Huat Teo and William Wai Lun Leong

Temasek Laboratories, National University of Singapore, T-Lab Building, 5A, Engineering Drive 1, Unit 09-02, Singapore 117411, Singapore; tsltshr@nus.edu.sg (R.S.H.T.); tsllwl@nus.edu.sg (W.W.L.L.)
* Correspondence: elehsn@gmail.com; Tel.: +65-98678943

Abstract: Multiple unmanned multirotor (MUM) systems are becoming a reality. They have a wide range of applications such as for surveillance, search and rescue, monitoring operations in hazardous environments and providing communication coverage services. Currently, an important issue in MUM is coverage control. In this paper, an existing coverage control algorithm has been extended to incorporate a new sensor model, which is downward facing and allows pan-tilt-zoom (PTZ). Two new constraints, namely view angle and collision avoidance, have also been included. Mobile network coverage among the MUMs is studied. Finally, the proposed scheme is tested in computer simulations.

Keywords: coverage control; distributed control; unmanned aerial vehicles

Citation: Huang, S.; Teo, R.S.H.; Leong, W.W.L. Multi-Camera Networks for Coverage Control of Drones. *Drones* **2022**, *6*, 67. https://doi.org/10.3390/drones6030067

Academic Editors: Arianna Pesci, Giordano Teza and Massimo Fabris

Received: 27 January 2022
Accepted: 25 February 2022
Published: 3 March 2022

Publisher's Note: MDPI stays neutral with regard to jurisdictional claims in published maps and institutional affiliations.

1. Introduction

Unmanned aerial vehicle (UAV) technology is currently a growing area. It offers many potential civil applications, inspiring scientists to undertake the development of new algorithms to automate UAV systems. Employing multiple unmanned multirotors (MUMs) [1–8] is rapidly becoming possible due to the development of computer hardware and communication technology. The use of UAVs is advantageous when compared to a single one. For example, when a task is very difficult, a single UAV may take a long time or may not be able to accomplish it effectively. The research challenge is then to develop the appropriate cooperation logic so that the UAVs work together to complete missions effectively and efficiently.

Coverage control is attracting research interest in MUMs [9]. The basic principle is to drive MUMs to optimal coverage for a given environment. The early works in coverage control address the visibility problem [10] or the Watchmen Tour Problem (WTP) [11], which determines the optimal number of guards and their routes, respectively, to observe a given area. These approaches form the primary research in this area, but they are not suitable for real applications. It is necessary to real world constraints in the development of coverage algorithms for MUMs. Cortés et al. [12,13] have proposed some approaches for coverage control, which are based on the Voronoi algorithm for MUMs. Here, the Voronoi algorithm is a decentralized iterative scheme to partition a 2D plane into several cells. Thereafter, the coverage problem received greater attention and many methods [4,14–16] have been proposed. Here, we give several examples. Schwager et al. [14,17] presented a distributed algorithm for dealing with the coverage, which controls MUMs to implement their coverage by configuring position and pan and tilt parameters. Piciarelli et al. [15] considered a fixed camera network and proposed a camera PTZ reconfiguration for coverage control with considering a relevance parameter, where an ellipse sensing field of view (FOV) is assumed. Parapari et al. [4] presented a distributed collision avoidance control for MUMs, but no PTZ configuration. Wang and Guo [16] handled the coverage problem by deriving a distributed control from a potential function, but no PTZ configuration. It should be noted that the result of Schwager et al. [17] is significant since the sensor used is the downward facing camera which is a practical configuration. The difficulty in this approach is that the sensing field of view is an arbitrary convex polygon of four sides. Most results do not consider the case of MUMs with downward facing cameras and arbitrary trapezoidal FOV flying in the sky. However, the camera focal parameter was not considered for configuration in that paper. In order to minimize mobility and hovering MUMs over the monitoring area, a simultaneous configuration of PTZ parameters is an important topic in coverage control. In addition, sometime a sensor would face an obstacle (e.g., a tree). Without the consideration of this situation in the coverage control, the sensor view may be occluded by dynamic or static obstacles. Thus, the view angle parameter is also important and should be considered. In a later work, Arslan et al. [18] presented a circular image sensor and developed an algorithm of configuring PTZ parameters. Unfortunately, the result of Arslan et al. [18] is based on a circular FOV, but the actual FOV is a trapezoidal one. In addition, Arslan et al. [18] use a fixed camera network to configure PTZ parameters without tuning the cameras' positions.

In this paper, we develop a distributed coverage control approach for MUMs with downward facing cameras. The proposed approach is based on the result of Schwager et al. [17]. A modified camera model is adopted in this paper. Based on this model, we extend Schwager et al.'s method to include more control variables in coverage control. The following features are improved compared with Schwager et al.'s method: (1) camera sensor model close to actual one; (2) configured rotation and PTZ parameters simultaneously; (3) imposed view angle constraints; (4) controller that can avoid collision among UAVs; (5) network convergence with more control variables and guaranteed collision avoidance.

Originally, a primary version of this paper was published in ICUAS 2018 [19]. We extend [19] to more contents: (1) the stability analysis is given for the view angle control; (2) we extend the coverage algorithm by incorporating with collision avoidance and the stability analysis is given; (3) more examples are tested in the simulation.

The paper is structured as follows: in Section 2, problem formulation and our research objectives are briefly described. In Section 3, the solution of a distributed coverage control with consideration of the unmanned aerial vehicle (UAV) position, rotation and PTZ parameters, and view angle constraint, is given. The simulation study is given in Section 4. Finally, the conclusion is given in Section 5.

2. Problem Statements

In this section, we describe the problem of coverage control of MUMs with downward facing cameras. We first define the environment and the camera model and then give the coverage control objective.

2.1. Environment

Let Q be a bounded environment $Q \subset R^2$. The degree of interest of the area in its interior is represented by a density function $\varphi(q)$ where $q \in Q$.

2.2. Camera Sensor Model

We consider n UAVs in an environment Q. Each UAV with a downward facing camera is located in position:

$$p_i = [x_i, y_i, z_i]^T.$$

Figure 1 shows the camera model, where the center of the camera is at the center of the lens; f is the focal length of the lens; L_l is the length of the camera image; L_w is the width of the camera image; each camera has a rectangular field of view (FOV), which is the intersection of the cone of the camera lens with the environment Q. There are two coordinate systems in the coverage problem: one is the camera coordinates (CC) of UAV i, and the other one is the global coordinates (GC). The CC is at the center of the camera lens, with the z-axis pointing downward facing to the ground through the lens, while the GC is fixed on the ground, with the z-axis points upward normal to the ground. In CC, the camera view is like a pyramid object whose base is a rectangle shape and whose sides are triangles uniting at a common apex, the center of the camera lens. The sides of the pyramid have the outward normal vectors e_1, e_2, e_3, e_4. Here, we derive the first outward normal vector e_1. The side view of the camera with pyramid FOV is shown in Figure 2. From the figure, it is known that two triangles $\Delta v_1 o v_2$ and $\Delta v_3 o v_4$ are similar since the vector e_1 and the line $o v_3$ are perpendicular. Thus, the projection of the normal vector e_1 on the y-axis is $o v_2$ and its value is f, while the rejection of the normal vector e_1 from the y-axis is $v_1 v_2$ and its value is $-\frac{L_l}{2}$. Therefore, the normal vector e_1 in the CC system is $[0 \ \ f \ \ -\frac{L_l}{2}]$. We are interested in the unit normal vector and thus it follows that:

$$e_1 = [0 \quad \frac{2f}{\sqrt{L_l^2+4f^2}} \quad -\frac{L_l}{\sqrt{L_l^2+4f^2}}]^T \tag{1}$$

In a similar way, we can obtain the other remained three outward normal vectors:

$$e_2 = [\frac{2f}{\sqrt{L_w^2 + 4f^2}} \quad 0 \quad -\frac{L_w}{\sqrt{L_w^2 + 4f^2}}]^T \tag{2}$$

$$e_3 = [0 \quad -\frac{2f}{\sqrt{L_l^2 + 4f^2}} \quad -\frac{L_l}{\sqrt{L_l^2 + 4f^2}}]^T \tag{3}$$

$$e_4 = [\frac{2f}{\sqrt{L_w^2 + 4f^2}} \quad 0 \quad -\frac{L_w}{\sqrt{L_w^2 + 4f^2}}]^T \tag{4}$$

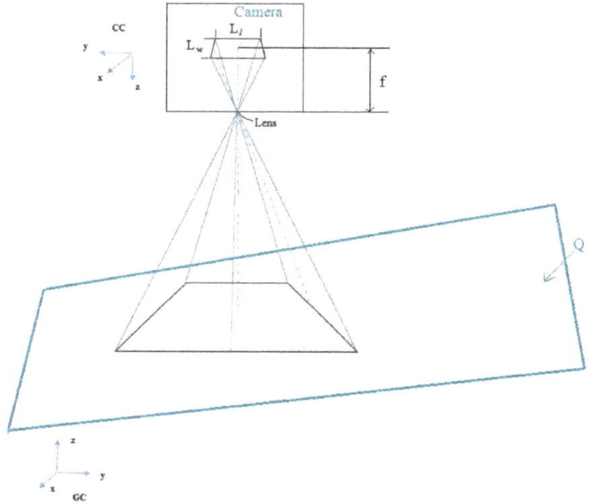

Figure 1. Camera model.

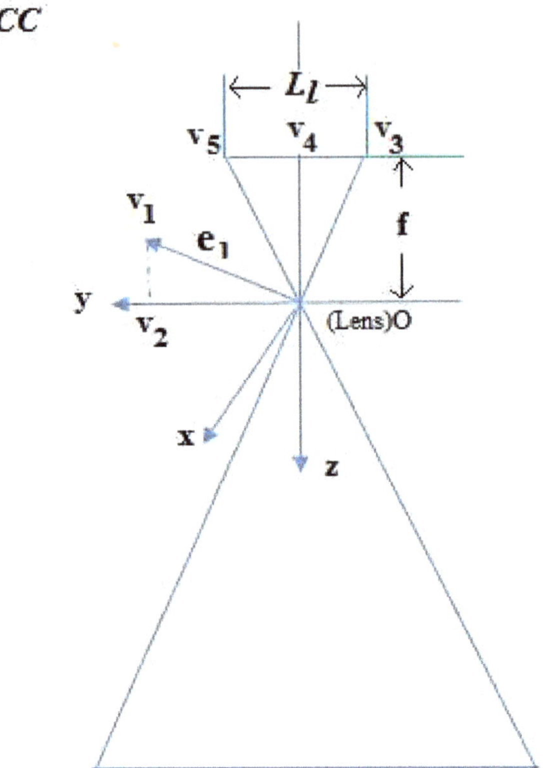

Figure 2. Side view of camera with pyramid FOV.

For the GC system, we want to transform it to get a CC system, which can be represented in 3D rotation matrix form as R. This matrix can be implemented by using two elementary transforms, each matrix realizes a rotation. This is achieved by rotating the z-axis by $\pi/2$ whose transformation is given by:

$$R_i(\frac{\pi}{2}) = \begin{bmatrix} cos(\pi/2) & sin(\pi/2) & 0 \\ -sin(\pi/2) & cos(\pi/2) & 0 \\ 0 & 0 & 1 \end{bmatrix},$$
(5)

and then turning the x-axis over by π whose transformation is given by:

$$R_i(\pi) = \begin{bmatrix} 1 & 0 & 0 \\ 0 & cos(\pi) & sin(\pi) \\ 0 & -sin(\pi) & cos(\pi) \end{bmatrix},$$
(6)

We can express this transformation by:

$$R_i^{\frac{\pi}{2} \to \pi} = R_i(\pi)R_i(\frac{\pi}{2}) = \begin{bmatrix} 0 & 1 & 0 \\ 1 & 0 & 0 \\ 0 & 0 & -1 \end{bmatrix},$$
(7)

Finally, for a 3D point v in GC, we can transform the vector $(v - p_i)$ into the CC system by the transformation $R_i^{\frac{\pi}{2} \to \pi}$, that is $R_i^{\frac{\pi}{2} \to \pi}(v - p_i)$. For a $q \in Q$, the vector $I_{32}q - p$ (where

$$I_{32} = \begin{bmatrix} 1 & 0 \\ 0 & 1 \\ 0 & 0 \end{bmatrix}$$) expressed in CC is given by $R_i^{\frac{\pi}{2} \to \pi}(I_{32}q - p)$.

Now, we are concerned about the pyramid FOV, as shown in Figure 1. This pyramid has four sides and they intersect with the ground to get four edges of the camera FOV. Each edge is defined as $l_k, k = 1, 2, 3, 4$. If the point q is on the edge l_k of the camera FOV, then:

$$e_k^T R_i^{\frac{\pi}{2} \to \pi}(I_{32}q - p_i) = 0, k = 1, 2, 3, 4$$

since the outward normal vector e_i and the vector $R_i^{\frac{\pi}{2} \to \pi}(I_{32}q - p_i)$ are perpendicular. Thus, for a given $q \in Q$, the camera FOV is represented by:

$$B_i = \left\{ q : \begin{bmatrix} e_{1i}^T R_i^{\frac{\pi}{2} \to \pi}(I_{32}q - p_i) \leq 0 \\ e_{2i}^T R_i^{\frac{\pi}{2} \to \pi}(I_{32}q - p_i) \leq 0 \\ e_{3i}^T R_i^{\frac{\pi}{2} \to \pi}(I_{32}q - p_i) \leq 0 \\ e_{4i}^T R_i^{\frac{\pi}{2} \to \pi}(I_{32}q - p_i) \leq 0 \end{bmatrix} \right\}. \tag{8}$$

The observed area A from the camera, which is a rectangle form is given by:

$$A = z^2 \frac{L_l L_w}{f^2}$$

which is obtained directly from Figure 1. The number of pixels are given by N_l and N_w. The area/pixel is given by:

$$\frac{area\,A}{pixel} = \frac{L_l L_w z^2}{N_l N_w f^2}. \tag{9}$$

Thus, we have a function:

$$g(p_i, q) = \begin{cases} \frac{L_{li} L_{wi} z_i^2}{N_{li} N_{wi} f_i^2}, & \text{for } q \in B_i \\ \infty, & \text{otherwise.} \end{cases} \tag{10}$$

Remark 1. *It should be noted that the area/pixel is different from the result of Schwager et al. [17], where Schwager et al. used a magnification factor to represent the area observed from the camera, while our expression is more practical by using area and pixel resolution. The proposed sensor model also differs from that of Schwager et al.'s work. The center of the UAV (or camera) is defined at the center of the camera lens in our model, while the center of the camera in Schwager et al.'s work was defined at the center of the camera focal. If the zoom parameter f is configured based on Schwager et al.'s work, this will result in the changes of the UAV altitude (it is redundant to adjust the zoom parameter in this situation). Whereas, the configuration based on our sensor model will not affect the UAV altitude. In addition, Schwager et al.'s work does not consider the size of the camera image, which affects the FOV.*

2.3. Wireless Camera Networks

In this paper, the coverage algorithm is applicable to the wireless camera networks formed of cameras with PTZ. This implies that each UAV is communicated with other UAVs. The main assumptions of this network are described below.

- Each camera can obtain its position and velocity by its GPS built in UAV.
- Each camera can obtain its FOV information by its camera image.

- The camera network is connected; therefore, each camera has a sensing and communication capability or multi-hop transmission capacity to transmit related information to neighbor cameras.

  It should be noted that in the present wireless camera network, (a) we need to transmit relevant FOV information to adjacent cameras to find FOV overlap, it is not necessary to transmit all FOVs of cameras in the environment. Therefore, the information broadcast can be set like two or three hops before it stops rebroadcasting (UDP protocol can be used). (b) It needs to communicate with all of the cameras to obtain only the position information of other UAVs.

2.4. Optimization Coverage Problem

For the multi-UAV coverage control, we have to design a cost function that can be used for evaluating the coverage performance. It is a natural way to use the information, as discussed in the previous section. Defining the cost function (see [17]) as our control objective is given by:

$$\mathcal{J} = \int_Q \left(\sum_{i=1}^{n} g(p_i, q)^{-1} + Y^{-1} \right)^{-1} dq \tag{11}$$

where $Y > 0$ is a positive constant. This constant is used to avoid the integration part as zero due to $g(p_i, q)^{-1}$. It may be interpreted as baseline information of the environment. The optimization coverage problem of multi-UAVs is to minimize the objective function (11).

3. Distributed Coverage with Configuration of Position, Yaw and PTZ

Our goal is to develop a distributed control for the coverage problem, where each UAV exchanges information with its neighboring UAVs. Thus, the area can be represented by:

$$J_{Nq} = \left(\sum_{i \in N_q} g(p_i, q)^{-1} + Y^{-1} \right)^{-1} \tag{12}$$

where N_q is the camera set described by:

$$N_q = \{ i : q \in B_i, i = 1, 2, ...n \}.$$

The objective function becomes:

$$\mathcal{J} = \int_Q J_{Nq} \varphi(q) dq \tag{13}$$

When the UAV heading yaw and pan-tilt-zoom (PTZ) cameras are considered in UAV camera networks, many problems arise. One of the problems is how to achieve the coverage by adjusting position, yaw (rotation) and PTZ parameters in a MUM system. Since the camera FOV affects coverage, any strategy for coverage control must be associated with cameras PTZ parameters. In this section, we develop the control law to consider such a general case, the configuration of position, yaw (rotation) and PTZ cameras. Consider the state of the UAV i with gimbal camera to be given by:

$$\chi_i = [x_i, y_i, z_i, \psi_i^r, \psi_i^p, \psi_i^t, f_i]^T,$$

where ψ_i^r, ψ_i^p, and ψ_i^t are the rotation (or yaw) , pan and tilt angles, respectively, and f_i is the focal length of the camera i. Usually, the rotation (yaw) is controlled by UAV, while the PTZ parameters are controlled by a gimbal camera system. We assume that both the UAV and camera are considered as one particle (camera lens representing this point) and their positions are the same one represented by this point. As explained in Section 2, the center of the UAV position is the camera lens. Consider the state equation of the camera i, that is:

$$\dot{\chi}_{ii} = u_i \tag{14}$$

where:

$$u_i = [u_{xi}, u_{yi}, u_{zi}, u_{\psi_i^r}, u_{\psi_i^p}, u_{\psi_i^t}, u_{f_i}]^T.$$

Let $p_i = [x_i, y_i, z_i]^T$ be the position of the camera i at the center of the lens (it is also the position of UAV). It should be noticed that for each UAV their dynamic is independent. For neighboring UAVs, the wireless communication network is assumed to be connected during the mission.

As shown in Section 2, a rectangular FOV on the GC coordinates is converted into one which is based on the CC system, by rotating a matrix $R_i^{\frac{\pi}{2} \to \pi}$. Sometimes we have to configure the yaw angle such that the coverage is achieved optimally. In this situation, a rotation matrix for rotating the yaw (rotation) angle is introduced as described below:

$$Rotation\ about\ z - axis\ (yaw\ angle)$$

$$R_{yaw}(\psi_i^r) = \begin{bmatrix} \cos\psi_i^r & \sin\psi_i^r & 0 \\ -\sin\psi_i^r & \cos\psi_i^r & 0 \\ 0 & 0 & 1 \end{bmatrix}. \tag{15}$$

Similarly, we can tune pan and tilt angles to control the camera to get a better FOV on the ground. The rotation matrices corresponding to the pan and tilt transformation are given by:

$$Rotation\ about\ x - axis\ (pan\ angle)$$

$$R_{pan}(\psi_i^p) = \begin{bmatrix} 1 & 0 & 0 \\ 0 & \cos\psi_i^p & \sin\psi_i^p \\ 0 & -\sin\psi_i^r & \cos\psi_i^r \end{bmatrix}, \tag{16}$$

and:

$$Rotation\ about\ y - axis\ (tilt\ angle)$$

$$R_{tilt}(\psi_i^t) = \begin{bmatrix} \cos\psi_i^t & 0 & -\sin\psi_i^t \\ 0 & 1 & 0 \\ \sin\psi_i^t & 0 & \cos\psi_i^t \end{bmatrix}, \tag{17}$$

respectively. Finally, for a 3D point v in a GC system, we can transform the vector $(v - p_i)$ into the CC system by rotating R_i, which is given by:

$$R_i = R_{tilt}(\psi_i^t) R_{pan}(\psi_i^p) R_{yaw}(\psi_i^r) R_i^{\frac{\pi}{2} \to \pi} \tag{18}$$

For a point q inside FOV, we can express the vector $(I_{32}q - g_i)$ in the CC system by $R_i(I_{32}q - p_i)$. Next, we are concerned about the FOV cone, as shown in Figure 1. By the transformation, the camera FOV shown in (8) in the CC system is changed to:

$$B_i = \{q : \begin{bmatrix} e_{1i}^T R_i(I_{32}q - p_i) \\ e_{2i}^T R_i(I_{32}q - p_i) \\ e_{3i}^T R_i(I_{32}q - p_i) \\ e_{4i}^T R_i(I_{32}q - p_i) \end{bmatrix} \leq 0\} \tag{19}$$

It should be noted that when tuning yaw or PTZ parameters, FOV is not a rectangular shape, as shown in Section 2, but it is a trapezium or trapezoid in this transformation

R_i. Thus, the area of such FOV shape can be viewed as a mapping defined by C_{fov} from the image area $L_l L_w$ to an area of a trapezium FOV on the ground, that is expressed as $C_{fov} : L_l L_w \rightarrow FOV$. Therefore, the $\frac{area}{pixel}$ in (9) should be changed to:

$$\frac{area}{pixel} = \frac{C_{fov}}{N_l N_w f^2} \parallel I_{32} q - p_i \parallel^2, \tag{20}$$

where $\parallel I_{32} q - p_i \parallel$ is the the distance between the camera and the point q inside FOV. Now, due to the use of yaw and PTZ cameras, the function $g(p_i, q)$ is changed to:

$$g(p_i, q) = \begin{cases} \frac{C_{fov}(L_{li} L_{wi})}{N_{li} N_{wi} f_i^2} \parallel I_{32} q - p_i \parallel^2, & \text{for } q \in B_i \\ \infty, & \text{otherwise.} \end{cases} \tag{21}$$

Consider the state equation of UAV i with the PTZ camera, that is:

$$\dot{x}_{ii} = u_i \tag{22}$$
$$u_i = -\rho \frac{\partial \mathcal{J}}{\partial \chi_i}, \chi_i = [p_i, \psi_i^r, \psi_i^p, \psi_i^t, f_i] \tag{23}$$

where $\rho > 0$ is a factor, and the gradient $\frac{\partial \mathcal{J}}{\partial \chi_i}$ (see Appendix) is given by:

$$\frac{\partial \mathcal{J}}{\partial p_i} = \sum_{k=1}^{4} \int_{Q \cap l_{ki}} (J_{Nq} - J_{Nq \setminus \{i\}}) \varphi(q) \frac{R_i^T e_{ki}}{\parallel I_{23} R_i^T e_{ki} \parallel} dq$$
$$- \int_{Q \cap B_i} \frac{2 N_{li} N_{wi} f_i^2 J_{Nq}^2 (I_{32} q - p_i)}{L_{li} L_{wi} \parallel I_{32} q - p_i \parallel^4} \varphi(q) dq, \tag{24}$$

$$\frac{\partial \mathcal{J}}{\partial \psi_i^w} = \sum_{k=1}^{4} \int_{Q \cap l_{ki}} (J_{Nq} - J_{Nq \setminus \{i\}}) \frac{\varphi(q) e_{ki}^T \frac{\partial R_i}{\partial \psi_i^w} (p_i - I_{32} q)}{\parallel I_{23} R_i^T e_{ki} \parallel} dq$$

$$w \in \{r, p, t\} \tag{25}$$

where $\frac{\partial R_i}{\partial \psi_i^w}$ can be obtained by differentiating R_i with respect to ψ_i^w directly, and:

$$\frac{\partial \mathcal{J}}{\partial f_i} = \sum_{k=1}^{4} \int_{Q \cap l_{ki}} (J_{Nq} - J_{Nq \setminus \{i\}}) \frac{\varphi(q) \frac{\partial e_{ki}^T}{\partial f_i} R_i (p_i - I_{32} q)}{\parallel I_{23} R_i^T e_{ki} \parallel} dq$$
$$- \int_{Q \cap B_i} \frac{2 J_{Nq}^2 f_i N_l N_w}{L_l L_w \parallel I_{32} q - p_i \parallel^2} \varphi(q) dq. \tag{26}$$

It is observed from (26) that if we fix the parameters x, y, z, the updating parameter f will not affect the position of UAV. This implies that we can hover the UAV at a certain position and tune PTZ parameters to get the optimal coverage. The factor ρ is used to affect the convergence to the coverage. Large values of ρ can speed up the convergence. However, it cannot be increased to a very large value due to the control limits.

The following theorem is given for configuring the parameters $x_i, y_i, z_i, \psi_i^r, \psi_i^p, \psi_i^t, f_i$ simultaneously.

Theorem 1. *In a network of n UAVs governed by dynamics (22) with downward facing cameras, the gradient of the cost function $\mathcal{J}$ with respect to the state variables $\chi_i = [x_i, y_i, z_i, \psi_i^r, \psi_i^p, \psi_i^t, f_i]^T$ is given by (24)–(26).*

The proposed coverage control is a decentralized form. Based on the gradient descent, the configuration of the control variables is an iterative process, as shown in Figure 3. The iteration is terminated when the desired coverage rate is reached, where the coverage rate is $\dfrac{multiple\ UAVs\ coverage\ of\ the\ area\ of\ interest}{area\ of\ interest}$.

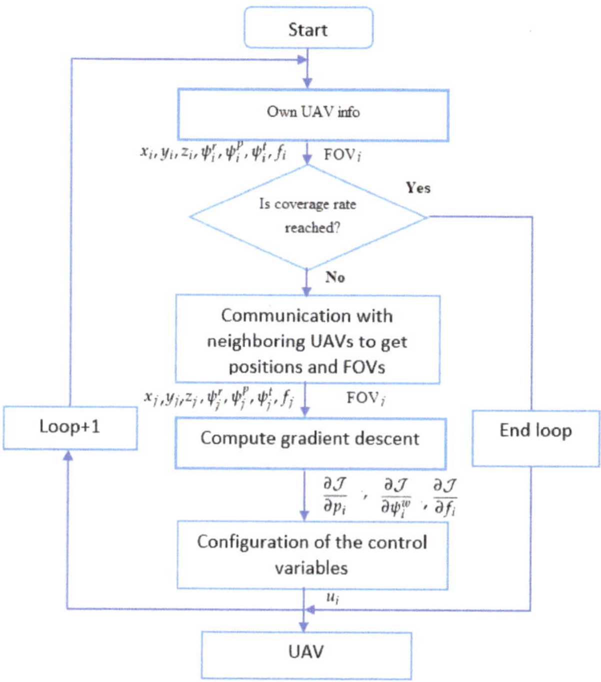

Figure 3. Chart block of decentralized coverage control scheme.

Remark 2. *In [17], the configuration of the focal parameter f is not considered. In the proposed algorithm, we can configure the position (x, y, z), yaw and PTZ parameters. This improves the results of [17].*

3.1. Incorporating View Angle

In a practical situation, sometimes a sensor (e.g., solid-state LiDAR) faces an obstacle (e.g., a tree), which may be between the sensor and the target and it can not capture the object image even if the target is within the view angle coverage of the sensor; this is known as an occlusion. If we consider the view angle as a target feature, this implies that the coverage must include this feature. Without the consideration of this feature in the coverage control, the target may be occluded by dynamic or static objects. For example, we hope the camera i to rotate its angle to 30 degrees so that we can see the observed region clearly. This is a reference following control problems.

Let us consider this situation in coverage control. Assume that we require the UAV to follow a desired view angle ψ_i^{wD}, $w \in \{r, p, t\}$. The controller design should be to achieve this objective, i.e., ψ_i^w approaches the desired ψ_i^{wD} as soon as possible. Let $\tilde{\psi}_i^w = \psi_i^{wD} - \psi_i^w$. We propose the following controller:

$$u_{\psi_i^w} = \dot{\psi}_i^{wD} + k_{pi}^w \tilde{\psi}_i^w \tag{27}$$

where $k_{pi}^w > 0$ is the constant control gain, which can be determined by users.

Assumption 1. *The desired angle $\psi_i^{wD}, w \in \{r, p, t\}$ is smooth and bounded, and its derivative $\dot{\psi}_i^{wD}$ is also bounded.*

Rearranging the state equations of the camera i, we get:

$$\dot{\eta}_i = u_{\eta i} \tag{28}$$

$$\dot{\psi}_i^w = u_{\psi_i^w}, w \in \{r, p, t\}. \tag{29}$$

where $\eta_i = [x_i, y_i, z_i, f_i]^T, u_{\eta i} = [u_{xi}, u_{yi}, u_{zi}, u_{fi}]^T, \psi_i^w = [\psi_i^r, \psi_i^p, \psi_i^t]^T$ *and* $u_{\psi_i^w} = [u_{\psi_i^r}, u_{\psi_i^p}, u_{\psi_i^t}]^T$.

We have the following theorem to ensure the stability.

Theorem 2. *In a network of n UAVs governed by dynamics (22) with downward facing cameras, the control laws $u_{xi}, u_{yi}, u_{zi}, u_{fi}$ are designed by (23), while the control laws $u_{\psi_i^w}, w \in \{r, p, t\}$ are designed by (27). If Assumption 1 holds, then:*
(1) the proposed control laws lead to a bounded coverage of $\mathcal{J}$;
(2) $\lim_{t\to\infty} \tilde{\psi}_i^w = 0$.

Proof. See Appendix A. $\square$

Remark 3. *For a camera network, it is possible that some UAVs use the position and PTZ control with/without considering view angle. In this situation, those UAVs without considering view angle are referred to as zero view angle control, i.e., the desired view angle is set to $\psi_i^{wD} = 0$. Thus, we have a similar result as in Theorem 2.*

Remark 4. *The stability in Theorem 2 ensures that (1) the coverage is bounded when applying the proposed coverage control with view angle; (2) it is converged to the desired view angle.*

3.2. Incorporating Collision Avoidance

Collision avoidance is a practical problem in multi-UAV networks. So far, the control design does not consider collisions with other UAVs when implementing the coverage objective. In what follows, we incorporate a potential function to design a coverage control with autonomously avoiding collisions.

Define a distance as $d_{ij} = ||p_i - p_j||, i, j = 1, 2, ...n, i \neq j$. Let Γ_i be the set of indices of cameras of the ith UAV for which d_{ij} is within certain regions; that is, $\Gamma_i = \{j : r < d_{ij} \leq R, j = 1, 2, ...n, j \neq i\}$. The view of $r < d_{ij} \leq R$ is shown in Figure 4, where r and R ($R > r > 0$) are the radii of the avoidance and detection regions. The potential function [20] based on n UAVs is given by:

$$V_{P_i} = \sum_{j=1, j\neq i}^{n} v_{P_{ij}} \tag{30}$$

where:

$$v_{P_{ij}} = \left(\min\left\{0, \frac{d_{ij}^2 - R^2}{d_{ij}^2 - r^2}\right\}\right)^2 \tag{31}$$

This is a repulsive potential function. The partial derivative of $v_{P_{ij}}$ with respect to $p_i = [x_i, y_i, z_i]^T$ is given by:

$$\frac{\partial v_{P_{ij}}^T}{\partial p_i} = \begin{cases} 0, & \text{if } d_{ij} \geq R \\ 4\frac{(R^2-r^2)(d_{ij}^2-R^2)}{(d_{ij}^2-r^2)^3}(p_i - p_j)^T, & \text{if } r < d_{ij} < R \\ 0, & \text{if } d_{ij} < r \end{cases} \tag{32}$$

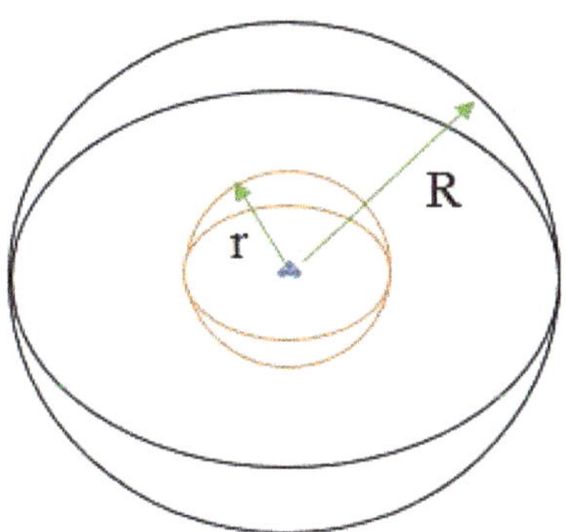

Figure 4. 3D view of avoidance function. Avoidance region with radius r and detection region with radius R.

Rearrange the state equations of the UAV i as:

$$\dot{p}_i = u_{pi} \tag{33}$$
$$\dot{\beta}_i = u_{\beta_i}, \tag{34}$$

where $\beta_i = [\psi_i^r, \psi_i^p, \psi_i^t, f_i]^T$.

Consider the following performance criterion:

$$V = \mathcal{J} + \frac{1}{2}\sum_{i=1}^{n} V_{P_i} \tag{35}$$

The following theorem is given to establish a stability result.

Theorem 3. *In a network of n UAVs governed by dynamics (22) with downward facing cameras, the control laws are given by:*

$$u_{pi} = -\rho\left(\frac{\partial \mathcal{J}}{\partial p_i} + \sum_{j=1, j\neq i}^{n} \frac{\partial v_{P_{ij}}}{\partial p_j}\right) \tag{36}$$

$$u_{\beta_i} = -\rho\frac{\partial \mathcal{J}}{\partial \beta_i}, \tag{37}$$

with (32), (A9), (A17) and (A18). If the initial configuration such that $d_{ij}(0) > r$, then:
(1) The proposed control laws lead to a bounded coverage of $\mathcal{J}$;
(2) The collision among UAVs is avoided.

Proof. See Appendix A. □

Remark 5. *If we incorporate the view angle control, in this situation, we have a similar conclusion as in Theorem 3.*

4. Simulation Studies

Simulation tests of the proposed coverage controls are given in this section. The HP computer with intel (R) Core(TM) i7-4770 CPU was used. The operation system was window 10. MATLAB software was used for simulations. In all case studies, the environment Q was the same as in the rectangle form $[-200, 200] \times [-200, 200] \subset R^2$. All UAVs were identical with the camera model, as shown in Figure 1. The parameters of the camera image sensor were $L_l = 3$ mm, $L_w = 5$ mm, $N_l = 640$ pixels, $N_v = 480$ pixels and Y was chosen as 1000.

4.1. Case 1

We considered 2 UAVs for coverage control. The sensor model in Section 2.2 was used in a rectangle form. The configuration of the camera network included UAV positions without rotation and PTZ tuning. The proposed algorithm (23) was used to control UAVs such that the coverage cost function was minimized. We have considered a density function $\phi(q) = e^{-(\frac{x-20}{20})^2 - (\frac{y-40}{40})^2}$ which is an area of interest (convex type). Two UAVs try to cover the area of interest by tuning the UAV position (x,y,z) parameters. Simulation results are illustrated in three parts: initial stage, after the initial stage and final stage. Initially, two UAVs started as seen in Figure 5. Figure 6 shows a representation of the coverage improvement after the initial stage. The final stage is shown in Figure 7. It was observed that the camera network coverage increased significantly through multiple rounds and the area was almost covered in the final stage.

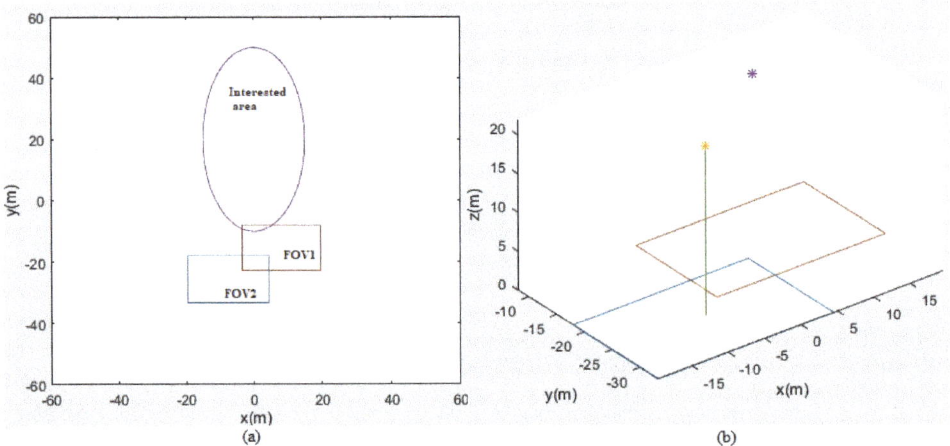

Figure 5. Test 1. Camera movement and sensing FOV (Initial configuration): (**a**) 2D coverage, (**b**) cameras' locations and their corresponding sensing regions.

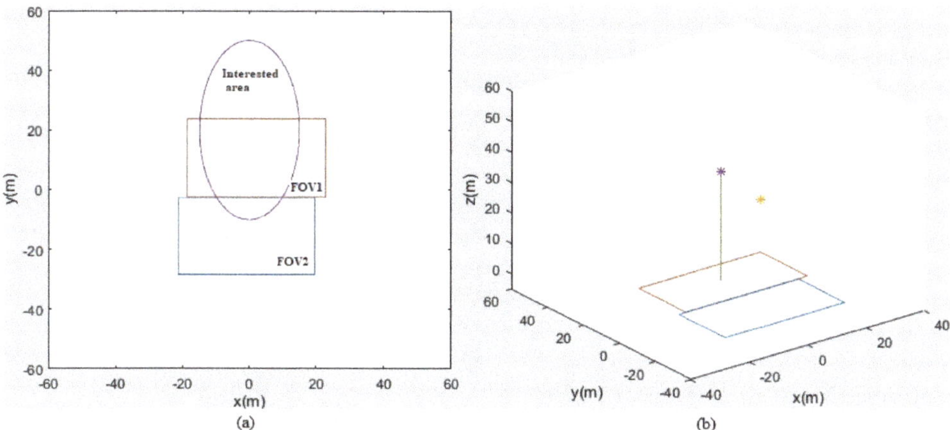

Figure 6. Test 1. Camera movement and sensing FOV (Middle configuration): (**a**) 2D coverage, (**b**) cameras' locations and their corresponding sensing regions.

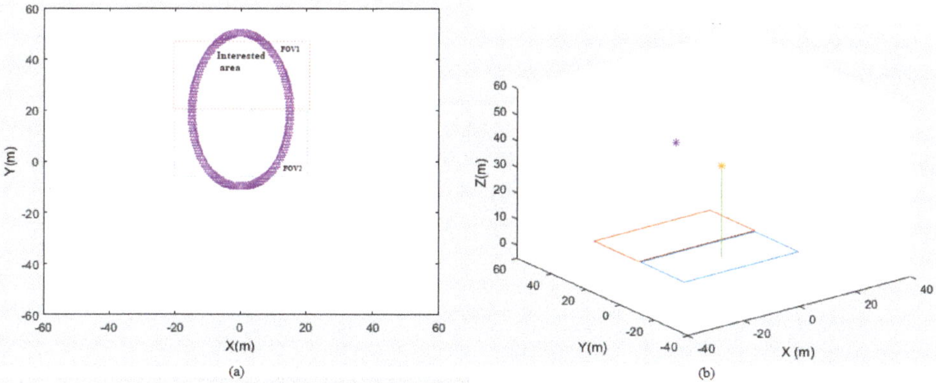

Figure 7. Test 1. Camera movement and sensing FOV (Final configuration): (**a**) 2D coverage, (**b**) cameras' locations and their corresponding sensing regions.

We consider configuring more parameters such as the UAV position (x,y,z), rotation, pan and tilt angles, where the proposed controls (23) were used. However, one UAV is required to configure the view angle (rotation), where the proposed control (27) was used. A density function $\phi(q) = e^{-(\frac{x}{20})^2 - (\frac{y-20}{40})^2} + e^{-(\frac{x-50}{50})^2 - (\frac{y-50}{20})^2}$ is a region of interest, which is a non-convex type, as shown in Figure 8a. Three UAVs try to cover the area of interest as soon as possible. Simulation results are illustrated in three parts: initial stage, after the initial stage and final stage. The first UAV, which is marked in blue color, configures the (x,y,z), rotation, pan and tilt parameters, but the rotation angle is required to follow the desired rotation angle 90 degree, starting at the x-axis of the CC system, which is right-handed. The other two UAVs use the coverage control with the configuration of the camera position (x, y, z), rotation, pan and tilt parameters. Initially, three UAVs start grouped, as seen in Figure 8 and the coverage is poor. Later, Figure 9 shows a representation of the coverage improvement after the initial stage and the 1/2 coverage of the area of interest is achieved. It is observed from Figures 8 and 9 that the view angle of the first UAV (marked in blue color) is controlled in 90 degree. The final stage is shown in Figure 10, where the rotation angle of the first UAV is still controlled in 90 degrees. It is observed from Figures 8–10 that the camera network coverage is increased significantly through multiple rounds and 93% of the area of interest is almost covered in the final stage.

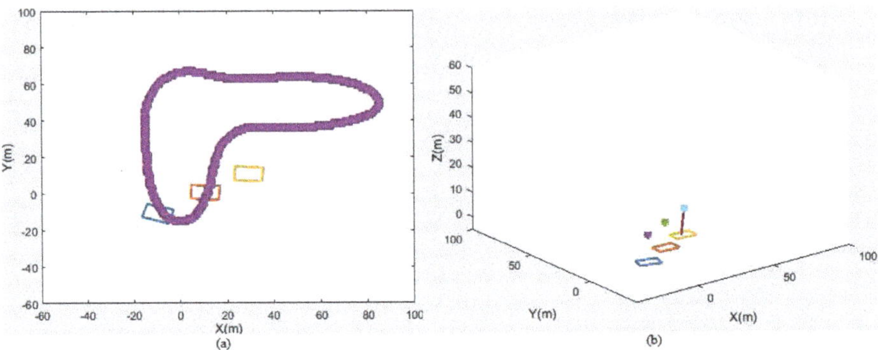

Figure 8. Test 2. Camera movement and sensing FOV (Initial configuration): (**a**) 2D coverage, (**b**) cameras' locations and their corresponding sensing regions.

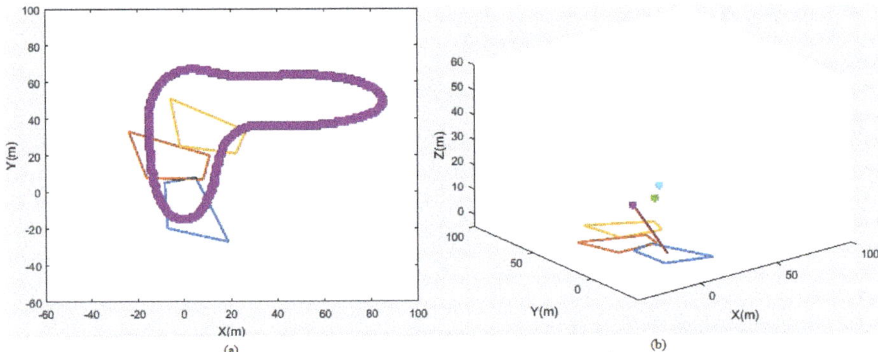

Figure 9. Test 2. Camera movement and sensing FOV (Middle configuration): (**a**) 2D coverage, (**b**) cameras' locations and their corresponding sensing regions.

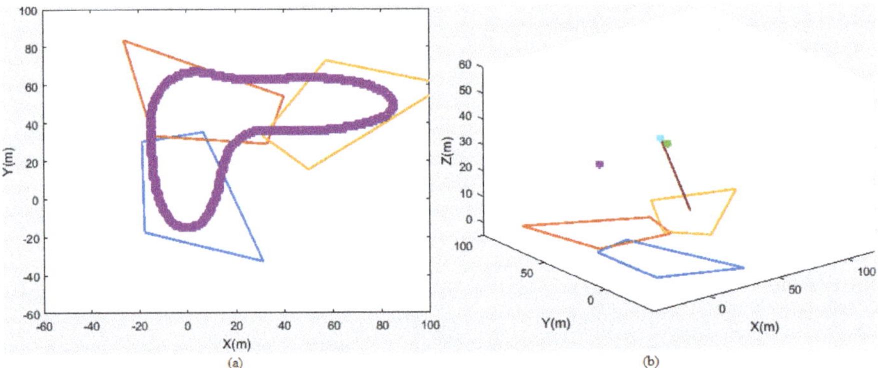

Figure 10. Test 2. Camera movement and sensing FOV (Final configuration): (**a**) 2D coverage, (**b**) cameras' locations and their corresponding sensing regions.

4.2. Case 3

We configured the camera focal length (zoom parameter), which was not controlled in Cases 1-2. The altitude (z) was fixed at 35 meters high so that the camera's FOV was observed as updating the camera's zoom parameter. Thus, the configuration parameters include the camera's position (x,y), rotation and PTZ. We used the coverage control laws, as proposed in (23). The density function $\phi(q)$ was the same as in Case 2. Four UAVs were intended to be used in the coverage control laws to cover the area of interest automatically. Simulation results are illustrated in three parts: initial stage, after the initial stage and final stage. Initially, four UAVs grouped, as seen in Figure 11, where the initial focal length was 8 for all cameras. It is observed that at the initial stage, the multi-UAVs were poor in coverage. After that, the coverage control was used to drive all UAVs to improve the coverage by configuring the (x,y), rotation and PTZ. Figure 12 shows a representation of the coverage improvement after the initial stage and almost 1/2 the coverage of the area of interest was achieved. It is observed that the the FOV of each UAV is changing such that the coverage performance is improved, especially in the size of the FOV of each UAV. This is because the zoom parameter was configured. The final stage is shown in Figure 13. It is observed that even though the altitude of all UAVs was fixed, the camera network coverage was achieved as about 97 percent by controlling four UAVs by configuring the zoom and other parameters.

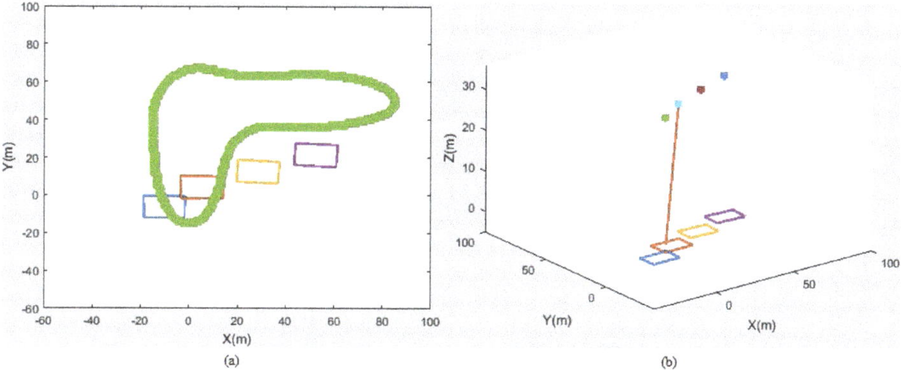

Figure 11. Test 3. Camera movement and sensing FOV (Initial configuration): (**a**) 2D coverage, (**b**) cameras' locations and their corresponding sensing regions.

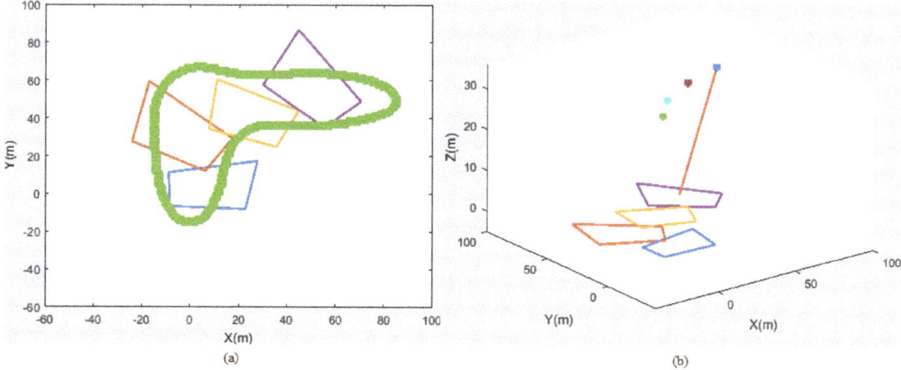

Figure 12. Test 3. Camera movement and sensing FOV (Middle configuration): (**a**) 2D coverage, (**b**) cameras' locations and their corresponding sensing regions.

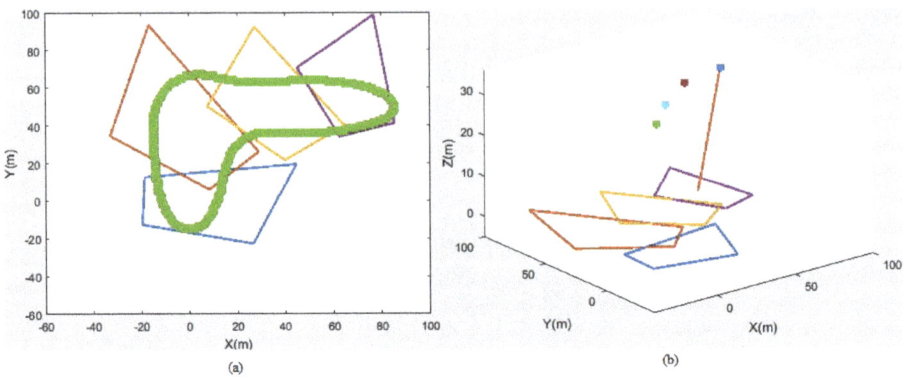

Figure 13. Test 3. Camera movement and sensing FOV (Final configuration): (**a**) 2D coverage, (**b**) cameras' locations and their corresponding sensing regions.

The previous results did not consider collision avoidance when configuring the camera parameters. In a practical situation, it is quite dangerous without collision avoidance mechanisms in controlling UAVs. In this case, we show the configuration of the parameters (x,y,z), rotation, pan and tilt with the collision avoidance proposed in Section 3.2. The density function $\phi(q)$ was the same as in Case 2. Four UAVs try to cover the area of interest area as soon as possible. It is required to have a coverage rate of 85 percent. Assume that the radii of the avoidance and detection were $r = 30$ and $R = 35$, respectively. Figure 14 shows a representation of the coverage without considering the collision avoidance, while Figure 15 shows a representation of the coverage with the collision avoidance function. The horizontal distance profile between the ith UAV and jth UAV ($i \geq j$) is shown in Figure 16. It is observed from Figure 16 that the minimum separation distance was about 25 m (<the radius of the avoidance) when, without considering the collision avoidance, it was greater than 30 m (>the radius of the avoidance) when considering the collision avoidance. This verifies that the proposed collision avoidance can work well. Nearly 85 percent of the camera network coverage rate was achieved. In this case, the coverage rate was less than in Cases 1–3, since we intend to demonstrate the collision avoidance and use a large radius of avoidance, i.e., 30.

The proposed decentralized control requires knowing the position and FOV information of the neighboring UAVs. As discussed in Section 2.3, a communication network should be built for exchanging information from each other. The proposed coverage control is an iterative process based on the gradient descent. Simulation results have shown that the desired coverage rate can be reached by multi-UAV configurations. Even if during the initial stage, UAVs are poor in coverage (see Cases 2 and 3), the desired coverage rate can still be achieved by configuration of the variables. Moreover, as a collision avoidance term is added into the control, it is not necessary to worry about avoiding issues when conducting a coverage control, which was observed in Case 4. The drawback of the proposed algorithm is that the FOV of the ith UAV may be outside of the area of interest, which is observed from Figure 13.

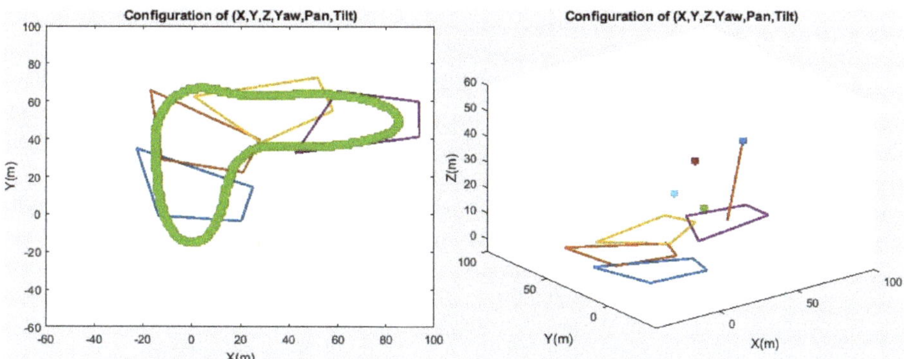

Figure 14. Test 4. Camera movement and sensing FOV without considering collision avoidance: (**a**) 2D coverage, (**b**) cameras' locations and their corresponding sensing regions.

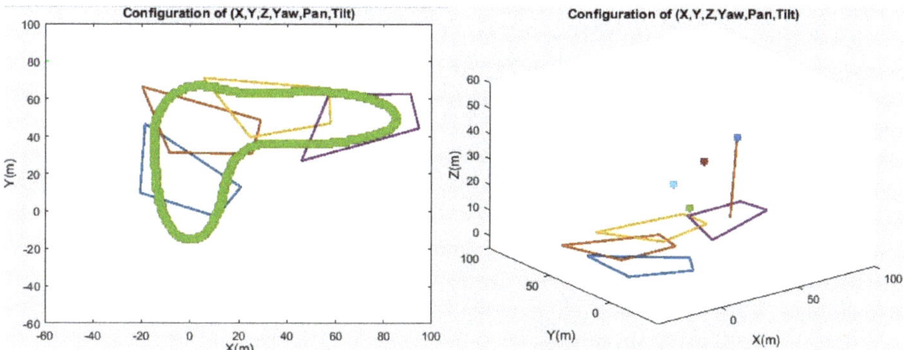

Figure 15. Test 4. Camera movement and sensing FOV with considering collision avoidance: (**a**) 2D coverage, (**b**) cameras' locations and their corresponding sensing regions.

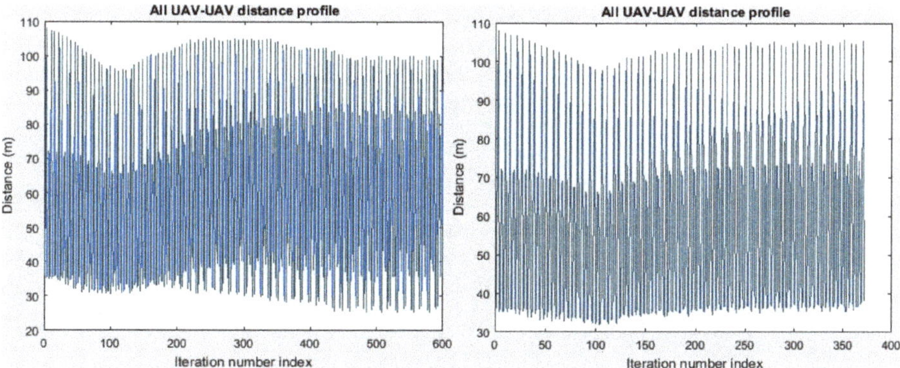

Figure 16. Test 4. Separation distance along horizontal direction: (**a**) configuration without considering collision avoidance, (**b**) configuration with considering collision avoidance.

5. Discussion and Conclusions

The development of coverage strategies is the main topic in multi-UAV systems. Technical considerations in camera networks consisting of multi-UAVs are still a challenge. Little work has been done in the design of the coverage control that involves the configuration of multiple parameters and view angle and collision avoidance.

This paper has presented a solution for coverage control of the multi-UAV system with downward facing cameras. We have extended the result of [19] to more contents. For example, we consider more parameters for tuning and enhancing control freedom. The view angle is discussed when designing the coverage issue. In addition, the proposed coverage control involves the collision avoidance issue. This is quite important when multiple UAVs are configuring the variables to achieve the coverage objective. The theoretical analysis is also given. The simulation tests have shown that the multi-UAVs with cameras can cover the whole interesting area automatically and verified that the proposed coverage control scheme is successful. The drawback of the proposed algorithm is that the FOV of the ith UAV may be outside of the area of interest. The current version also cannot handle invisible and obscured areas. This is a 3D coverage issue.

In future research we will consider the deformation of reality that may arise in the image due to the differentiation of the tilt angle or rotation. Another objective is to implement the proposed method on a real multi-UAV system.

Author Contributions: Conceptualization, S.H. and R.S.H.T.; Methodology, S.H. and R.S.H.T; software, W.W.L.L.; formal analysis,W.W.L.L. All authors have read and agreed to the published version of the manuscript.

Funding: This research received no external funding.

Institutional Review Board Statement: This paper has been approved by Temasek Lab. on Feb.15,2022 (Number is 2720).

Conflicts of Interest: The authors declare no conflict of interest.

Abbreviations

The following abbreviations are used in this manuscript:

Symbols

q	point belongs to a known environment
Q	environment
$\varphi(q)$	the density function of the area of interest
p_i	the ith UAV's position
e_i	the ith outward normal vectors
f	the focal length of the lens
L_l	the length of the camera image
L_w	the width of the camera image
R_i	the rotation matrix of the ith UAV
A	the observed area by camera
B_i	the field of view of the ith camera
$g(p_i, q)$	the area/pixel information
n	the total number of UAVs
J_{Nq}	the cost index
ψ_i^w	the angle of the ith camera,w={r,p,t}
u_i	the controlled variable of the ith UAV
N_l	the pixel number of the length of image
N_w	the pixel number of the width of image

List of abbreviations

MUM	Multiple unmanned multirotor
PTZ	Pan-tilt-zoom
UAV	Unmanned aerial vehicle
WTP	Watchmen Tour Problem
FOV	Field of view
CC	Camera coordinates
GC	Global coordinates
GPS	Global positioning system
LiDAR	Light detection and ranging

Appendix A

Proof of Theorem 1.

Taking a similar procedure as in [17] (see Equation (37) of [17]), we have:

$$\frac{\partial H}{\partial p_i} = \int\limits_{Q \cap \partial B_i} (J_{Nq} - J_{Nq \setminus \{i\}}) \phi(q) \frac{\partial q_{Q \cap \partial B_i}^T}{\partial p_i} n_{Q \cap \partial B_i} dq$$

$$+ \int\limits_{Q \cap B_i} \frac{\partial J_{Nq}}{\partial p_i} \phi(q) dq \tag{A1}$$

where $q_{Q \cap \partial B_i}$ and $n_{Q \cap \partial B_i}$ are denoted as the point q and the outward normal vector n are along the boundary $Q \cap \partial B_i$ of UAV i, respectively. We have to calculate:

$$\frac{\partial q_{Q \cap \partial B_i}^T}{\partial p_i} n_{Q \cap \partial B_i} \tag{A2}$$

and:

$$\frac{\partial J_{Nq}}{\partial p_i}. \tag{A3}$$

Since $n_{Q \cap \partial B_i}$ is composed of four edges of the camera FOV on the ground, it is decomposed into n_{1i}, n_{2i}, n_{3i} and n_{4i} along the edges l_{1i}, l_{2i}, l_{3i} and l_{4i}, respectively. It should be noted that each n_{ki} is in the GC system and thus n_{ki} can be obtained through an inverse transformation R_i to e_{ki}, that is $I_{23} R_i^{-1} e_{ki} = I_{23} R_i^T e_{ki}$. Considering a unit vector, we have:

$$n_{ki} = \frac{I_{23} R_i^T e_{ki}}{\| I_{23} R_i^T e_{ki} \|}. \tag{A4}$$

When q lies on the boundary $Q \cap \partial B_i$, the condition $e_{ki}^T R_i (I_{32} q - p_i) = 0$ is satisfied. Thus, by differentiating this condition, we can obtain:

$$\frac{\partial q_{Q \cap \partial B_i}^T}{\partial p_i} n_{ki} = \frac{R_i^T e_{ki}}{\| I_{23} R_i^T e_{ki} \|} \tag{A5}$$

The term of $\frac{\partial J_{Nq}}{\partial p_i}$ can be calculated as:

$$\frac{\partial J_{Nq}}{\partial p_i} = \frac{\partial}{\partial g_i} \left(\sum_{j=1}^{n} g_j^{-1} + Y^{-1} \right)^{-1} \frac{\partial g_i}{\partial p_i} \tag{A6}$$

$$= J_{Nq}^2 \frac{1}{g_i^2} \frac{\partial g_i}{\partial p_i} \tag{A7}$$

$$= -\frac{2 N_{li} N_{wi} f_i^2 J_{Nq}^2 (I_{32} q - p_i)}{L_{li} L_{wi} \| I_{32} q - p_i \|^4} \tag{A8}$$

Thus, we have:

$$\frac{\partial \mathcal{J}}{\partial p_i} = \sum_{k=1}^{4} \int\limits_{Q \cap l_{ki}} (J_{Nq} - J_{Nq \setminus \{i\}}) \varphi(q) \frac{R_i^T e_{ki}}{\| I_{23} R_i^T e_{ki} \|} dq$$

$$- \int\limits_{Q \cap B_i} \frac{2 N_{li} N_{wi} f_i^2 J_{Nq}^2 (I_{32} q - p_i)}{L_{li} L_{wi} \| I_{32} q - p_i \|^4} \varphi(q) dq \tag{A9}$$

For calculating the gradient of $\partial\mathcal{J}$ with respect to f_i, we have:

$$\frac{\partial\mathcal{J}}{\partial f_i} = \int_{Q\cap\partial B_i} (J_{Nq} - J_{Nq\setminus\{i\}})\varphi(q)\frac{\partial q_{Q\cap\partial B_i}^T}{\partial f_i}n_{Q\cap\partial B_i}dq$$
$$+ \int_{Q\cap B_i}\frac{\partial J_{Nq}}{\partial f_i}\varphi(q)dq \tag{A10}$$

Differentiating the boundary $e_{ki}^T R_i(I_{32}q - p_i) = 0$, we have:

$$\frac{\partial e_{ki}^T}{\partial f_i}R_i(I_{32}q - p_i) + e_{ki}^T R_i I_{32}\frac{\partial q}{\partial f_i} = 0 \tag{A11}$$

and:

$$\frac{\partial q^T}{\partial f_i}n_{ki} = -\frac{\frac{\partial e_{ki}^T}{\partial f_i}R_i(I_{32}q - p_i)}{\parallel I_{23}R_i^T e_{ki}\parallel} \tag{A12}$$

For the gradient $\frac{\partial e_{ki}^T}{\partial f_i}$, from Equations (1)–(4), we have:

$$\frac{\partial e_{1i}^T}{\partial f_i} = [0 \quad 2L_{li}^2 \quad 4f_i L_{li}]/(4f_i^2 + L_{li}^2)^{3/2} \tag{A13}$$

$$\frac{\partial e_{2i}^T}{\partial f_i} = [2L_{wi}^2 \quad 0 \quad 4f_i L_{wi}]/(4f_i^2 + L_{wi}^2)^{3/2} \tag{A14}$$

$$\frac{\partial e_{3i}^T}{\partial f_i} = [0 \quad -2L_{wi}^2 \quad 4f_i L_{wi}]/(4f_i^2 + L_{wi}^2)^{3/2} \tag{A15}$$

$$\frac{\partial e_{4i}^T}{\partial f_i} = [-2L_{wi}^2 \quad 0 \quad 4f_i L_{li}]/(4f_i^2 + L_{li}^2)^{3/2} \tag{A16}$$

Thus, it follows that:

$$\frac{\partial\mathcal{J}}{\partial f_i} = \sum_{k=1}^4 \int_{Q\cap l_{ki}} (J_{Nq} - J_{Nq\setminus\{i\}})\frac{\varphi(q)\frac{\partial e_{ki}^T}{\partial f_i}R_i(p_i - I_{32}q)}{\parallel I_{23}R_i^T e_{ki}\parallel}dq$$
$$- \int_{Q\cap B_i}\frac{2J_{Nq}^2 f_i N_l N_w}{L_l L_w \parallel I_{32}q - p_i\parallel^2}\varphi(q)dq \tag{A17}$$

Similarly, we have:

$$\frac{\partial\mathcal{J}}{\partial\psi_i^w} =$$
$$\sum_{k=1}^4 \int_{Q\cap l_{ki}} (J_{Nq} - J_{Nq\setminus\{i\}})\frac{\varphi(q)e_{ki}^T\frac{\partial R_i}{\partial\psi_i^w}(p_i - I_{32}q)}{\parallel I_{23}R_i^T e_{ki}\parallel}dq$$

$$w \in \{r, p, t\} \tag{A18}$$

where $\frac{\partial R_i}{\partial\psi_i^w}$ can be obtained by differentiating R_i with respect to ψ_i^w directly. $\quad\square$

Proof of Theorem 2.

Consider the Lyapunov-like function:

$$V = \mathcal{J} + \sum_{w\in\{r,p,t\}}\frac{1}{2}\tilde{\psi}_i^{w2} \tag{A19}$$

Its time derivative is given by:

$$\dot{V} = \sum_{i=1}^{n} \left[\frac{\partial \mathcal{J}^T}{\partial \eta_i} \dot{\eta}_i + \sum_{w \in \{r,p,t\}} \tilde{\psi}_i^w \dot{\tilde{\psi}}_i^w \right]$$

$$= \sum_{i=1}^{n} \left[\frac{\partial \mathcal{J}^T}{\partial \eta_i} \dot{\eta}_i + \sum_{w \in \{r,p,t\}} \tilde{\psi}_i^w (\dot{\psi}_i^{wD} - \dot{\psi}_i^w) \right] \tag{A20}$$

Substituting the state Equations (28) and (29) with the control laws (23) and (27) yields:

$$\dot{V} = -\sum_{i=1}^{n} \left[\rho \frac{\partial \mathcal{J}^T}{\partial \eta_i} \frac{\partial \mathcal{J}}{\partial \eta_i} + \sum_{w \in \{r,p,t\}} k_{pi}^w (\tilde{\psi}_i^w)^2 \right]$$

$$\leq 0 \tag{A21}$$

Since $\dot{V} \leq 0$, the camera network is bounded, i.e.,

$$\mathcal{J} + \sum_{w \in \{r,p,t\}} \frac{1}{2} \tilde{\psi}_i^{w2} \leq H(0) + \sum_{w \in \{r,p,t\}} \frac{1}{2} \tilde{\psi}_i^w(0)^2. \tag{A22}$$

This implies that $\mathcal{J}, \tilde{\psi}_i^w$ are bounded. The conclusion (1) is proved.

Next, we prove that conclusion (2) holds. The view angle controls result in $\dot{\tilde{\psi}}_i^w = -k_{pi}^w \tilde{\psi}_i^w$. Since $\tilde{\psi}_i^w$ is bounded, this implies that $\dot{\tilde{\psi}}_i^w$ is also bounded. Equation (A21) and the positive definiteness of V imply that:

$$V(t) - V(0) = -\int_0^t \sum_{i=1}^{n} \left[\rho \frac{\partial \mathcal{J}^T}{\partial \eta_i} \frac{\partial \mathcal{J}}{\partial \eta_i} + \sum_{w \in \{r,p,t\}} k_{pi}^w (\tilde{\psi}_i^w)^2 \right]$$

This implies that:

$$\int_0^t k_{pi}^w (\tilde{\psi}_i^w)^2 \leq V(0) - V(t) - \int_0^t \sum_{i=1}^{n} \rho \frac{\partial \mathcal{J}^T}{\partial \eta_i} \frac{\partial \mathcal{J}}{\partial \eta_i}$$

$$\leq V(0) \tag{A23}$$

Thus, $\int_0^t k_{pi}^w (\tilde{\psi}_i^w)^2$ is bounded. By virtue of Barbalat's lemma, we have:

$$\lim_{t \to \infty} \tilde{\psi}_i^w = 0 \tag{A24}$$

where we have used the fact that $V(t) > 0$. This completes the proof. $\square$

Proof of Theorem 3.

We consider the Lyapunov-like function (35) and its time derivative is given by:

$$
\begin{aligned}
\dot{V} &= \sum_{i=1}^{n}\left[\frac{\partial \mathcal{J}^T}{\partial p_i}\dot{p}_i + \frac{\partial \mathcal{J}^T}{\partial \beta_i}\dot{\beta}_i + \frac{1}{2}\left(\frac{\partial V_{P_i}^T}{\partial p_i}\dot{p}_i + \frac{\partial V_{P_i}^T}{\partial p_j}\dot{p}_j\right)\right] \\
&= \sum_{i=1}^{n}\left(\frac{\partial \mathcal{J}^T}{\partial p_i}\dot{p}_i + \frac{\partial \mathcal{J}^T}{\partial \beta_i}\dot{\beta}_i\right) \\
&\quad + \frac{1}{2}\sum_{i=1}^{n}\sum_{j=1,j\neq i}^{n}\left(\frac{\partial v_{P_{ij}}^T}{\partial p_i}\dot{p}_i + \frac{\partial v_{P_{ij}}^T}{\partial p_j}\dot{p}_j\right)] \\
&= \sum_{i=1}^{n}\left(\frac{\partial \mathcal{J}^T}{\partial p_i}\dot{p}_i + \frac{\partial \mathcal{J}^T}{\partial \beta_i}\dot{\beta}_i\right) \\
&\quad + \frac{1}{2}\sum_{i=1}^{n}\sum_{j=1,j\neq i}^{n}\frac{\partial v_{P_{ij}}^T}{\partial p_i}\dot{p}_i + \frac{1}{2}\sum_{i=1}^{n}\sum_{j=1,j\neq i}^{n}\frac{\partial v_{P_{ij}}^T}{\partial p_j}\dot{p}_j
\end{aligned}
\tag{A25}
$$

Exchanging i, j in the last term yields:

$$
\begin{aligned}
\dot{V} &= \sum_{i=1}^{n}\left(\frac{\partial \mathcal{J}^T}{\partial p_i}\dot{p}_i + \frac{\partial \mathcal{J}^T}{\partial \beta_i}\dot{\beta}_i\right) \\
&\quad + \frac{1}{2}\sum_{i=1}^{n}\sum_{j=1,j\neq i}^{n}\frac{\partial v_{P_{ij}}^T}{\partial p_i}\dot{p}_i + \frac{1}{2}\sum_{i=1}^{n}\sum_{j=1,j\neq i}^{n}\frac{\partial v_{P_{ji}}^T}{\partial p_i}\dot{p}_i
\end{aligned}
\tag{A26}
$$

Notice that $\frac{\partial v_{P_{ji}}^T}{\partial p_i} = \frac{\partial v_{P_{ij}}^T}{\partial p_i}$. Thus, we have:

$$
\begin{aligned}
\dot{V} &= \sum_{i=1}^{n}\left(\frac{\partial \mathcal{J}^T}{\partial p_i}\dot{p}_i + \frac{\partial \mathcal{J}^T}{\partial \beta_i}\dot{\beta}_i\right) + \sum_{i=1}^{n}\sum_{j=1,j\neq i}^{n}\frac{\partial v_{P_{ij}}^T}{\partial p_i}\dot{p}_i \\
&= \sum_{i=1}^{n}\left[\left(\frac{\partial \mathcal{J}^T}{\partial p_i} + \sum_{j=1,j\neq i}^{n}\frac{\partial v_{P_{ij}}^T}{\partial p_i}\right)\dot{p}_i + \frac{\partial \mathcal{J}^T}{\partial \beta_i}\dot{\beta}_i\right]
\end{aligned}
\tag{A27}
$$

Substituting the control laws (36) and (37) into the above equation yields:

$$
\begin{aligned}
\dot{V} &= -\sum_{i=1}^{n}\left[\rho\left(\frac{\partial \mathcal{J}^T}{\partial p_i} + \sum_{j=1,j\neq i}^{n}\frac{\partial v_{P_{ij}}^T}{\partial p_i}\right)\left(\frac{\partial \mathcal{J}}{\partial p_i} + \sum_{j=1,j\neq i}^{n}\frac{\partial v_{P_{ij}}}{\partial p_i}\right)\right. \\
&\quad \left. + \rho\left(\frac{\partial \mathcal{J}^T}{\partial \beta_i}\right)\left(\frac{\partial \mathcal{J}}{\partial \beta_i}\right)\right] \\
&\leq 0
\end{aligned}
\tag{A28}
$$

Since $V > 0$, $\dot{V} \leq 0$ and this proves that H is bounded.

Next, we prove the collision avoidance of the proposed control. With the initial condition $d_{ij}(0) > r$, we show that if the safe distance, i.e., $d_{ij}(t) > r$, is violated, then:

$$
\lim_{d_{ij}\to r^+} v_{P_{ij}} = +\infty
\tag{A29}
$$

according to (31). However, this results in the Lyapunov-like function $V \to \infty$. From $V > 0$ and $\dot{V} \leq 0$, it is known that V is bounded. Thus, $V \to \infty$ contradicts the bounded V. Therefore, the safe distance $d_{ij} > r$ holds over the whole coverage control and the collision among UAVs is avoided. $\square$

References

1. Huang, S.; Teo, R.S.H.; Tan, K.K. Collision avoidance of multi unmanned aerial vehicles: A review. *Annu. Rev. Control.* **2019**, *48*, 147–164 [CrossRef]
2. Huang, S.; Teo, R.S.H.; Leong, W.L.; Martinel, N.; Forest, G.L.; Micheloni, C. Coverage Control of Multiple Unmanned Aerial Vehicles: A Short Review. *Unmanned Syst.* **2018**, *6*, 131–144 [CrossRef]
3. Tan, C.Y.; Huang, S.; Tan, K.K.; Teo,R.S.H.; Liu, W.; Lin, F. Collision Avoidance Design on Unmanned Aerial Vehicle in 3D Space. *Unmanned Syst.* **2018**, *6*, 277–295 [CrossRef]
4. Parapari, H.F.; Abdollahi, F.; Menhaj, M.B. Distributed coverage control for mobile robots with limited-range sector sensors. 2 In Proceedings of the 016 IEEE International Conference on Advanced Intelligent Mechatronics, Banff, AB, Canada, 12–15 July 2016; pp. 1079–1085
5. Bhattacharya, S.; Ghrist, R.; Kumar, V. Multi-robot coverage and exploration in non-Euclidean metric spaces. In *Algorithmic Foundations of Robotics X*; Frazzoli, E., Lozano-Perez, T., Roy, N., Rus, D., Eds.; Springer: Berlin/Heidelberg, Germany, 2013; pp. 245–262
6. Bullo, F.; Carli, R.; Frasca, P. Gossip coverage control for robotic networks: Dynamical systems on the space of partitions. *SIAM J. Control. Optim.*, **2012**, *50*, 419–447 [CrossRef]
7. Daingade, S.; Sinha, A.; Borkar, A.; Arya, H. Multi UAV Formation Control for Target Monitoring. In Proceedings of the 2015 Indian Control Conference Indian Institute of Technology Madras, Chennai, India, 5–7 January 2015; pp. 25–30.
8. Smith, S.L.; Broucke, M.E.; Francis, B.A. Stabilizing a multi-agent system to an equilibrium polygon formation. In Proceedings of the 17th International Symposium on Mathematical Theory of Networks and Systems, Kyoto, Japan, 24–28 July 2006; pp. 2415–2424.
9. Huang, S.; Teo, R.S.H.; Leong, W.L.; Martinel, N.; Forest, G.L.; Micheloni, C. Coverage Control of Multi Unmanned Aerial Vehicles: A Short Review, Unmanned Systems. **2018**, *6*,131-144 Huang,S.; Teo, R.; Leong, W.L. Review of coverage control of multi unmanned aerial vehicles. Unmanned Systems in Proc. 2017 11th Asian Control Conference (ASCC), 228–232. 2017.
10. Chvatal, V. A combinatorial theorem in plane geometry. *J. Comb. Theory (B)* **1975**, *18*, 39–41. [CrossRef]
11. Carlsson, S.; Nilsson, B.J.; Ntafos, S.C. Optimum guard covers and m-watchmen routes for restricted polygons. *Int. J. Comput. Geom. Appl.* **1993**, *3*, 85–105. [CrossRef]
12. Cortés, J.; Martínez, S.; Karatas, T.; Bullo, F. Coverage control for mobile sensing networks. Berlin/Heidelberg, Germany, IEEE International Conference on Robotics and Automation, Washington, DC, USA, May 11 –15, 2002; pp. 1327–1332.
13. Cortés, J.; Martínez, S.; Karatas, T.; Bullo, F. Coverage control for mobile sensing networks. *IEEE Trans. Robot. Autom.* **2004**, *20*, 243–255.
14. Schwager, M.; Julian, B.J.; Rus, D. Optimal coverage for multiple hovering robots with downward facing cameras. In Proceedings of the IEEE International Conference on Robotics and Automation, Kobe, Japan, 12–17 May 2009, pp.3515–3522.
15. Piciarelli, C.; Micheloni, C.; Foresti, G.L. PTZ Camera Network Reconfiguration. In Proceedings of the IEEE/ACM Intl. Conf. on Distributed Smart Cameras, Como, Italy, 30 August–2 September 2009; pp. 1–8
16. Wang, H.; Guo,Y. A decentralized control for mobile sensor network effective coverage. In Proceedings of the 7th World Congress on Intelligent Control and Automation, Chongqing, China, 25–27 June 2008; pp.473–478.
17. Schwager, M.; Julian, B.J.; Angermann, M.; Rus, D. Eyes in the sky: Decentralized control for the deployment of robotic camera networks. *Proc. IEEE* **2011**, *99*, 1541–1561. [CrossRef]
18. Arslan, O.; Min, H.; Koditschek, D.E. Voronoi-based coverage control of pan/tilt/zoom camera networks. In Proceedings of the 2018 IEEE International Conference on Robotics and Automation (ICRA), Brisbane, Australia, 21–25 May 2018
19. Huang, S.; Teo,R.S.H.; Leong,W.L. Distributed Coverage Control for Multiple Unmanned Multirotors with Downward Facing Pan-Tilt-Zoom-Cameras. In Proceedings of the 2018 International Conference on Unmanned Aircraft Systems (ICUAS) Dallas, TX, USA, 12–15 June 2018; pp. 744–751. 2004
20. Mastellone, S.; Stipanović, D.M.; Graunke, C.R.; Intlekofer, K.A.; Spong, M.W. Formation Control and Collision Avoidance for Multi-agent Non-holonomic Systems: Theory and Experiments. *Int. J. Robot. Res.* **2008**, *27*, 107–126 [CrossRef]

drones

Article

An Experimental Apparatus for Icing Tests of Low Altitude Hovering Drones

Eric Villeneuve [1],*, Abdallah Samad [1], Christophe Volat [1], Mathieu Béland [2] and Maxime Lapalme [2]

[1] Department of Applied Sciences, University of Québec in Chicoutimi, 555 Boulevard de l'Université, Chicoutimi, QC G7H 2B1, Canada; abdallah.samad1@uqac.ca (A.S.); christophe_volat@uqac.ca (C.V.)
[2] Bell Textron Canada Limited, 12 800 rue de l'Avenir, Mirabel, QC J7J 1R4, Canada; mbeland01@bellflight.com (M.B.); mlapalme01@bellflight.com (M.L.)
* Correspondence: eric_villeneuve@uqac.ca

Abstract: The icing facilities of the Anti-Icing Materials International Laboratory AMIL have been adapted to reproduce icing conditions on a Bell APT70 drone rotor, typical of small-to-medium UAV models. As part of an extensive icing test campaign, this paper presents the design and preliminary testing of the experimental setup and representative icing conditions calibration in the laboratory's cold chamber. The drone rotor used has four blades with a diameter of 0.66 m and a maximum tip speed of 208 m/s. For the icing conditions, freezing rain and freezing drizzle were selected. A Liquid Water Content (*LWC*) calculation methodology for a rotor in hover was developed, and procedures to determine experimental *LWC* in the facility are presented in this paper. For the test setup, the cold chamber test section was adapted to fit the rotor and to control its ground clearance. Testing was aimed at studying the effect of rotor height h on aerodynamic performance, both with and without icing conditions. Results show no significant effect on the ground effect between h = 2 m and h = 4 m in dry runs, while the icing behavior can be largely influenced for certain conditions by the proximity of the precipitation source, which depend on the height of the rotor in these experiments.

Keywords: aerospace; icing; drones; UAV; experimental setup; cold room; hover flight

Citation: Villeneuve, E.; Samad, A.; Volat, C.; Béland, M.; Lapalme, M. An Experimental Apparatus for Icing Tests of Low Altitude Hovering Drones. *Drones* **2022**, *6*, 68. https://doi.org/10.3390/drones6030068

Academic Editors: Arianna Pesci, Giordano Teza and Massimo Fabris

Received: 17 January 2022
Accepted: 3 March 2022
Published: 6 March 2022

Publisher's Note: MDPI stays neutral with regard to jurisdictional claims in published maps and institutional affiliations.

1. Introduction

The last decade saw a huge influx towards the use of Unmanned Aerial Vehicles (UAVs) and drones for a wide variety of military, commercial and recreational applications [1]. Up until 2019, close to a million drones had already been registered by the FAA for personal and commercial use, and the number keeps growing [1]; more than one-third of registered drones are designated for commercial applications.

The APT70 is the newest addition to the fully autonomous and fully electric family of drones developed by Bell [2]. Designed with Vertical TakeOff Landing (VTOL) and wing-borne flight capabilities, the APT70 is capable of reaching speeds of 86 knots, handling loads up to 31 kg, and has a range of up to 56 km per flight (or 30 min flight time). The drone offers commercial, medical, and military solutions; however, when operated in some areas of North America or other cold countries, it will be vulnerable to the dangers of both ground and in-flight icing. Icing has been long shown to be a notorious and critical phenomenon for the operations of all sorts of aircraft, and historical records indicate many accidents that are directly linked to ice accumulation on both airplanes and rotorcraft [3].

For rotorcraft, in particular, icing increases drag and rotor torque, reduces lift and rotor thrust and induces severe vibrations to the rotor system [4]. A study by Liu et al. [5] found that ice accumulation could cause up to 70% thrust loss and increase power consumption up to 250%, compared to operation prior to icing. If exposed to icing effects during flight, most UAVs risk a potential loss of control and dangerous accidents. The literature shows that icing studies have been vigorously conducted in the past for fixed-wing aircraft [6–12] and helicopter rotors [6,13–20]. For example, Narducci and Kreeger developed a high-fidelity method to evaluate the ice accumulation for a helicopter flying through an icing cloud in hover [19] as well as in forward flight [20]. Chen et al. performed CFD numerical simulations and optimization analyses for rotor anti-icing based on big data analytics [21]. Xi and Qi-Jun proposed a new three-dimensional icing model capable of simulating ice accretion on rotors [22]. On the other hand, solutions for protection from ice accumulation during flight have also been studied and proposed, from electro-thermal heaters to coatings and piezoelectric devices [4,15,23,24]. Yet, icing and de-icing studies for smaller-sized UAVs and drones are still rare and government agencies are pushing towards more research in this field [5,25,26].

For take-off/landing and hovering rotor experiments, cold chambers are more advantageous compared to icing wind tunnels due to the absence of upstream flow. However, a proper assessment and calibration of the icing parameters, particularly the Liquid Water Content (*LWC*), for a rotor in hover is made difficult by the absence of an airstream velocity that is needed for the *LWC* measuring equipment to function [27]. Experiments were done at the AERTS cold chamber and a process to calculate the *LWC* experimentally based upon icing wind tunnel calibration procedures was developed using a numerical iterative procedure described in [14]. Comparisons were made for the ice thickness, impingement limits and ice shape, especially at inboard stations. Recently, Brouwers et al. [27] developed a new icing model to predict the ice accretions and shedding for a rotor in hover. The model was correlated with previously published ice shapes for both small- and full-scale ice accretion results. Wang et al. [17] experimented on a hovering rotor subjected to icing conditions in a cold chamber. The Median Volumetric Diameter (MVD) was obtained by applying silicone oil on the surface of a glass sample, which was then subjected to the spray. The glass sample was then put under a microscope and a statistical approach allowed the calculation of the droplet diameter. The *LWC* was calculated based on the method of [14]. They studied the effects of temperature, rotation speed, *LWC*, icing time, number of blades and blade material on the resulting ice shape.

The Anti-Icing Materials International Laboratory (AMIL) has a track record for conducting icing experiments with helicopter rotors [18,23,28–32]. With the objective of studying the icing effects on the take-off/hover mode of the Bell APT70 rotor, a new and innovative test setup was developed. This paper presents the preparation and development of the experimental apparatus, including the calibration of the icing parameters, used to conduct the icing tests on the rotor. Analysis and procedures to quantify MVD, precipitation rate λ and *LWC* in the chamber are also presented. A new method for the determination of an *LWC* equivalency for ground and low altitude icing where no significant airflow is present is also proposed in this paper. Furthermore, an evaluation of the rotor height and the influence on the ground effect, as well as the influence of distance of the nozzles to the rotor, is performed to determine the requirements for proper rotor positioning for icing tests. The laboratory's nine-meter-high (9 m) cold chamber was selected for this investigation because of its high ceiling as well as its large test section.

2. Materials and Methods: Icing Precipitation

2.1. Cold Chamber Characteristics

The rotor icing test setup was installed in the AMIL cold chamber, identified as the 9M-Chamber (Figure 1), in which the test rotor was positioned at the center of the icing test section. The 9M chamber is 9.10 m high, 5.50 m long, and 3.50 m wide. Air temperature can be controlled between $-32\,°C$ and $5\,°C \pm 0.5\,°C$. Measurement of the air temperature in the test section was done using calibrated Omega RTDs. Two heat exchangers connected to a compressor were installed in the upper section of the chamber. Four ventilators were located in front of the exchangers and provided the airflow toward the test section. The air was distributed uniformly in the climatic chamber through a perforated ceiling and can circulate either in ascending or descending directions. Test conditions can be indefinitely maintained for continuous operation of the rotor. Inside the chamber and surrounding the rotor, the icing test section has an area of 2 by 2 m, where the icing precipitation is obtained and is delimited by both the chambers walls and Plexiglas movable walls.

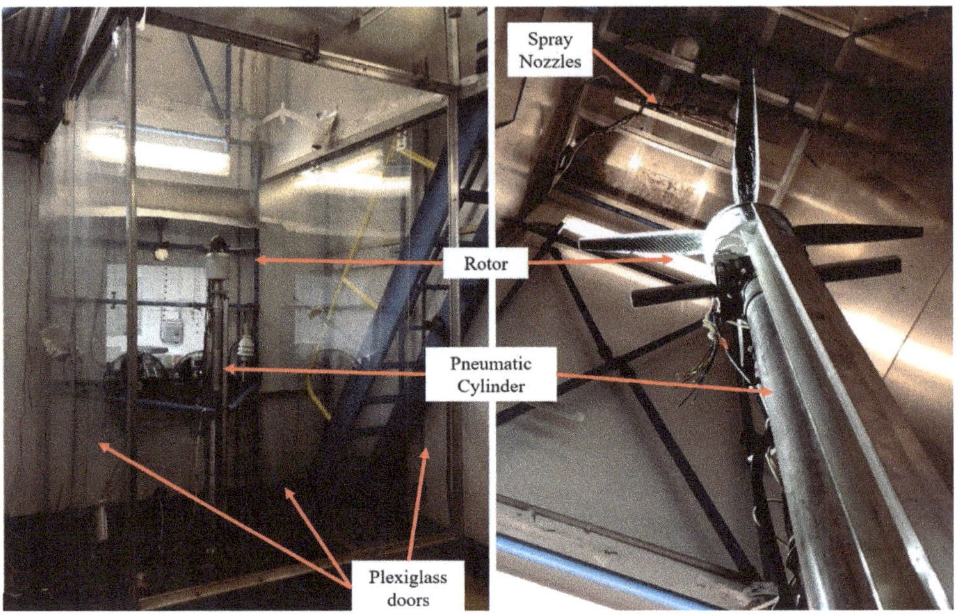

Figure 1. Photo of the experimented rotor system placed inside the cold chamber.

One of the purposes of the study is to determine the effects of the rotor height on the results during a dry run (without water or ice) as well as under icing conditions. The distance from the ground must be evaluated to limit the ground effect influence on the results. The distance from the nozzles to the rotor must also be evaluated to make sure that the icing cloud reaches equilibrium before impacting the rotor in order to be representative of atmospheric conditions. The rotor was installed on top of a pneumatic cylinder, as shown in Figure 1, which can be extended vertically using compressed air, with a maximum height of 5 m.

2.2. Icing Nozzle Array

Two icing nozzles were located on the chamber ceiling to generate the icing cloud. A photograph of the spray rig in the ceiling with the two installed nozzles is shown in Figure 2. The nozzles were located at the edges of the array and directly above the center of the rotor shaft to distribute the spray evenly in the chamber. Hydraulic sprinklers produced gravity-fed icing with pressurized nozzle sprayers. Different nozzle heads can be installed to generate different droplet sizes, and heating elements were installed in the system to prevent nozzles from freezing between testing. The droplet speed corresponds to their freefall values in the vertical airflow (see Section 2.6). Distilled water maintained in a refrigerator was used to produce the precipitation. Equal air pressure and water pressure were delivered to each nozzle. Measuring the water and air pressures at the input of the water and airlines to the nozzles ensures precise readings of the pressure differential controlling the droplet size.

Figure 2. The cold chamber's icing nozzle array.

2.3. MVD Measurement

To measure the Median Volumetric Diameter (MVD), a thin layer of silicone oil was applied on the surface of a glass slide, as shown in Figure 3. This method, described and recommended in AIR4906 [33], used a glass slide coated with silicone oil, which is passed through the spray range of the nozzles, to collect a certain number of droplets. Due to the incompatibility between oil and water, the droplets were wrapped in silicone oil and kept their original shapes for a short time. The droplets on the glass slide were observed and photographed under a camera; the observation results are shown in Figure 3.

Figure 3. A photograph of the collected water droplets on the glass slide with silicone oil.

The MVD can be obtained by statistical calculation of droplet diameters. This technique is a commonly used and accepted method in the industry [33]. To obtain different MVD in the cold chamber, different nozzle heads are installed on the nozzle array. In this study, two models are used, a first one to obtain a smaller droplet size representing freezing drizzle precipitation and a second one to obtain significantly larger droplets representative of freezing rain precipitation. This method has been validated using a cloud imaging probe from Droplet Measurement Technologies.

2.4. Precipitation Rate

The Precipitation Rate λ, or Icing Intensity, is a commonly used measure of in-ground and low-altitude icing. This measure represents the mass of water, or frozen precipitation, that falls over an area per unit of time. To measure the precipitation rate in the chamber, a catch pan technique was used, as required in aircraft ground icing standards [34,35] and used in other studies [36]. After choosing and installing the desired spray head, 24 catch pans were placed in the center of the chamber at 2 m, as seen in Figure 4. As shown in the figure, the catch pans (blue interior) were placed on a table at the center of the test section and directly under the icing nozzle array. Each catch pan was individually weighed empty, and its weight was recorded. The spray was activated for a period of up to 30 min and the pans were again weighed together with the collected amount of water. The difference between the two sets of recorded weights is the weight of the collected water, which is then used to calculate λ with the Equation (1). M_W is the mass of collected water in g and t_{prec} is the precipitation collection time in min. This measurement setup respects the requirements of the industry standards [34,35]. In addition, multiple measurements were repeated throughout the test campaign to ensure the stability of the precipitation rate.

$$\lambda = \frac{M_W}{t_{prec}} \times 60 \tag{1}$$

Figure 4. A photo of the catch pans used to collect water for water precipitation calculation.

2.5. LWC Estimation and Droplet Velocities

The Liquid Water Content (*LWC*) is a measure of the mass of water contained in a volume of air. This value is commonly used for in-flight icing but also for ground and low altitude icing. *LWC* sensors, like the King probe or rotating cylinder methods [37], can be readily used to measure the *LWC* in an icing cloud. However, those sensors either require a certain flow velocity (usually 10–15 m/s and higher) to function properly or must use the flow velocity in the calculation to determine the *LWC* value from their measurements. In-ground icing and low altitude hover mode, unlike in wind tunnel, no airflow is generated during the calibration of the icing conditions, making those sensors unreliable since the airflow value is close to 0. This makes the use of the *LWC* as a parameter impractical for this kind of testing and very difficult to perform a direct comparison between λ and *LWC*. Another method [27] required measurement of ice accretion on the rotor in order to calculate the *LWC*, but it is dependent on blade shape, which is not suitable for an easy universal *LWC* measurement, as well as the particular blade profile used in this study. *LWC* measurement is a common and recurrent problem in the domain, especially since most standards use *LWC* for their requirements for in-flight and low altitude icing like the FAA/AR-09/45 [38] and AC29-2C [39], while most standards use λ [34,40] for ground icing.

A new methodology is developed to settle this problem and calculate *LWC* in the cold chamber for hovering rotor tests and perform a direct comparison to λ to match all the different industry standards. The new methodology calculates the terminal velocity of the water droplets based on the freefall equations and mass balance of the droplet. The velocity is then used together with λ to determine the *LWC* in the chamber.

In this work, the water droplets are assumed to have the shape of a sphere while dropping from the nozzles downwards. The data for the drag coefficient C_D of a sphere are obtained as a function of *Re* from the literature. The Re for a sphere is calculated by Equation (2). For simplicity, the C_D is calculated using a linear interpolation presented at Equation (3), of the well-established drag coefficient and Reynold's number relationship between $0 \leq Re \leq 1000$ [41]. It is acceptable to use that approximation in that first section of the relationship, as demonstrated by the correlation coefficient of $R^2 = 0.9851$. Past this region, for $1000 \leq Re \leq 100{,}000$, the C_D is constant at 0.3. During this test campaign, it is not expected to test with water droplets with an MVD higher than 800 μm. At this size, the sphere Re is approximately 200, which is well below the higher boundary of the Re correlation. V_T is the terminal velocity of the water droplet, and $\vartheta_{air} = 0.00001328$ m^2/s is the kinematic viscosity of air.

$$Re = \frac{V_T D}{\vartheta_{air}}, \tag{2}$$

$$C_D = 28.651 * Re^{-0.722}, \tag{3}$$

A force balance on the freefalling water droplet results in the gravitational and drag force (D) to be present on the droplet, as expressed in Equation (4). $M_{Droplet}$ is the mass of the droplet in kg, and a is the acceleration in m^2/s. The force from the droplet mass can be expressed by Equation (5), relating the water density ($\rho_w = 997$ kg/m^3) to the sphere volume ($V_{Droplet}$) and the gravitational acceleration ($g = 9.81$ m/s^2). r is the radius of the water droplet in meters.

$$F = (M - D)_{Droplet} = M_{Droplet} * a, \tag{4}$$

$$M_{Droplet} = \rho_w * V_{Droplet} * g = \rho_w * \frac{3}{4}\pi r^3 * g, \tag{5}$$

The drag force, which is linked to the sphere drag coefficient C_D, is obtained from Equation (6). C_D can be obtained from Equation (3) once the Re for the droplet is obtained. The density of air used is $\rho_{air} = 1.2922$ kg/m^3, while the density of air will be different slightly different at lower temperatures, up to 1.367 at $-15\,°C$; calculations indicate that the final terminal velocity calculations would only be impacted by less than 2.2%.

$$D_{Droplet} = \frac{1}{2}\rho_{air} * V_T^2 * A * C_D = \frac{1}{2}\rho_{air} * V_T^2 * \pi r^2 * C_D, \tag{6}$$

Once all the terms are rearranged and considering zero acceleration of the droplet, the terminal velocity of the droplet is expressed with Equation (7). V_T can then be used together with the precipitation λ (g dm^{-2} h^{-1}) to calculate the liquid water content *LWC* (g/m^3), as described in Equation (8). The conversions of λ are outlined in Equation (9).

$$V_T = \sqrt{\frac{8}{3}r * \left(\frac{\rho_w}{\rho_{air}}\right) * \left(\frac{g}{C_D}\right)}, \tag{7}$$

$$LWC\left(\frac{g}{m^3}\right) = \frac{\lambda\left(g * m^{-2} * s^{-1}\right)}{V_T\left(m * s^{-1}\right)}, \tag{8}$$

$$\lambda\left(g * m^{-2} * s^{-1}\right) * 60 * 60 * 10^{-1} * 10^{-1} = \lambda\left(g * dm^{-2} * h^{-1}\right), \tag{9}$$

To calculate the distance required for a droplet to reach its V_T, a time-dependent approach is taken. For every time step Δt, the droplet velocity V_{i+1} is calculated by using Equation (10) where V_i is the droplet velocity at the previous timestep and a_i is the droplet acceleration. a_i is calculated using Equation (11) by applying the force balance shown in Equation (4). The distance traveled between the two timesteps Δh is then calculated using Equation (12). The procedure is looped until the droplet acceleration is zero, where the terminal velocity is reached.

$$V_{i+1} = V_i + a_i * \Delta t, \tag{10}$$

$$a_i = \frac{M_W * g - D_{Droplet}}{M_W}, \tag{11}$$

$$\Delta h = (V_{i+1} + V_i) * 0.5 * \Delta t, \tag{12}$$

The distance traveled by a droplet before reaching its V_T, calculated as a function of droplet size, is presented in Figure 5. The figure shows droplet sizes between 100 μm and 1500 μm. For small droplets, V_T is reached for heights as low as 0.05 m, whereas bigger droplets require a higher distance to reach their V_T. For a droplet size of 1500 μm, a height of almost 8 m is required.

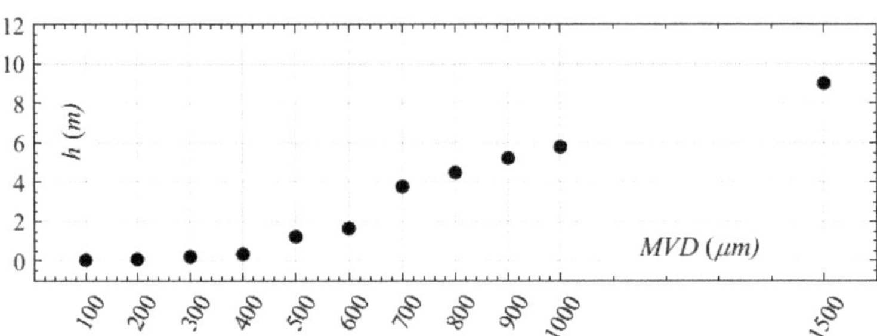

Figure 5. Height (m) Required to Achieve Terminal Velocity in (m/s) for a Range of Droplet Sizes between 100 μm and 1500 μm.

In this research, the droplets used have an MVD of either 120 μm or 800 μm. According to Figure 5, the droplets at 120 μm need 0.04 m to reach their V_T whereas those at 800 μm need at least 4 m. Provided that the nozzles are at a height of 9 m, and that the rotor is tested between 2 m and 4 m off the ground, then the distance between the rotor and nozzles during tests will be between 5 m and 7 m. This range then satisfies the height requirement for the used droplets and therefore agrees with the methodology of *LWC* calculation.

Based on the above, and with the proper height requirement to reach the V_T for each droplet size, the λ needed to produce an *LWC* = 0.5 g/m^3 can now be calculated. A *LWC* of 0.5 g/m^3 is typical for freezing rain and freezing drizzle outside clouds, according to FAA/AR-09/45 [38]. Figure 6 shows the variation of λ needed to produce a *LWC* = 0.5 g/m^3 for the variety of droplet sizes. As the figure shows, a higher λ is needed when the droplet size increases. A maximum of 100 g dm^{-2} h^{-1} was estimated for a droplet size of 1500 μm. For the MVD selected for testing, the λ for a *LWC* = 0.5 g/m^3 is around 5 g dm^{-2} h^{-1} for 120 μm and around 73 g dm^{-2} h^{-1} for 800 μm.

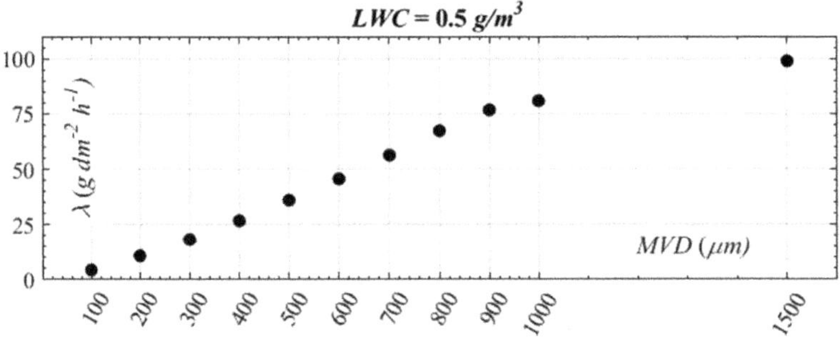

Figure 6. Variation of the Precipitation Rate in g dm^{-2} h^{-1} Required to Achieve an *LWC* of 0.5 g/m^3 Versus Droplet Size in (μm).

In a similar analogy to the preceding section, the *LWC* produced based on a λ of 25 g dm^{-2} h^{-1} is investigated which is typical for ground icing. Figure 7 shows the variation of the *LWC* versus the droplet size that is obtained for a λ of 25 g dm^{-2} h^{-1}. For the fixed λ, the *LWC* produced decreases with increasing droplet size, varying between 3 g/m^3 at 100 μm down to almost 0.1 g/m^3 for a droplet size of 1500 μm. For the MVD selected, this corresponds to a *LWC* of 2.53 g/m^3 for droplets of 120 μm and a *LWC* of 0.19 g/m^3 for droplets of 800 μm.

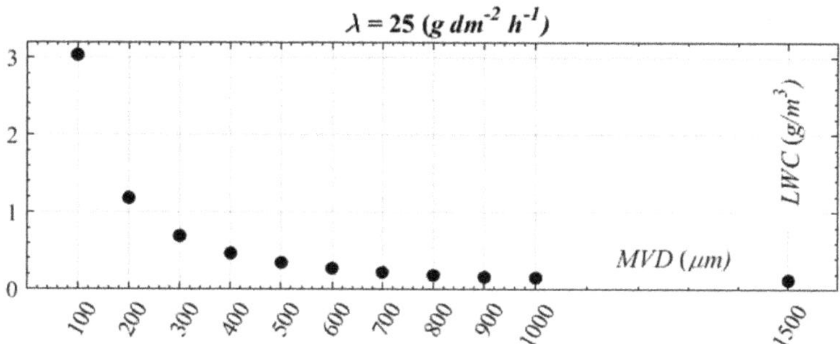

Figure 7. Variation of the *LWC* in (g/m^3) Required to Achieve λ = 25 g dm^{-2} h^{-1} Versus Droplet Size in (μm).

2.6. Tracking the Droplets' Terminal Velocity

An experiment was conducted to validate the water droplets terminal velocity after falling from the chamber's ceiling, making sure the calculations from the previous section were correct and to confirm that the droplets had reached terminal velocity at rotor height. For the experiment, the chamber was darkened, and a light source was placed behind the falling stream of water. A camera was used to take photos of the falling water droplets at a rate of 200 photos per second (a picture every 5 ms). Using a ruler as an established scale, the distance traveled by a single droplet between two photos was then measured using digitizing software. A cover with a single slit is installed on top of the camera to obtain a single slice of droplets falling at a single and known distance from the camera. Knowing the time difference between the photos from the rate of the camera, the velocity of the droplet could be calculated. This was done for many droplets, of both similar and different sizes, to confirm the consistency between the velocities of the different droplets and validate the calculation for the whole range of droplet size generated. This simple setup was developed to strengthen confidence in the theoretical calculations done at the previous section, using simple and widely accepted theory, and for this reason the precision of the method was deemed sufficient.

2.7. Ice Shape Documentation & Measurements

An apparatus and a methodology were developed to document the ice shapes and its measurements in this experimental campaign. A database of photographic images, thickness measurements and 3D scans was created. For the 2D photographs, a custom-built rig was fabricated to hold a high-resolution camera in place, at a focal distance from the blade. The rig has three positions for the camera that enabled taking photos of the blade and ice from the front, side and top of the blade in a consistent manner between tests, similar as those presented in [5,36].

As for the ice thickness measurements, one blade was selected as a control blade where straight lines were drawn using a water-resistant ink at 9 different radial locations between the root and tip of the blade, as shown in Figure 8. A digital caliper was used to measure the chord at those locations without ice and then used again after each test to measure the new chord with the accumulated ice. A subtraction of the former from the latter will result in the chordwise ice thickness at the leading edge found at the marked radial locations. Those thicknesses will serve as a simple indication of the severity of the ice accumulation along and in addition to the pictures and scan gathered.

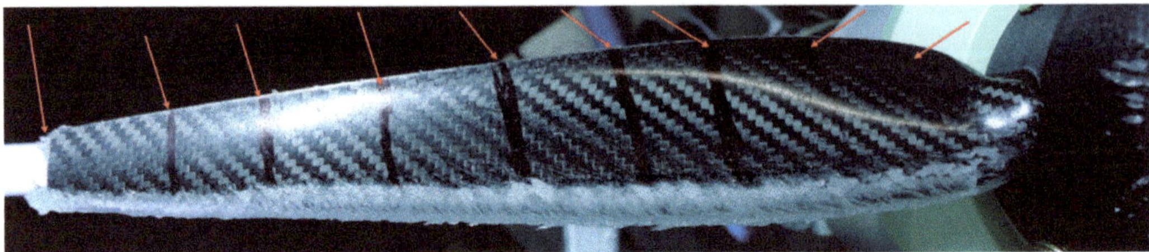

Figure 8. A Photograph Showing the Markings of the Control Blade Used for Ice Thickness Measurements.

Finally, multiple 3D scans of the ice accumulation were also gathered. Figure 9a shows a sample scan of the clean rotor blade, prior to any icing exposure. The result of a scan with accumulated ice is shown in Figure 9b. Three different models of 3D scanners from the company Creaform were used to scan the blade with its accumulated ice: the GoSCAN 50, using white light technology, the HandySCAN 307/700, using a red laser, and the HandySCAN Black, using a blue laser. The white and opaque rime ice was easier to scan due to its better reflection of the light than the clear glaze ice. To help with clear glaze ice, very fine clay powder was brushed on the ice layer. Not all types of 3D scanners were able to provide a satisfactory scan of glaze ice, and the red-light model was deemed the most suitable one for this testing. The VXelements software was used to obtain the 3d images.

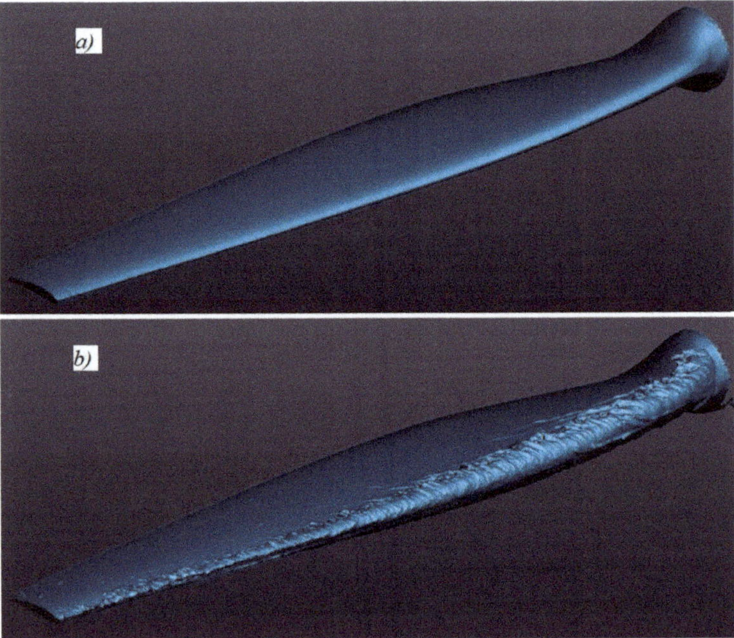

Figure 9. Results of Scan with Creaform HandySCAN 307/700 (red laser) 3D Scanner for a: (**a**) Clean Blade and (**b**) Blade with Accumulated Ice.

2.8. Icing Conditions

To select the icing conditions for testing, two main standards are used: the FAA/AR-09/45 [38] and the SAE standards ARP 5485 [34] and AS5901 [40]. Those documents determine conditions for ground and low altitude icing. The two types of precipitation selected are freezing drizzle and freezing rain. In [34,40], the droplet size for freezing drizzle is 300 μm and for the freezing rain is 1000 μm, while in FAA/AR-09/45, it is 120 μm and 800 μm respectively. The two droplet sizes selected are 120 μm and 800 μm, corresponding to nozzle heads readily available at the laboratory. The SAE documents specify the precipitation rate for the icing precipitation with the more severe condition set at 25 g dm^{-2} h^{-1}. In the FAA document, however, the icing precipitation requirements are expressed in terms of *LWC*, with values of 0.4 to 0.5 g/m^3 for both freezing drizzle and freezing rain. For simplicity, a single value of 0.5 g/m^3 is considered in this test campaign. The technique and calculations developed and detailed in the previous section are used to compare and convert *LWC* to λ. In addition to those two requirements, an additional requirement of 0.25 in of water per hour is also defined by the industrial partner itself. The comprehensive grid of icing conditions selected for the whole test campaign and the rationale for each of those conditions is presented in Table 1. For the preliminary testing performed in this paper, only the conditions in bold were used.

Table 1. Details of icing parameters used for tests along with their rationale.

λ (g dm^{-2} h^{-1})	MVD (μm)	*LWC* (g/m^3)	Rational
5	120	0.47	0.5 g/m^3 (FAA/AR-09/45)
25	120	2.35	Typical Ground Icing + Light Rain (2 L/h/m^2)
25	800	0.19	Typical Ground Icing + Light Rain (2 L/h/m^2)
67	120	6.31	0.25 in. Water/hr for APT70 Requirement + Moderate Rain (6 L/h/m^2)
67	800	0.50	0.25 in. Water/hr for APT70 Requirement + 0.5 g/m^3 (FAA/AR-09/45) + Moderate Rain (6 L/h/m^2)
80	**120**	**7.53**	**Typical Ground Icing + Moderate Rain (8 L/h/m^2)**
80	**800**	**0.59**	**Typical Ground Icing + Moderate Rain (8 L/h/m^2)**

Aside from the water spray parameters, the air temperature in the cold chamber plays a crucial role in the type of accumulated ice. To make sure rotor icing is studied for the glaze, rime, and mixed ice conditions during the experimental campaign, three air temperatures are used. These are T_α = −5 °C, −12 °C, and −15 °C. The highest T_α used was −5°C since ice accumulation was poor or not possible when temperatures of −2 °C or 0 °C were used. For the preliminary testing of this study, only results at T_α = −5 °C and −15 °C are presented.

3. Materials and Methods: Drone Rotor Setup

3.1. Drone Rotor Assembly

In the center of the chamber, a 12-kW Hacker Q150-45-4 motor is used to power the drone's rotor. The motor and hub system is installed in the center of the chamber's test section on a pneumatic cylinder, as shown in Figure 10, to test different rotor heights. Compressed air is supplied to the motor via an external feeder to increase operation time before overheating the motor. To measure thrust, torque and mechanical power consumption of the rotor, a two-axis Futek MBA-FSH04262 load cell was added between the motor and the holding post. Data acquisition, as well as control of the rotor system, is performed using a custom Labview interface. This measurement and control system has been extensively tested and used by the industrial partner for other aerodynamical studies.

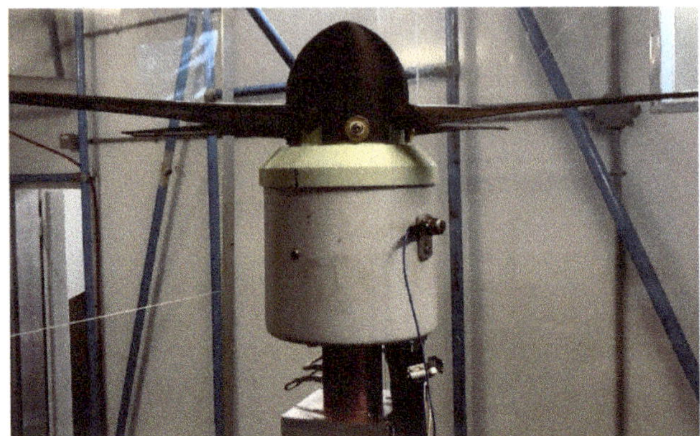

Figure 10. A Photograph of the Installed Rotor Showing the Blades and Load Cell.

The rotor studied in this experiment is the slightly modified 81% scaled version of the Bell APT70 drone rotor. The modifications are related to the root attachment, which is circular on the subscale rather than rectangular. For both rotors, the blade roots are mostly covered by a spinner cone which should not make a significant difference. It has four blades (presented in Figure 11) with a diameter of 0.66 m and a NACA 4412 aerodynamic profile. Each blade is formed of two carbon fiber parts that are glued together to form one hollow blade 25.4 cm long. The blades are twisted and have a variable chord from root to tip. Equations (13) and (14) are used to describe the curve fitting result based on the non-dimensional radial location r for the twist φ (°) and chord c (m), respectively. The configuration can reach 6000 RPM, reproducing the full-scale drone maximum tip speeds.

$$\varphi = -65.187r^3 + 169.58r^2 - 161.36r + 68.5, \tag{13}$$

$$c = \left(-10.632r^4 + 30.997r^3 - 33.185r^2 + 13.566r + 0.0397\right) * \frac{2.54}{100}, \tag{14}$$

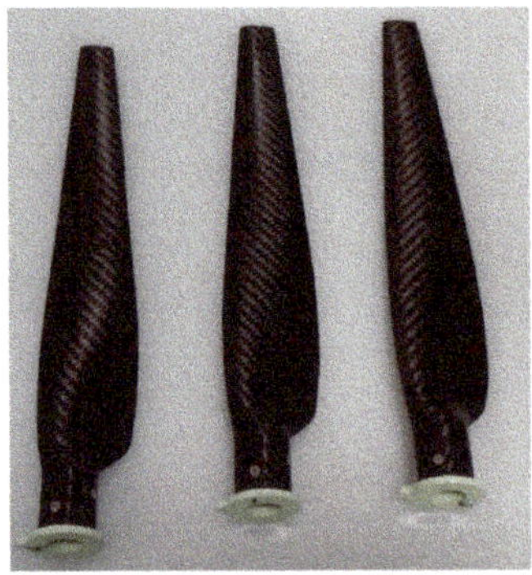

Figure 11. A Photo of the Rotor Blades Used for the Experimental Icing Tests.

3.2. Rotor Test Parameters

In order to better understand the impact of ice accretion on rotor performance in hover conditions for a drone like the Bell APT70, icing tests are to be conducted at different speeds, air temperatures, droplet sizes, and precipitation rates are required. Three levels of RPM are targeted for a comprehensive investigation: low, medium and high. Since the rotor studied in this experiment is a scaled-down version, it is important to set similarity parameters with the full-scale APT70 rotor. Similar tip speeds are expected to produce similar aerodynamic performances and flow behavior. Similar centrifugal forces are expected to produce similar ice accretion and ice shedding. It is not possible to match both parameters at the same time, so for each level of RPM investigated for the full-scale rotor, two RPM values are obtained for the scaled-down model, one to match the tip-speed and one to match the centrifugal force of each level, resulting in a total of six RPM values for experimental testing (Table 2).

Table 2. Calculated RPM Values for the Scaled-Down Rotor.

Thrust Level	Scaling Rule	RPM
Low	Same centrifugal force	3880
	Same tip-speed	4300
Medium	Same centrifugal force	4440
	Same tip-speed	4950
High	Same centrifugal force	4950
	Same tip-speed	5540

4300 RPM is close to 4440 RPM and is then rejected to minimize the number of tests. Only 3880, 4440, 4950 and 5540 are kept. However, during preliminary testing with icing spray, it was remarked that the electrical power consumption for tests at Ω = 4950 RPM reached around 11 kW. Therefore, to avoid exceeding the rated motor power of 12 kW, rotor speeds for icing tests were limited to 4950 RPM. Table 3 lists the different Ω and rotor heights h used for the icing tests at the different icing conditions selected (II.H).

Table 3. List of Tested Rotor Speeds and Rotor Heights.

Height (m)	Ω; Icing Tests (RPM)
2	3880
	4400
4	4950
	4950

3.3. Testing Procedure

Icing tests were conducted in the cold chamber following a specific protocol that ensures the safety of the personnel, maintains the healthy operation of equipment and provides reliable and consistent results. The procedure is outlined step-by-step as follows.

- The motor cooling air is first turned on, followed by the power supply for the motor. The rotor spin safety switch is then deactivated, and the test begins. The test is initiated in the software and data acquisition starts recording.
- The rotor speed is incrementally increased from 0 to the desired steady-state speed. After a short stabilization period at the target rotation speed, the water spray is activated, and ice accumulation begins where the torque and electrical power consumed continuously increase and thrust decreases. The vibration levels, motor temperature and electrical power consumed are closely monitored throughout the spray time.
- The test is stopped whenever one of those conditions happens; (listed in the order of actual occurrence during tests): 1- ice sheds from the blade, causing severe vibrations (>2 Inches Per Second (*ips*)) OR when the vibration levels reach 2 *ips* even without ice shedding; 2- the electrical power consumed approaches the rated power of the motor (12 kw); 3- the test duration exceeds 20 min and; 4- the motor temperature reaches 100 °C.
- After the test is concluded, the motor alimentation is switched off and the safety switch of the rotor is activated. A visual inspection of the equipment is first done to check for any damage.
- The photography platform is then installed on the rotor pole, and photos of the ice shape are taken from the front, side and upper directions. A scan is performed of the

ice accumulation with the 3D scanner and a digital caliper is then used to measure the ice thickness at 9 different and pre-marked blade locations.

- Finally, the ice is then melted off the rotor using a heat gun and cleaned using industrial grade paper towel to prepare it for the next test.

3.4. Post-Processing and Non-Dimensional Coefficients

To analyze the data and determine the severity of each test, non-dimensional coefficients are used for test comparison. The non-dimensional thrust coefficient C_T as well as the torque coefficient C_Q for each test are calculated as shown in Equation (15), where ρ is the density of air (kg/m^3), Ω is the rotor speed (rev/s) and d is the rotor diameter (m).

$$C_T = \frac{T}{\rho\Omega^2 d^4}, \quad C_Q = \frac{Q}{\rho\Omega^2 d^5}, \tag{15}$$

For each test, average values for each of the recorded parameters during the stabilization period, prior to water initiation, must be established. Figure 12 shows an example of an icing test where the stabilization period is highlighted in orange for all the curves of rotor speed and electrical power (Figure 12a), as well as for the torque and thrust (Figure 12b). The period begins once the rotor reaches the target speed and ends at the beginning of the water spray, which is directly followed by an increase of torque and decrease of thrust.

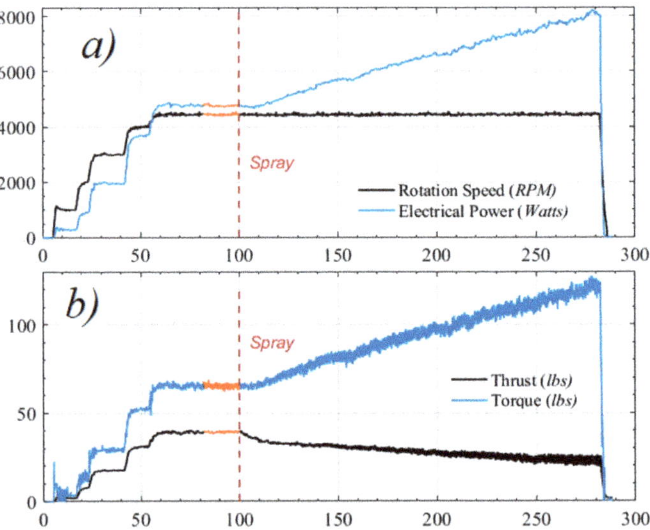

Figure 12. Example of variation of rotor performance data with time for (**a**) rotation speed and electrical power; and (**b**) thrust and torque.

To observe the effect of ice accretion on rotor performance, the measurements taken during icing are compared to their calculated average values prior to water spray initiation. This is how Figure 13 is obtained, which shows the percentage decline of the measured C_T with time (Figure 13a) as well as the parallel percentage increase of C_Q (Figure 13b), based on their respective average values without ice.

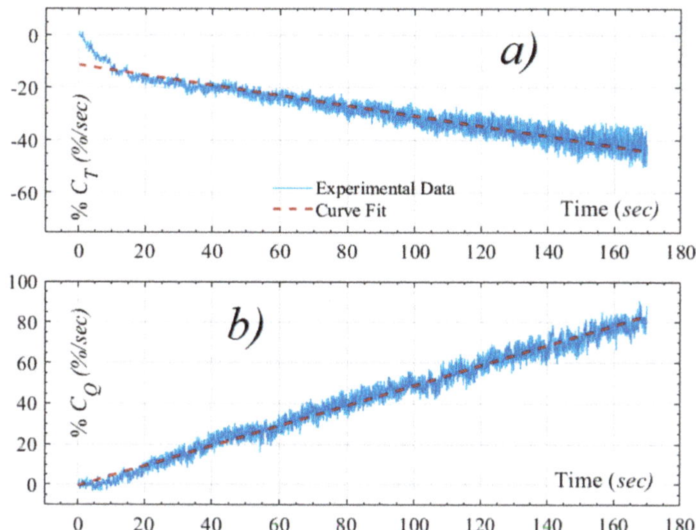

Figure 13. Variation of non-dimensional parameters with time during icing for (**a**) thrust coefficient C_T; and (**b**) torque coefficient C_Q.

Testing showed that for the absolute majority of cases, the variations shown in Figure 13 are linear or could be curve fitted by a linear variation as shown by the red dashed lines of the figure. Therefore, the severity of a test is then directly related to the C_T slope (referred to as C_T^* (%/s)) as well as the C_Q slope (C_Q^* (%/s)). In other words, a test is considered more severe if C_T^* is lower and C_Q^* is higher. It should be noted that the C_T slope during icing is always negative, while the C_Q slope is always positive.

It is also interesting to determine the required increase of mechanical power (P^+) to maintain the same initial thrust without ice in order to use only one output parameter to sort tests by severity. This logic requires the calculation of the required increase of rotor speed (Ω^+ (%/s)) using Equation (16). The increase of torque required to maintain initial thrust (C_Q^+ (%/s)) is then calculated using Equation (17), and P^+ is finally obtained using Equation (18).

$$\Omega^+ = \frac{1}{\sqrt{1+C_T^*}}, \tag{16}$$

$$C_Q^+ = \frac{1}{1+C_T^*} * C_Q^*, \tag{17}$$

$$P^+ = \frac{1}{\sqrt{1+C_T^*}} * C_Q^+, \tag{18}$$

4. Results

4.1. Validation of LWC Estimation—Droplet Terminal Velocity

Figure 14 shows the theoretical results of V_T calculation obtained for different water droplet sizes, compared to the experimental data measured using the procedure described in Section 2.6. Numerical data indicate that V_T increases with the droplet size. This ranges from as low as 0.25 m/s at 100 µm to as high as almost 5.5 m/s when the droplet size is 1500 µm. Experimental data show good agreement with numerical predictions, especially when the error bars are considered, principally caused by the determination of the droplet diameters. This basic measurement technique allows for the validation of the theoretical calculations, but the improvement to the technique could be made to increase the accuracy of the measure, and thus a better resolution to measure the droplet diameters.

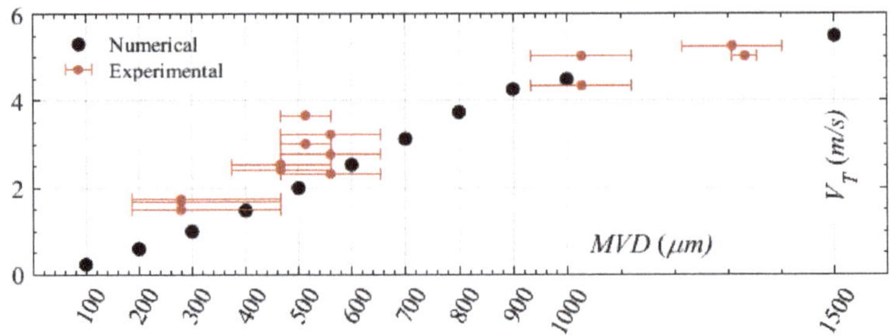

Figure 14. Comparison between the Calculated V_T Values for Water Droplets Obtained through Numerical Analysis and Experimental Measurements.

4.2. Assessment of Rotor Height on Ground Effect and Rotor Performance

To determine the minimum distance from the ground to the rotor, preliminary tests at different heights are performed. The test setup, as detailed in Section 2.1, allows testing at heights between h = 2 m and h = 5 m from the ground. For this work, testing is done at h = 2 m and at h = 4 m without blade pitch adjustment. Experimental data for C_T and C_Q were gathered during both dry runs as well as during the initial dry period of icing tests prior to water spray.

Figure 15 shows the variation of C_T versus rotation speed at h = 2 m and h = 4 m. Curve fitting was also applied to each set of data, and the variation obtained is also linear for both. The difference between the curve fitting results of C_T at 2 m, compared to those with the same rotation speed at 4 m, varies between 2.77% at the lowest tested RPM (3000 RPM) and up to 6.53% at the highest (4950 RPM).

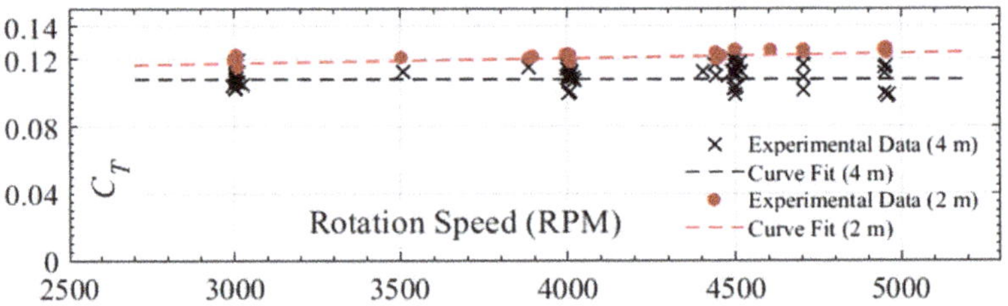

Figure 15. Comparison of the Measured C_T during Dry-Runs for Rotor at h = 2 m and h = 4 m.

Similarly, Figure 16 shows the C_Q variation with Ω for both rotor heights along with their corresponding curve fitting results. The difference between the curve fitting results of C_Q at h = 2 m, compared to those with the same rotation speed at h = 4 m, varies between 4.1% at the lowest tested RPM (3000 RPM) to 3.65% at the highest (4950 RPM).

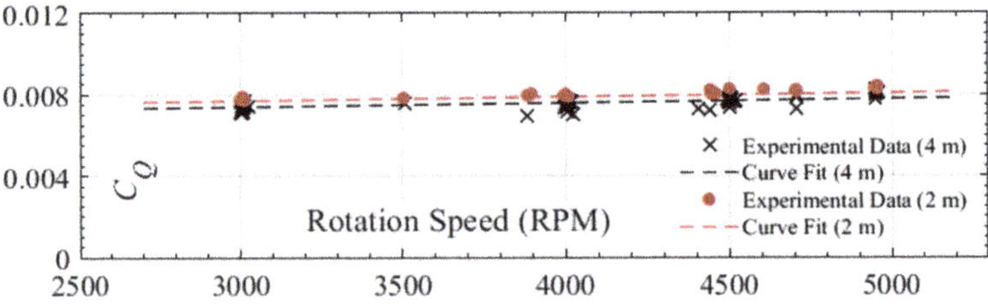

Figure 16. Comparison of the measured C_Q during dry runs for rotor at h = 2 m and h = 4 m.

Finally, the C_T vs. C_Q curve is compared for both sets of measured data at 2 m and 4 m, as shown in Figure 17. A curve fitting scheme is applied to measurement points, and results for both sets at 2 m and 4 m show good agreement. On average, the error between the two curve fittings was around 1.28%. The error is consistent with findings of previous works, where the ground effect is usually absent when h/d > 1 [42], such as in this case. Based on these results, it is possible to assess that there is no significant ground effect at h = 2 m and no significant differences between the rotor performances at h = 2 m and h = 4 m.

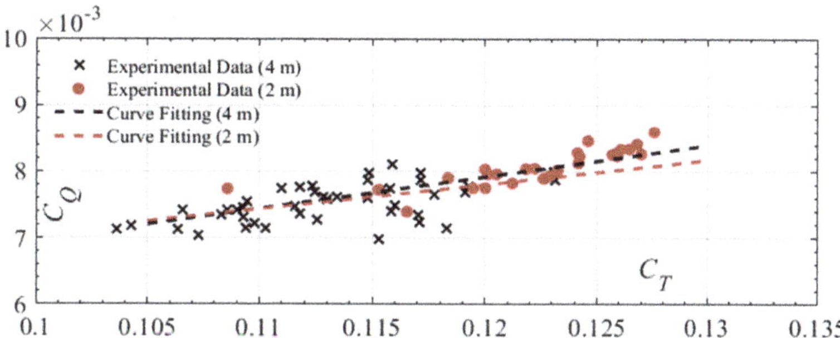

Figure 17. C_T vs. C_Q comparison during dry runs for rotor at h = 2 m and h = 4 m.

4.3. Aerodynamic Parameters at Different Heights—Icing Tests

Table 4 presents the results of the calculations for the preliminary icing tests done at h = 2 m and h = 4 m. The table compares tests performed with the exact same cold room and spray nozzles configuration, with the only difference being the rotor height. The main reason for this comparison is to study the effect of rotor proximity to the nozzles. The tests in the table were all done at Ω = 4950 RPM, θ = 11.7° and λ = 80 g dm^{-2} h^{-1}, with the MVD and T_∞ also specified in the table.

Table 4. Comparison between calculated aerodynamic parameters for rotor at h = 2 m and h = 4 m; (Tests at Ω = 4950 RPM, θ = 11.7° and λ = 80 g dm^{-2} h^{-1}.

MVD (µm)	T_∞ (°C)	Height (m)	C_T^* (%/s)	C_Q^* (%/s)	C_Q^+ (%/s)	P^+ (%/s)	Icing Time (s)
	−5	2	−0.124	0.430	0.491	0.524	169
		4	−0.195	0.565	0.703	0.785	114
120		Ratio	1.57	1.31	1.43	1.50	0.67
	−15	2	−0.136	0.317	0.367	0.394	162
		4	−0.226	0.400	0.548	0.642	106
		Ratio	1.66	1.27	1.50	1.62	0.65
	−5	2	−0.036	0.061	0.063	0.065	321
		4	−0.012	0.047	0.048	0.048	761
800		Ratio	0.33	0.78	0.76	0.75	2.37
	−15	2	−0.049	0.081	0.085	0.087	326
		4	−0.081	0.121	0.132	0.138	220
		Ratio	1.65	1.49	1.55	1.58	0.67

For all the cases tested, except for MVD = 800 µm at T_∞ = −5 °C, the performances of the rotor at h = 4 m degrades between 1.27 to 1.66 faster than at h = 2 m. This is also reflected in the shorter icing time at h = 4 m before testing has to be stopped. The measurements of the icing precipitation intensity (λ) was done at a h = 2 m. To explain this difference, new measurements of the precipitation intensity are repeated at h = 4 m, with the nozzle settings defined during calibration at h = 2 m. Due to the fact that the icing cloud is not fully dispersed in the test section area and is more concentrated, the precipitation intensities obtained at h = 4 m are 1.3 to 1.4 times higher compared to those obtained for the same nozzle settings at h = 2 m for both droplet sizes. The water spray angle and area of coverage are smaller at h = 4 m than at h = 2 m, so a higher λ is obtained, explaining the difference in the results.

Results differ when the larger droplets (MVD = 800 µm) are used at T_∞ = −5 °C. The aerodynamic performances decreased more slowly at h = 4 m than at h = 2 m in opposition to what is obtained at the other test conditions. This can be explained by the fact that some of the larger water droplets of the distribution spectrum are not reaching freezing temperature while falling from the nozzles. The larger droplets are expected to take a longer time during the free fall to reach freezing temperature, and the shorter distance between the rotor and nozzles at h = 4 m seems to lower ice accretion at a warmer temperature like T_∞ = −5 °C. At h = 2 m the hub and blades are 7 m away from the nozzles, while at h = 4 m they are only 5 m away. The droplets are released around 0 to 1°C and cool down during their freefall. With the aerodynamic heating, those very large droplets after a freefall of only 5 m take much longer to freeze on the blade due to their higher temperature when they reach the rotor, and most of them are expulsed from the rotating blades. The expulsion of droplets from the blades could be observed visually during testing. This could also be observed by the lower vibration rates during the test, as well as with the lower amount of ice that was accreted when the test was finalized and longer test durations.

Keeping in mind that the performances of the rotor without ice were similar between h = 4 m and h = 2 m, and since the icing parameters are calibrated at h = 2 m, it is decided that the test campaign will be performed at h = 2 m. On the other hand, the differences brought by the rotor height during icing pose an interesting topic to be elaborated in future studies, especially on the large droplets' temperature equilibrium.

5. Conclusions

This paper presented the design process and methodology used to develop an innovative experimental test rig to study the effects of icing on the rotor of a Bell APT70 drone in take-off/hover flying mode. The test rig was installed in the 9-m-high AMIL cold chamber whose unique capabilities allowed to study the ground effect and rotor proximity to the icing nozzle array at different rotor heights. Techniques for the measurement of MVD and λ in the chamber are presented, followed by a new methodology to calculate the *LWC* for a rotor in hover flight, based on the MVD and λ.

A preliminary experimental comparison of aerodynamic parameters between rotor heights at h = 2 m and h = 4 m showed no difference in the results, demonstrating that no ground effect is obtained at h = 2 m, and further icing tests were therefore decided to be carried out at that lower height, with the advantage of experimental convenience and extra distance from the nozzles. When icing tests were carried out at a distance of 7 m from the nozzles to the rotor (h = 2 m) and at 5 m (h = 4 m), the icing accumulation was changed at warm temperatures for the larger droplets. The larger droplets at that warmer temperature did not cool enough, and very few droplets froze on the blades after a freefall of only 5 m (h = 4 m). The larger droplets of the spectrum did not totally reach terminal velocity either. For those reasons, the decision to position the setup at h = 2 m (7 m from the nozzles) was taken.

Following this important study on the development of a unique test right, a comprehensive investigation on the impact of drone rotor icing can now be undertaken. The effects of Pitch Angles θ, Liquid Water Content (*LWC*), Median Volume Diameter (MVD), air temperatures T_∞, rotor speeds Ω and precipitation rates λ on rotor performance and resulting ice accumulation will be investigated. Then additional testing will be done to study different active and passive solutions to the icing problematic. These studies pave the way for further investigation of the impact of ice on UAV performances and the development of ice protection systems for drone applications.

Author Contributions: Conceptualization, E.V., A.S., M.B., M.L. and C.V.; methodology, E.V., M.B. and M.L.; software, E.V. and A.S.; validation, E.V., A.S., M.B. and M.L.; formal analysis, E.V. and A.S.; investigation, E.V., A.S. and M.B.; writing—original draft preparation, E.V. and A.S.; writing—review and editing, E.V. and A.S.; supervision, C.V.; project administration, E.V.; funding acquisition, C.V. All authors have read and agreed to the published version of the manuscript.

Funding: This research was funded by Bell Textron Canada Ltd.

Acknowledgments: We acknowledge the support provided by Bell Textron Canada Ltd.

Conflicts of Interest: The authors declare no conflict of interest.

References

1. Administration, F.A. UAS by the Numbers. 2021. Available online: https://www.faa.gov/uas/resources/by_the_numbers/ (accessed on 23 November 2021).
2. Flight, B. BELL APT—Autonomous Pod Transport. 2021. Available online: https://www.bellflight.com/products/bell-apt (accessed on 26 November 2021).
3. Cao, Y.; Tan, W.; Wu, Z. Aircraft icing: An ongoing threat to aviation safety. *Aerosp. Sci. Technol.* **2018**, *75*, 353–385. [CrossRef]
4. Yamazaki, M.; Jemcov, A.; Sakaue, H. A Review on the Current Status of Icing Physics and Mitigation in Aviation. *Aerospace* **2021**, *8*, 188. [CrossRef]
5. Liu, Y.; Li, L.; Ning, Z.; Tian, W.; Hu, H. Experimental Investigation on the Dynamic Icing Process over a Rotating Propeller Model. *J. Propuls. Power* **2018**, *34*, 933–946. [CrossRef]
6. Potapczuk, M.G. Aircraft icing research at NASA Glenn research center. *J. Aerosp. Eng.* **2013**, *26*, 260–276. [CrossRef]
7. Dukhan, N.; De Witt, K.J.; Masiulaniec, K.C.; Van Fossen, G.J. Experimental Frossling Numbers for Ice-Roughened NACA 0012 Airfoils. *J. Aircr.* **2003**, *40*, 1161–1167. [CrossRef]
8. Fortin, G.; Laforte, J.-L.; Beisswenger, A. Prediction of ice shapes on NACA0012 2D airfoil. In Proceedings of the FAA In-Flight Icing/Ground De-Icing International Conference & Exhibition, 16–20 June 2003; SAE International United States: Warrendale, PA, USA, 2003.
9. Özgen, S.; Canıbek, M. Ice accretion simulation on multi-element airfoils using extended Messinger model. *J. Heat Mass Transf.* **2009**, *45*, 305–322. [CrossRef]
10. Tsao, J.-C.; Lee, S. *Evaluation of Icing Scaling on Swept NACA 0012 Airfoil Models*; NASA Glenn Research Center: Cleveland, OH, USA, 2012.
11. Tsao, J.-C. Further Evaluation of Swept Wing Icing Scaling with Maximum Combined Cross Section Ice Shape Profiles. In Proceedings of the 2018 Atmospheric and Space Environments Conference, Atlanta, GA, USA, 25–29 June 2018; American Institute of Aeronautics and Astronautics: Atlanta, GA, USA, 2018; p. 3183.
12. Liu, Y.; Zhang, K.; Tian, W.; Hu, H. An experimental investigation on the dynamic ice accretion and unsteady heat transfer over an airfoil surface with embedded initial ice roughness. *Int. J. Heat Mass Transf.* **2020**, *146*, 118900. [CrossRef]
13. Tsao, J.-C.; Kreeger, R. *Further Evaluation of Scaling Methods for Rotorcraft Icing*; SAE Technical Paper; SAE: Warrendale, PA, USA, 2011.
14. Palacios, J.L.; Han, Y.; Brouwers, E.W.; Smith, E.C. Icing Environment Rotor Test Stand Liquid Water Content Measurement Procedures and Ice Shape Correlation. *J. Am. Helicopter Soc.* **2012**, *57*, 29–40. [CrossRef]
15. Wright, J.; Aubert, R. Icing wind tunnel test of a full scale heated tail rotor model. In Proceedings of the AHS 70th Annual Forum, Montreal, QC, Canada, 20–22 May 2014; American Helicopter Society (AHS): Montreal, QC, Canada, 2014; pp. 20–22.
16. Li, Y.; Sun, C.; Jiang, Y.; Feng, F. Scaling Method of the Rotating Blade of a Wind Turbine for a Rime Ice Wind Tunnel Test. *Energies* **2019**, *12*, 627. [CrossRef]
17. Wang, Z.; Zhu, C.; Zhao, N. Experimental Study on the Effect of Different Parameters on Rotor Blade Icing in a Cold Chamber. *Appl. Sci.* **2020**, *10*, 5884. [CrossRef]
18. Fortin, G.; Perron, J. Spinning rotor blade tests in icing wind tunnel. In Proceedings of the 1st AIAA Atmospheric and Space Environments Conference, San Antonio, TX, USA, 22–25 June 2009; American Institute Aeronautics and Astronautics: San Antonio, TX, USA, 2009; p. 4260.
19. Narducci, R.; Kreeger, R.E. *Analysis of a Hovering Rotor in Icing Conditions*; NASA: Phoenix, AZ, USA, 2012.
20. Narducci, R.; Kreeger, R.E. *Application of a High-Fidelity Icing Analysis Method to a Model-Scale Rotor in Forward Flight*; NASA: Virginia Beach, VA, USA, 2012.
21. Chen, L.; Zhang, Y.; Wu, Q.; Chen, Z.; Peng, Y. Numerical Simulation and Optimization Analysis of Anti-/De-Icing Component of Helicopter Rotor Based on Big Data Analytics. In *Theory, Methodology, Tools and Applications for Modeling and Simulation of Complex Systems, Proceedings of the AsiaSim 2016, SCS AutumnSim 2016, Beijing, China, 8–11 October 2016*; Springer: Singapore, 2016; pp. 585–601.
22. Xi, C.; Qi-Jun, Z. Numerical Simulations for Ice Accretion on Rotors Using New Three-Dimensional Icing Model. *J. Aircr.* **2017**, *54*, 1428–1442. [CrossRef]

23. Villeneuve, E.; Harvey, D.; Zimcik, D.; Aubert, R.; Perron, J. Piezoelectric Deicing System for Rotorcraft. *J. Am. Helicopter Soc.* **2015**, *60*, 1–12. [CrossRef]
24. Liu, Y.; Li, L.; Hu, H. An Experimental Study on the Effects of Surface Wettability on the Ice Accretion over a Rotating UAS Propeller. In Proceedings of the 9th AIAA Atmospheric and Space Environments Conference, AIAA, Denver, CO, USA, 5–9 June 2017.
25. Canada, T. *Transport Canada's Drone Strategy to 2025*; Transport Canada: Ottawa, ON, Canada, 2021.
26. Liu, Y.; Li, L.; Li, H.; Hu, H. An experimental study of surface wettability effects on dynamic ice accretion process over an UAS propeller model. *Aerosp. Sci. Technol.* **2018**, *73*, 164–172. [CrossRef]
27. Brouwers, E.W.; Palacios, J.L.; Smith, E.C.; Peterson, A.A. The experimental investigation of a rotor hover icing model with shedding. In Proceedings of the American Helicopter Society 66th Annual Forum, Phoenix, AZ, USA, 11–13 May 2010.
28. Samad, A.; Villeneuve, E.; Blackburn, C.; Morency, F.; Volat, C. An Experimental Investigation of the Convective Heat Transfer on a Small Helicopter Rotor with Anti-Icing and De-Icing Test Setups. *Aerospace* **2021**, *8*, 96. [CrossRef]
29. Samad, A.; Villeneuve, E.; Morency, F.; Volat, C. A Numerical and Experimental Investigation of the Convective Heat Transfer on a Small Helicopter Rotor Test Setup. *Aerospace* **2021**, *8*, 53. [CrossRef]
30. Villeneuve, E.; Volat, C.; Ghinet, S. Numerical and Experimental Investigation of the Design of a Piezoelectric De-Icing System for Small Rotorcraft Part 1/3: Development of a Flat Plate Numerical Model with Experimental Validation. *Aerospace* **2020**, *7*, 62. [CrossRef]
31. Villeneuve, E.; Blackburn, C.; Volat, C. Design and Development of an Experimental Setup of Electrically Powered Spinning Rotor Blades in Icing Wind Tunnel and Preliminary Testing with Surface Coatings as Hybrid Protection Solution. *Aerospace* **2021**, *8*, 98. [CrossRef]
32. Villeneuve, E.; Ghinet, S.; Volat, C. Experimental Study of a Piezoelectric De-Icing System Implemented to Rotorcraft Blades. *Appl. Sci.* **2021**, *11*, 9869. [CrossRef]
33. SAE. *Droplet Sizing Instrumentation Used in Icing Facilities—Aerospace Standard AIR4906*; AC-9C Aircraft Icing Technology Committee: Warrendale, PA, USA, 1995.
34. SAE. *Endurance Time Tests for Aircraft Deicing/Anti-Icing Fluids SAE Type II, III, and IV—Aerospace Standard ARP5485*; G-12HOT Holdover Time Committee: Warrendale, PA, USA, 2004.
35. SAE. *Endurance Time Tests for Aircraft Deicing/Anti-icing Fluids SAE Type I—Aerospace Standard ARP5945*; G-12HOT Holdover Time Committee: Warrendale, PA, USA, 2007.
36. Orchard, D. Investigation of tolerance for icing of UAV rotors/propellers: Phase 3 test results. In *Laboratory Memorandum (National Research Council of Canada. Aerospace. Aerodynamics Laboratory)*; no. LM-AL-2021-0051; National Research Council of Canada. Aerospace: Ottawa, ON, Canada, 2021.
37. SAE. *Calibration and Acceptance of Icing Wind Tunnels—Aerospace Standard ARP5905*; AC-9C Aircraft Icing Technology Committee: Warrendale, PA, USA, 2003.
38. Jeck, R.K. *FAA/AR-09/45: Models And Characteristics of Freezing Rain And Freezing Drizzle For Aircraft Icing Applications*; Federal Aviation Administration, Department of Transportation: Springfield, VA, USA, 2010.
39. Advisory Circular 29-2C, C.o.T.C.R. *29-2C, Certification of Transport Category Rotorcraft*; Federal Aviation Administration, Department of Transportation: Washington, DC, USA, 2008.
40. SAE. *Water Spray and High Humidity Endurance Test Methods for AMS1424 and AMS1428 Aircraft Deicing/Anti-Icing Fluids—Aerospace Standard ARP5901*; G-12ADF Aircraft Deicing Fluids: Warrendale, PA, USA, 2019.
41. Timmerman, P.; Van Der Weele, J.P. On the rise and fall of a ball with linear or quadratic drag. *Am. J. Phys.* **1999**, *67*, 538–546. [CrossRef]
42. Johnson, W. *Rotorcraft Aeromechanics*; Cambridge University Press: Cambridge, UK, 2013; Volume 36.

 *drones*

Article

A Time-Efficient Method to Avoid Collisions for Collision Cones: An Implementation for UAVs Navigating in Dynamic Environments

Manaram Gnanasekera *[iD] and Jay Katupitiya

School of Mechanical and Manufacturing Engineering, University of New South Wales, Sydney, NSW 2052, Australia; j.katupitiya@unsw.edu.au
* Correspondence: m.gnanasekera@unsw.edu.au

Abstract: This paper presents a methodology that can be used to avoid collisions of aerial drones. Even though there are many collision avoidance methods available in literature, collision cone is a proven method that can be used to predict a collision beforehand. In this research, we propose an algorithm to avoid a collision in a time-efficient manner for collision cone based aerial collision avoidance approaches. Furthermore, the paper has considered all possible scenarios including heading change, speed change and combined heading and speed change, to avoid a collision. The heading-based method was mathematically proven to be the most time-efficient method out of the three. The proposed heading-based method was compared with other work presented in the literature and validated with both simulations and experiments. A Matrice 600 Pro hexacopter is used for the collision avoidance experiments.

Keywords: UAV; collision cones; collision avoidance; drones; time-efficiency

Citation: Gnanasekera, M.; Katupitiya, J. A Time-Efficient Method to Avoid Collisions for Collision Cones: An Implementation for UAVs Navigating in Dynamic Environments. *Drones* 2022, 6, 106. https://doi.org/10.3390/drones6050106

Academic Editors: Arianna Pesci, Giordano Teza and Massimo Fabris

Received: 29 March 2022
Accepted: 21 April 2022
Published: 25 April 2022

Publisher's Note: MDPI stays neutral with regard to jurisdictional claims in published maps and institutional affiliations.

1. Introduction

From high-tech surveillance vehicles used in the military to toys used by children, Unmanned Aerial Vehicles (UAV) have come a long way. A flying drone (UAV) above us has been almost a daily experience for a person living in this era of technology. The rapid commercialization pace will make the future skies congested with busy drones assisting us, humans, in many ways. However, there is a darker side to this technological advancement. A congested air space could lead to aerial collisions. The emerging risk of aerial vehicle collisions is substantial and the potential to cause injury to persons and damage to property is ever-increasing. More importantly, UAVs carry flammable materials which may cause fire due to a collision. Therefore, avoiding aerial collisions is important.

A vast array of solutions could be found on avoiding obstacles in past literature. Generally, collision avoidance methods are categorized as global planning based obstacle circumvention (global planners) and local planning based obstacle circumvention (local planners). Global planners use accumulated sensor data and *a priori* information when determining collision-free paths. Algorithms such as PRM [1] (Probabilistic Road Maps), A* [2] and D* [3] are global planners. Local planners, on the other hand, work in a given local search space. In general, local planners operate in highly dynamic environments. Local planners are also known as reactive methods. RRT methods [4], vector field histograms [5], dynamic window approach [6], artificial potential fields [7] and collision cones [8]/velocity obstacles [9] are typical examples for local planners. A sky full of UAVs is a dynamic environment, therefore the authors of this research have been biased towards local planners-based reactive collision avoidance.

Sampling-based methods such as RRT methods will possess a high computational cost when the environment becomes congested. The dynamic window approach takes the motion dynamics of the robot into account. Vector field histograms is a method that is used to find the displacement vectors, given the local range sensor data. Artificial potential field method is a vastly used algorithm which is based on two fields, a repulsive field and an attractive field. However, out of all the reactive methods up to date, the collision cone method possesses the ability to foretell the occurrence of a potential collision. On the other hand, the collision cone concept is based on a paradigm that is based on relative velocities of obstacles and the moving agents in the dynamic environment [8].

Many variants of the collision cone method could be found in the literature. In [9], the concept of collision cone is enhanced to avoid multiple obstacles with the aid of the Minkowski vector sum operator. The Method in [10] proposes a non-linear velocity obstacle algorithm which is composed of local obstacle avoidance and global motion planning. The dynamics of the obstacles and the perception limitations have been addressed by the approach proposed in [11] by applying the probabilistic velocity obstacle method [12] to a dynamic occupancy grid. Authors in [13] present a reciprocal velocity obstacle approach, which guarantees the safe and oscillation free navigation among a flock of agents. The method proposed in [14] incorporates the kinematics and the sensor uncertainty of the robots and avoids collisions by considering the reciprocity. Discrete optimization methods are used in [15] for agent motion planning. This method extends the velocity obstacle concept to a quadratic optimization problem to find collision-free paths.

In spite of the substantial profusion of literature available on collision cone, time-efficient collision avoidance has been scarcely examined. Time efficiency is a vital factor in many aerial applications. A concept in the name of "Time Scaled Collision Cone" proposed in [16,17] remains the sole contribution towards time optimality (time efficiency) in collision cone literature. In a nutshell, this approach avoids a collision through acceleration/de-acceleration based on a non-linear time scaling function by maintaining the original path. The authors of this script have compared the merits of the proposed method over the "Time Scaled Collision Cone" method in Section 6 and discussed them in detail.

It is important to note that the authors of this manuscript have published a previous paper [18] on heading-based collision avoidance by designing the collision cone with respect to the UAV. By understanding the disadvantages of this design, the authors have completely changed the idea and designed the collision cone with respect to the obstacle in the proposed work. The stages of the proposed method are discussed in the forthcoming sections.

2. System Description and Problem Definition

We consider the UAV to be a circle of radius R_u, traveling in a two dimensional plane by maintaining a constant height from the ground. Let $x_u(t), y_u(t)$ be the position of the UAV in form of Cartesian co-ordinates. It travels in a speed of $v_u(t)$ and a heading angle of $\theta_u(t)$ from the abscissa (X-axis). The two dimensional plane consists of multiple circular dynamic obstacles. All the obstacles travel in a much lesser speed compared to the speed of the UAV ($v_u(t)$). These obstacles travel at a constant speed and a constant heading. As described before the task of the UAV is to navigate to a goal location by being in a time-efficient collision-free path. Initially, the authors of the manuscript have presented a comprehensive examination on collision avoidance in the presence of a single obstacle by determining the best algorithm that guarantees time efficiency. Subsequently, the preferred algorithm has been enhanced for multiple obstacle avoidance.

Figure 1 presents a typical single obstacle scenario. The obstacle consists of a radius of R_o, a speed of v_o and a heading of θ_o. The standard first step of a collision cone procedure [8] is to re-arrange one of the objects (either the obstacle or the UAV) as a point. If the collision cone is drawn with respect to the obstacle as in Figure 1, the UAV will be drawn as a point by adding the UAV's radius to the obstacle's radius. As a result, the UAV will become a point with 0 radius and the obstacle will be enlarged to a bigger circle with a larger radius of $R_u + R_o$ (its own radius plus the UAV's radius). The UAV is labeled as U and the center of the obstacle circle is labeled as O in Figure 1. Thereafter, two tangents from U will be drawn to the circle O. The point of tangencies are marked as P_1 and P_2. At this point, the diagram will appear as a cone. Thereafter, a relative vector triangle with respect to $\vec{v}_o$

will be drawn at U as shown in Figure 1. Let, θ_{uo} be the angle of the $\vec{v}_{uo}$ with the abscissa. A potential collision between the UAV and the obstacle could be confirmed if the condition in Definition 1 is satisfied ($\angle P_1 UX$ and $\angle P_2 UX$ are given with the abscissa).

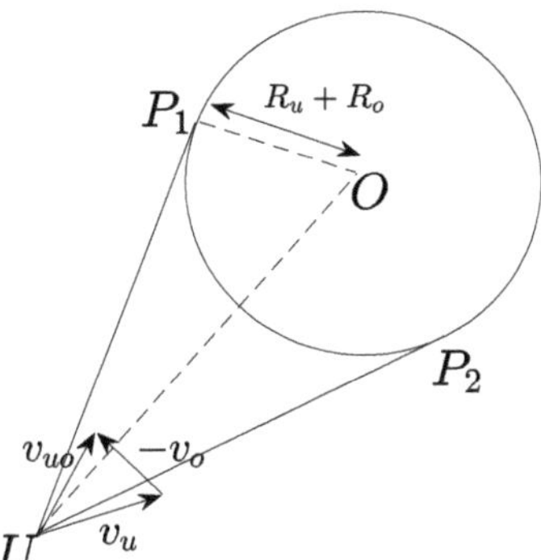

Figure 1. The collision cone is drawn with respect to the obstacle's velocity. The UAV is traveling in $\vec{v}_u$ and the obstacle is traveling in $\vec{v}_o$. The UAV has a radius of R_u and the obstacle has a radius of R_o. The UAV is drawn as a point and the obstacle is drawn with a radius of $R_u + R_o$.

Definition 1. *A potential collision could be guaranteed if,* $\angle P_1 UX > \theta_{uo} > \angle P_2 UX$ *(Figure 1).*

In an instance where the condition in Definition 1 is satisfied, the potential collision could be avoided in three different approaches. The UAV's speed and heading has been differed in the following ways to avoid a collision:

1. Varying the speed and maintaining a constant heading (Purely Speed Solution).
2. Varying the heading and maintaining a constant speed (Purely Heading Solution).
3. Varying both speed and heading (Hybrid Solution).

The forthcoming sections, Sections 3 and 4, provide an exhaustive discussion on purely speed solution, purely heading solution and hybrid solution, respectively. At the end of Section 4, the most time-efficient method is determined. Section 5 enhances the algorithm to handle multiple obstacles. Section 6 presents the results and Section 7 concludes with a few final remarks.

3. Purely Speed Solution(PSS)

As stated in Section 2 a potential collision between a UAV and an obstacle could be avoided by varying the speed of the UAV ($v_u(t)$) and maintaining the heading fixed at θ_u. To assure the time-efficiency criteria the UAV should maneuver at $v_u(t) = v_{max}$ in an obstacle-free environment. The translational model of the UAV is introduced in (1). Term $B_s v_u(t)$ in Equation (1) refers to the translational coulomb resistance, where $B_s \in \mathbb{R}$. Due to the fixed heading, the UAV will be translated forward by applying force F.

$$
\begin{aligned}
\dot{v}_u(t) &= F - B_s v_u(t), \\
\dot{y}_u(t) &= v_u(t)\sin(\theta_u), \\
\dot{x}_u(t) &= v_u(t)\cos(\theta_u).
\end{aligned}
\tag{1}
$$

We use a sliding mode controller based control strategy when calculating F of (1). Let $e(t)$ be the error between the calculated $v_a(t)$ and the current velocity $v_u(t)$ of the UAV. Calculation of $v_a(t)$ will be discussed in detail in the forthcoming Section 3.1.

$$
e(t) = v_a(t) - v_u(t).
\tag{2}
$$

With the aid of (2), we introduce the sliding surface of the sliding mode controller in (3). Where, $\lambda \in \mathbb{R}$.

$$\sigma_s(t) = \dot{e}(t) + \lambda e(t). \tag{3}$$

The navigation law based on the sliding mode control strategy is introduced in (4).

$$F = \text{sgn}(\sigma_s(t)). \tag{4}$$

3.1. Speed Calculation Methodology (Calculation of $V_A(t)$)

Our main objective is to calculate $v_a(t)$ in a given collision scenario. A typical collision scenario is shown in Figure 2a. Both $v_u(t)$ and v_o are presented in vector forms as $\vec{v}_u$, $\vec{v}_o$ at a certain time instance. The UAV labeled as U is traveling towards the goal location G in $\vec{v}_u$ velocity in a fixed θ_u direction. The UAV has detected an obstacle O traveling in $\vec{v}_o$ velocity and in a θ_o direction. The collision cone (Figure 2a) is with respect to the obstacle and let the radius of the base of the cone be R, where $R = R_u + R_o$. If the value of $\angle OUT_2$ and $\angle OUT_1$ is Y and the angle between UO and $\vec{v}_{uo}$ vector is χ ($\angle OUB$). Based on the collision cone, if $\chi < $ Y, the UAV will have a certain collision with the obstacle in time to come. The collision will occur at a point near C, if the UAV and the obstacle continue to move headings and velocities unchanged.

Given that, v_o, θ_o, θ_u are fixed and v_{max} is a maximum, reducing the speed of the UAV becomes the solitary option to avoid the collision.

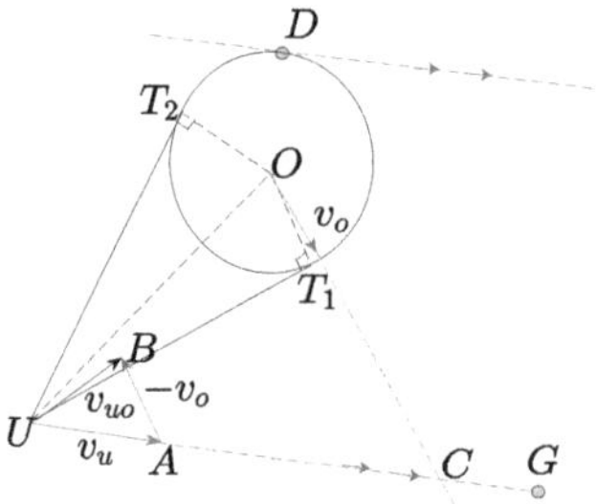

(a)

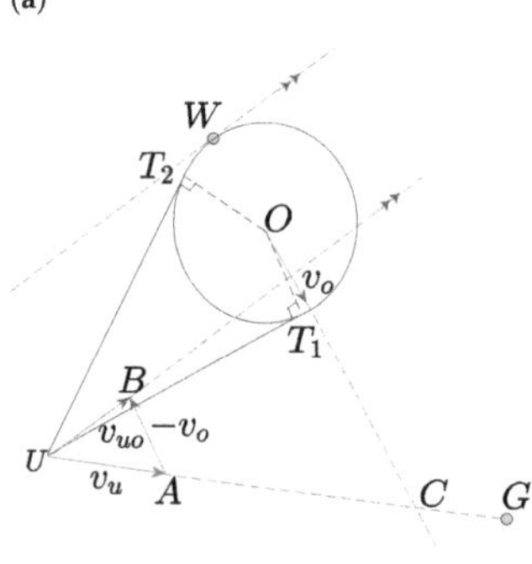

(b)

Figure 2. The figure shows a typical potential collision situation between a UAV and an obstacle. The two sub-figures show the real scenario (the actual situation) and the relative (with respect to the obstacle) scenario. (a) The actual travel path of the UAV is UG and the obstacle's trajectory intersects the path at C. (b) Shows the relative travel path. Point W is the replica of D in the relative scenario. When the UAV actually moves from U to G, relatively it travels in the $\vec{UB}$ direction.

Let $D(t)$ be the point of tangency to the line parallel to UG (Figure 2a). $D(t)$ is a point located on the safety boundary (periphery) of the moving obstacle. All the surface points including $D(t)$ of the obstacle will intersect the UG line at some point of time in future.

Remark 1. *Let $t = t_d$ be the time of intersection of the dynamic point $D(t)$ with the line UG. $D(t_d)$ becomes the farthest obstacle's surface point that the UAV could reach uncollided. An UAV that travels in a time-efficient speed should always arrive at $D(t_d)$ when $D(t_d) \cap UG$ at $t = t_d$.*

We have the independence of deciding a speed profile that satisfies Remark 1. UG is the actual traveling path irrespective of the speed changes made by the UAV during the time of travel. We examine the displacement of the UAV with respect to the obstacle as in Figure 2b. We have addressed the speed-based avoidance problem in a three-staged approach. Namely, hazard stage, intermediate stage and post-hazard stage. The UAV's travel path from the moment of sensing the obstacle till nearing the surface of the obstacle is discussed under the hazard stage. The intermediate stage highlights the UAV's navigation at close range to the obstacle's safety boundary until the line of sight to the goal becomes unimpeded. Thereafter, the UAV travels straight to the goal during the post-hazard stage. At each stage we calculate the time efficient $v_a(t)$.

Through out the manuscript we will be considering the Galilean relativity. Therefore, we can propose the following Remark 2.

Remark 2. *Let the completion time of a certain event be T in the real scenario. The same event will take the same T completion time in the relative scenario.*

When the UAV travels to $D(t_d)$ by satisfying Remark 1, the UAV's relative movement with respect to the obstacle should reach the point W of Figure 2b according to Remark 2. It is evident that there are multiple relative paths, which satisfies Remark 1. However, to suit any collision cone scenario, we select $U\widehat{T_2}W$ as the relative path, which satisfies Remark 1. Relative motion on UT_2 becomes the hazard stage and on $\widehat{T_2W}$ becomes the intermediate stage.

Calculation of the UAV's velocities in the different stages will de discussed in Sections 3.1.1–3.1.3.

3.1.1. Hazard Stage

Figure 3a shows the $UA'B'\triangle$ vector triangle after the speed adjustment of the UAV's speed vector. When measured in the counter clockwise direction from the abscissa, $\angle B'UX$ and $\angle B'A'X$ angles are known. Therefore, the value of $\angle UB'A'$ could be calculated and let the value be θ_a. Since, the heading direction of the UAV is fixed, $\angle A'UX$ will be a constant. Similarly, $\angle A'UB'$ could be calculated as θ_b. By applying the Law of Sines to $UA'B'\triangle$, the speed of the UAV could be calculated.

$$v_u = v_o\left(\frac{sin(\theta_a)}{sin(\theta_b)}\right). \tag{5}$$

The calculated v_u value in (5) can be applied to $v_a(t)$ of (2).

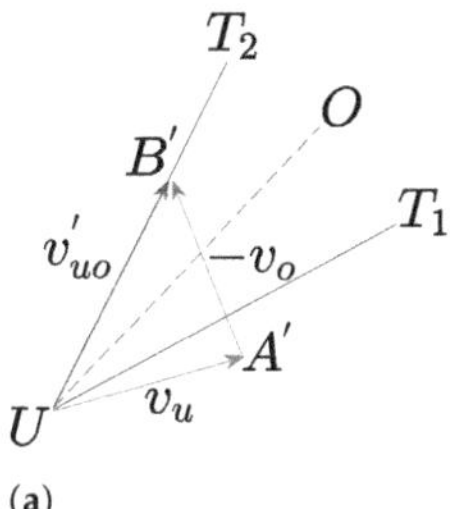

(**a**)

Figure 3. *Cont.*

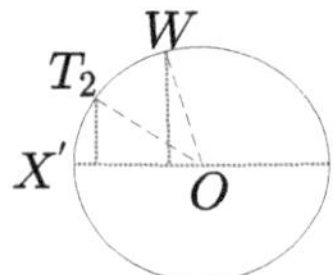

(**b**)

Figure 3. The subfigures show the corrections required during the hazard stage and the intermediate stage. The speed correction during the hazard stage and the intermediate stages. (**a**) The relative velocity vector triangles after the speed correction is shown. $\vec{v}'_{uo}$ is the is the relative velocity vector after the speed is corrected. $\vec{v}'_{uo}$ coincides with UT_2 line segment. (**b**) The intermediate stage when considering the UAV's relative motion. $X'O$ is parallel to the X-axis/abscissa. T_2 is the entrance point of the intermediate stage. W is the final point of the intermediate stage.

3.1.2. Intermediate Stage

Since the relative displacement of the UAV is through the $\overset{\frown}{T_2W}$ arc, the UAV has to maneuver in a variable speed. Let, $X'O$ be parallel to the X-axis (abscissa). Therefore, $\angle X'OW$ and $\angle X'OT_2$ become known angles (Figure 3b. Let, $\angle X'OW = \theta_2$ and $\angle X'OT_2 = \theta_1$. Let, the UAV's speed at T_2 be v_{u1}, which is calculated from (5). At W the UAV should travel in the maximum relative speed. We introduce (6) and (7) by considering the relative displacement of the UAV from T_2 to W in the X-direction (direction of abscissa).

$$v_{x1} = v_{u1}(cos(\theta_u)) - v_o(cos(\theta_o)). \tag{6}$$

$$v_{x2} = v_{max}(cos(\theta_u)) - v_o(cos(\theta_o)). \tag{7}$$

Assumption 1. *The UAV's relative displacement from T_2 to W in the X-direction happens with a constant jerk.*

We introduce (8), by applying the motion equation $v = v_0 + a_0t + \frac{1}{2}jt^2$ to the relative motion from T_2 to W. It is important to note that $a_0 = 0$ at T_2. This is in the X direction, and we will be using the (6) and (7) for the calculations.

$$j = \frac{2(v_{x2} - v_{x1})}{t^2}. \tag{8}$$

Similarly, we apply the motion equation $s = s_0 + v_0t + \frac{1}{2}a_0t^2 + \frac{1}{6}jt^3$ to the relative displacement from T_2 to W and obtain (9).

$$j = \frac{6(Rcos(\theta_1) - Rcos(\theta_2) - v_{x1}t)}{t^3}. \tag{9}$$

From (8) and (9) we can obtain a value for the jerk, $j = J_x$ (Assumption 1) and the total travel time T of the intermediate stage. If the sample time of the system is Δt, the total time could be introduced as in (10). Where, $N \in \mathbb{R}$.

$$T = N\Delta t. \tag{10}$$

With the aid of the motion equation $v = v_0 + a_0 t + \frac{1}{2} j t^2$ we introduce (11) to calculate the velocity of the UAV at any t_u time during the intermediate stage. Where, $t_u = M\Delta t$ and $M \leqslant N$. $M \in \mathbb{R}$.

$$v_u = \frac{v_{x1} + v_0(cos(\theta_o)) + \frac{1}{2} J_x t_u^2}{cos(\theta_u)}.$$ (11)

Similar to the hazard stage, the calculated v_u value in (11) is applied to $v_a(t)$ in (2).

3.1.3. Post-Hazard Stage

The UAV completely eliminates the risk of collision with the particular obstacle when entering the post-hazard stage from the intermediate stage (from W in the relative scenario). Therefore, it maintains the v_{max} speed throughout the post-hazard stage until the goal's position. In other words, $v_a(t) = v_{max}$ during the post-hazard stage.

4. Purely Heading Solution (PHS)

In this section, we consider a heading only navigation strategy by keeping the speed at a constant. Figure 4a manifests a typical collision cone scenario constructed with respect to the obstacle. UG line depicts the original path which the UAV should travel and UG' is the relative path corresponding to the UG original path. The UAV travels in v_{max} speed at all times. Put comprehensively, less deviation from the UG line would result a lesser completion time. Thereby, we introduce the cost function stated in (12). Let, $i(t)$ be the line connecting the moving UAV and the goal location. Let $\vartheta(t)$ be the angle between the UAV's travel direction and $i(t)$.

$$J(t) = \int v_{max} sin(\vartheta(t)) dt.$$ (12)

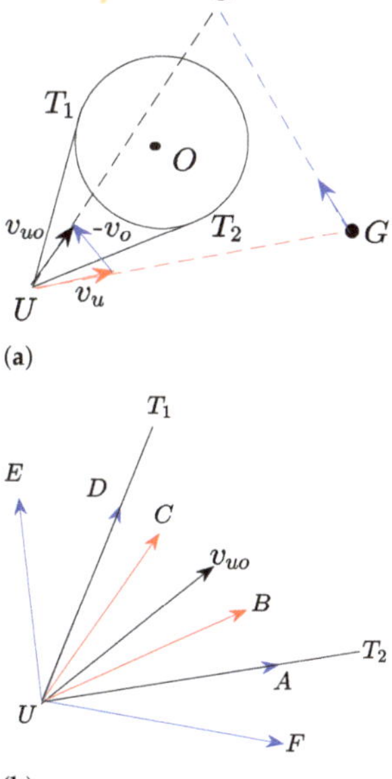

(a)

(b)

Figure 4. The collision scenario and the possible heading corrections. (**a**) A potential collision with UAV U and obstacle O. In the relative scenario, the UAV will meet the goal at G' in the presence of no collision. (**b**) Different directions that the v_{uo} vector could be directed at. Vectors $\vec{UC}$ and $\vec{UB}$ will navigate the UAV inside the collision cone until the safety boundary of the obstacle. On the other hand, vectors $\vec{UD}$, $\vec{UA}$, $\vec{UE}$, $\vec{UF}$ will immediately move the UAV out of the collision cone.

Figure 4b illustrates a number of directions $\vec{v}_{uo}$ can be directed at by varying the UAV's heading. Initially, the focus of this section will be laid upon a method on directing the UAV out of the collision cone immediately during the initial sensing of the obstacle (e.g., $\vec{UD}$, $\vec{UA}$, $\vec{UE}$, $\vec{UF}$). We name this method PHSO (Purely Heading Solution Outside Collision Cone). At the latter part of this section, approaches which make the UAV navigate inside the collision cone (e.g., $\vec{UC}$, $\vec{UB}$) will be examined (for ease of reference, we name the methods as PHSI (Purely Heading Solution Inside Collision Cone)) and compared with the former (PHSO with PHSI). The most time-efficient method out of the two will be determined with rigorous mathematical proofs at the latter part of the section.

4.1. System Description for the Heading Based Avoidance Task

As the UAV is traveling in $v_u = v_{max}$ constant speed, a force will not be required for the translational motion. The required heading changes are made by varying the torque of the system. The rotational and the translational motion is introduced in (13). I is the inertia of the UAV and β_h is a resistance constant ($\beta_h \in \mathbb{R}$). The term $(-\beta_h/I)\omega_u(t)$ is the Rotational Coulomb resistance component of the UAV.

$$
\begin{aligned}
\dot{x}_u(t) &= v_u \cos(\theta_u(t)), \\
\dot{y}_u(t) &= v_u \sin(\theta_u(t)), \\
\dot{\theta}_u(t) &= \omega_u(t), \\
\dot{\omega}_u(t) &= (-\beta/I)\omega_u(t) + \tau/I.
\end{aligned}
\tag{13}
$$

We introduce the error $\zeta(t)$ in (14). Where, $\theta_u(t)$ is the current heading of the UAV and $\theta_c(t)$ is the proposed algorithm's heading, which is discussed in Section 4.2.

$$
\zeta(t) = \theta_c(t) - \theta_u(t).
\tag{14}
$$

With the aid of (14), we introduce the sliding surface of the sliding mode controller in (15). Where, $\Lambda \in \mathbb{R}$.

$$
\sigma_h(t) = \dot{\zeta}(t) + \Lambda\zeta(t).
\tag{15}
$$

In (16), we introduce the navigation law.

$$
\tau = \mathrm{sgn}(\sigma_h(t)).
\tag{16}
$$

The calculation of $\theta_c(t)$ is comprehensively discussed in Section 4.2.

4.2. Introduction to Phso Method and the Calculation of $\theta_c(t)$

Similar to Section 3 we have addressed the heading based avoidance problem in the same three staged approach. Namely, hazard stage, intermediate stage and post-hazard stage. When considering the relative scenario, the UAV should navigate through UT_1 edge (Figure 4a) in order to have a low cost value for $J(t)$ and to still be out of the collision cone. The UAV's travel path from the moment of sensing the obstacle till nearing to the surface of the obstacle is discussed under the hazard stage. After nearing to the surface of the obstacle, it is evident that the UAV has to travel at a close range to the moving obstacle boundary to have the lowest cost value for $J(t)$. The UAV will travel in the intermediate stage until the line of sight between itself and the goal becomes unhindered by the obstacle's surface. When the line of sight becomes clear, the UAV will enter the post-hazard stage and navigate directly to the goal.

4.2.1. Hazard Stage

In accordance with the introduction, the collision cone concept [8] has been the basis of the avoidance notion particularity in this stage. Figure 4a presents a situation where an UAV has initially perceived an obstacle from a certain distance. The collision cone in Figure 4a is drawn with respect to the obstacle's velocity v_o. T_1 and T_2 are two points of tangencies. Let Y be the value of angles $\angle T_2 UO$ and $\angle T_1 UO$. Let χ be the angle between the vector $\vec{v}_{uo}$ and UO. With the introduction of χ and Y, Definition 1 can be re arranged as Definition 2.

Definition 2. *If the condition $\chi < Y$ satisfies for a given collision cone, the particular UAV and the obstacle will collide patently.*

Therefore, a suitable heading adjustment should be made by the UAV to dissatisfy the condition given in Definition 2 to circumvent the obstacle. However, as stated previously, the UAV should reach the safety boundary of the obstacle at a point of tangency. Thereby, the relative vector $\vec{v}_{ou}$ should point at T_1 (Figure 4a) throughout the course of the hazard stage.

Figure 5 shows two relative vector diagrams. UH_1N_1 $\triangle$, which is drawn prior to the heading correction and UH_2N_2 $\triangle$ subsequent to the UAV's heading correction. Let the angle of $-\vec{v}_o$ with the X axis be θ'_o and let the angle $\angle XUN_2$ with the X axis be β. It should be noted that, both θ'_o and β angles are known angles. Thereby, $\angle UN_2H_2$ could be found as λ. Let α be the $\angle H_2UN_2$ angle. With the Law of Sines applied to UN_2H_2 $\triangle$, we introduce (17).

$$\alpha = arcsin\left(\frac{v_o}{v_u}sin(\lambda)\right). \tag{17}$$

Now, $\theta_c(t)$ could be calculated with (18).

$$\theta_c(t) = \beta - \alpha. \tag{18}$$

$\theta_c(t)$ will be applied to (14).

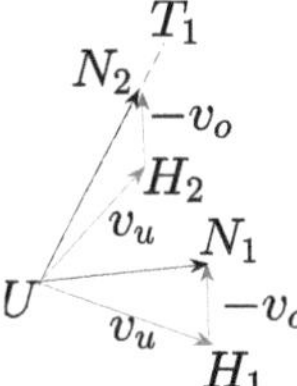

Figure 5. The relative velocity vector triangles needed for the angle calculations are shown in this figure. Heading correction during the hazard stage. UH_1N_1 $\triangle$ is the relative vector triangle prior to the heading correction. UH_2N_2 $\triangle$ is the relative vector triangle after the heading correction.

4.2.2. Intermediate Stage

Due to the switching behavior of the sliding mode controller we have introduced an additional safety boundary. The additional boundary with a radius of $R_o + 2\xi$ is known as the Exterior safety boundary. The safety boundary the is closest to the obstacle is called the Interior safety boundary (which has a radius of $R_o + \xi$). Where, $\xi \in \mathbb{R}$. During the intermediate stage, we will be considering the real scenario (the actual UAV movement).

Control law (16) should maneuver the UAV between exterior safety boundary and the obstacle surface throughout the intermediate stage. We propose Algorithm 1 to effectuate this requirement. Initially, let the position of the UAV be $x_u(t)$, $y_u(t)$ with a heading direction of $\theta_u(t)$. If the sample time of the UAV is δT, we introduce (19), (20) and (21) to predict the location of the UAV at $t = t + \delta T$, when, $\theta_u(t + \delta T) = \theta_u(t)$, $\theta_u(t + \delta T) = \theta_u(t) + \delta\theta$ and $\theta_u(t + \delta T) = \theta_u(t) - \delta\theta$. Where $\delta\theta \in \mathbb{R}$.

$$\hat{x}_{u1}(t + \delta T) = x_u(t) + v_u \cos(\theta_u - \delta\theta)\delta T,$$
$$\hat{y}_{u1}(t + \delta T) = y_u(t) + v_u \sin(\theta_u - \delta\theta)\delta T. \tag{19}$$

$$\hat{x}_{u2}(t + \delta T) = x_u(t) + v_u \cos(\theta_u)\delta T,$$
$$\hat{y}_{u2}(t + \delta T) = y_u(t) + v_u \sin(\theta_u)\delta T. \tag{20}$$

$$\hat{x}_{u3}(t + \delta T) = x_u(t) + v_u \cos(\theta_u + \delta\theta)\delta T,$$
$$\hat{y}_{u3}(t + \delta T) = y_u(t) + v_u \sin(\theta_u + \delta\theta)\delta T. \tag{21}$$

By assuming the obstacle's heading to be unchanged at θ_o, we introduce (22) to calculate the location of the obstacle when $t = t + \delta T$. Let the initial position of the obstacle be $x_o(t)$, $y_o(t)$.

$$\hat{x}_o(t + \delta T) = x_o(t) + v_o \cos(\theta_o)\delta T,$$
$$\hat{y}_o(t + \delta T) = y_o(t) + v_o \sin(\theta_o)\delta T. \tag{22}$$

We introduce R_1 in (23), R_2 in (24) and R_3 in (25).

$$R_1 = \sqrt{(\hat{x}_{u1}(t + \delta T) - \hat{x}_o(t + \delta T))^2 + (\hat{y}_{u1}(t + \delta T) - \hat{y}_o(t + \delta T))^2}. \tag{23}$$

$$R_2 = \sqrt{(\hat{x}_{u2}(t + \delta T) - \hat{x}_o(t + \delta T))^2 + (\hat{y}_{u2}(t + \delta T) - \hat{y}_o(t + \delta T))^2}. \tag{24}$$

$$R_3 = \sqrt{(\hat{x}_{u3}(t + \delta T) - \hat{x}_o(t + \delta T))^2 + (\hat{y}_{u3}(t + \delta T) - \hat{y}_o(t + \delta T))^2}. \tag{25}$$

With the aid of (23),(24) and (25) we introduce Δ_{R1} as in (26), Δ_{R2} as in (27) and Δ_{R3} as in (28).

$$\Delta_{R1} = |(R_1 - (R_o + \zeta))|. \tag{26}$$

$$\Delta_{R2} = |(R_2 - (R_o + \zeta))|. \tag{27}$$

$$\Delta_{R3} = |(R_3 - (R_o + \zeta))|. \tag{28}$$

We propose Algorithm 1, which will keep the UAV in between the exterior safety boundary and the actual obstacle surface throughout the intermediate stage.

Algorithm 1 Heading at $t = t + \delta T$

Input: Δ_{R1}, Δ_{R2} , Δ_{R3}
Output: $\theta_u(t + \delta T)$
1: **if** $(\Delta_{R1} < \Delta_{R2})$ **and** $(\Delta_{R1} < \Delta_{R3})$ **then**
2:    $\theta_u(t + \delta T) \leftarrow \theta_u(t) - \delta\theta$;
3: **else if** $(\Delta_{R2} < \Delta_{R1})$ **and** $(\Delta_{R2} < \Delta_{R3})$ **then**
4:    $\theta_u(t + \delta T) \leftarrow \theta_u(t)$;
5: **else**
6:    $\theta_u(t + \delta T) \leftarrow \theta_u(t) + \delta\theta$;
7: **end if**

4.2.3. Post-Hazard Stage

The line of sight from the UAV towards the goal will be de-linked by the obstacle during the hazard and the intermediate stages. The UAV will enter the post-hazard stage immediately after the line of sight gets unhindered. Then, move straight to the goal. We introduce (29), (30) and (31) to find the transition point from the hazard stage to the post-hazard stage.

$$m1(t) = \frac{y_u(t) - y_o(t)}{x_u(t) - x_o(t)}. \tag{29}$$

$$m2(t) = \frac{y_g - y_u(t)}{x_g - x_u(t)}. \tag{30}$$

The UAV will enter the post-hazard stage when the condition in (31) is satisfied at $t = t_{tr}$.

$$m1(t_{tr})m2(t_{tr}) = -1. \tag{31}$$

We introduce (32) to find $\theta_c(t)$ during the post-hazard stage.

$$\theta_c(t) = arctan(m2(t_{tr})). \tag{32}$$

4.3. Introduction to Phsi Method and the Time-Efficiency Comparison between Phso and Phsi

Even though Section 4.2 finds the time-efficient heading possibilities out of the collision cone, the possibilities shown in Figure 6a still remain unexamined. To begin with, we will extend the study conducted in Section 4.2 by investigating the possibilities inside the collision cone. The obstacle in Figure 6a has A, B_i, C, D, E, F, H, I labels on its periphery. Similar to navigating from U to A (over the edge of the cone), the UAV has the ability to travel directly towards the periphery of the obstacle and then to maneuver over the periphery of the obstacle. As an example the UAV could travel through UC and $\overset{\frown}{CB_i}$. Where, $i \in \mathbb{N}$. However, the time-efficiency when traveling inside the cone should be compared with the solution provided in Section 4.2.

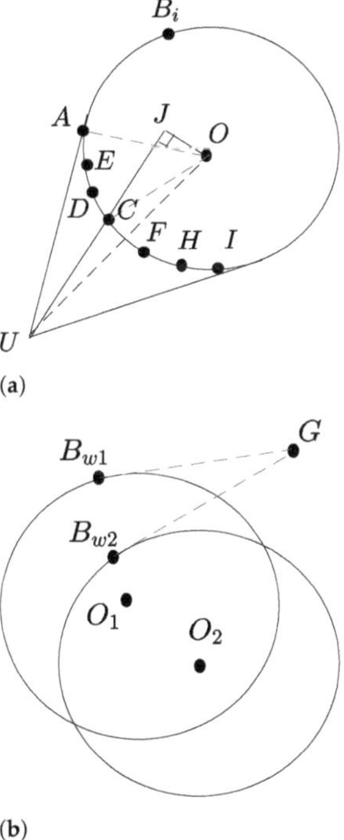

(a)

(b)

Figure 6. The figures explains the *PHSI* method. Especially the hazard stage and the intermediate stage. (a) The points on the obstacle's periphery (safety boundary) which are inside the collision cone are labeled as A, C, D, E, F, H, I. (b) Post-hazard stages of the two cases. B_{w1} is the final point of the intermediate stage of the case coming through A. B_{w2} is the final point of the case coming through C.

Theorem 1. *The point C in Figure 6a is at the periphery of the obstacle inside the collision cone. Let t_1 be the time taken for the UAV to travel through the path $UAB_1G(t_1)$, where $i = 1$ and t_2 be the time taken for the UAV to travel through the path $UCB_2G(t_2)$, where $i = 2$. B_1 and B_2 are the last points of the intermediate stage or the first points of the post hazard stage ($i \in \mathbb{N}$). At any given collision cone scenario $t_1 < t_2$.*

Proof of Theorem 1. Firstly, we introduce the following terms for the UAV's travel through the relative path $UCB_2G(t_2)$,

- v_a—Relative velocity of the UAV during the hazard stage.
- $v_b(t)$—Relative velocity of the UAV during the intermediate stage.
- d_r—The relative distance between the UAV and the obstacle when the obstacle was initially sensed (UO distance).
- r_r—The relative distance traveled by the UAV during the intermediate stage from C to A (Figure 6a).

□

With the availability of range sensors which could measure up to 300 m (e.g.,: LDM301, Sick DME5000-321) we make a reasonable assumption as in Assumption 2.

Assumption 2. *If the radius of the base of the collision cone is R. Then, $10R \leq d_r$.*

By considering the UAV's radius along with the secure safety boundary, we introduce a sensible assumption in Assumption 3.

Assumption 3. *For any given collision cone $\gamma + \psi \geq C$. Where, $\angle AUO = \gamma$, $\angle JUO = \psi$ and $C \in \mathbb{R}$.*

Assumption 4. *Let v_b be the maximum relative velocity during the intermediate stage (v_b is the maximum velocity, $v_b(t)$ could reach). For ease of analysis, we assume that the UAV travels at v_b during the intermediate stage.*

Let, $\angle OCJ = \phi$, $\angle AUO = \gamma$, $\angle JUO = \psi$, $\angle AOC = \theta$.
We introduce δ_r in Equation (33).

$$\delta_r = d_r \cos(\gamma) - (d_r \cos(\psi) - R \cos(\phi)). \tag{33}$$

Similar to δ_r and r_r in the relative scenario, we can also consider δ_w and r_w in the real scenario. With the aid of Remark 2 and Assumption 3, we can introduce (34) and (35). Where $k1, k2 \in \mathbb{R}$.

$$\delta_r = k_1 \delta_w \tag{34}$$

$$r_r = k_2 r_w \tag{35}$$

(36) could be obtained by (35)/(34).

$$\frac{r_w}{\delta_w} = \frac{k_1}{k_2} \frac{r_r}{\delta_r}. \tag{36}$$

(36) could be rearranged as $\dfrac{r_w}{\delta_w} \propto \dfrac{r_r}{\delta_r}$. We introduce (37) by applying $\phi = \dfrac{\pi}{2} - (\gamma + \theta - \psi)$ to (33),

$$\delta_r = d_r \cos(\gamma) - (d_r \cos(\psi) - R \cos(\frac{\pi}{2} - (\gamma + \theta - \psi))). \tag{37}$$

(37) could be simplified as (38).

$$\delta_r = d_r \cos(\gamma) - d_r \cos(\psi) + R sin(\gamma + \theta - \psi). \tag{38}$$

The terms in (38) could be rearranged in the form of (39).

$$\delta_r = -2d_r \sin(\frac{\gamma - \psi}{2}) \sin(\frac{\gamma + \psi}{2}) + R \sin(\theta) \cos(\gamma - \psi) + R \cos(\theta) \sin(\gamma - \psi). \tag{39}$$

The term $R \cos(\theta) \sin(\gamma - \psi)$ of (39) could be expanded as in (40).

$$\delta_r = -2d_r \sin(\frac{\gamma - \psi}{2}) \sin(\frac{\gamma + \psi}{2}) + R \sin(\theta) cos(\gamma - \psi) + 2R cos(\theta) \sin(\frac{\gamma - \psi}{2}) \cos(\frac{\gamma - \psi}{2}). \tag{40}$$

With the aid of Assumption 2 we apply $d_r = 10R$ to (40).

$$\delta_r = -20R\sin(\frac{\gamma - \psi}{2})\sin(\frac{\gamma + \psi}{2}) + R\sin(\theta)\cos(\gamma - \psi) + 2Rcos(\theta)\sin(\frac{\gamma - \psi}{2})\cos(\frac{\gamma - \psi}{2}). \tag{41}$$

A simplified version of (41) is presented in (42).

$$\delta_r = -2R\sin(\frac{\gamma - \psi}{2})\left\{10\sin(\frac{\gamma + \psi}{2}) - \cos(\theta)\cos(\frac{\gamma - \psi}{2})\right\} + R\sin(\theta)\cos(\gamma - \psi). \tag{42}$$

With the aid of Assumptions 2 and 3 we can state that $10\sin(\frac{\gamma + \psi}{2}) > 1$ and $0 < \cos(\theta)\cos(\frac{\gamma - \psi}{2}) < 1$. Therefore, we can confirm that $-2R\sin(\frac{\gamma - \psi}{2})\left\{10\sin(\frac{\gamma + \psi}{2}) - \cos(\theta)\cos(\frac{\gamma - \psi}{2})\right\} < 0$. Furthermore, we also can affirm that $R\sin(\theta)\cos(\gamma - \psi) < R\sin(\theta)$ as $0 < \cos(\gamma - \psi) < 1$. Thereby, $\delta_r < R\sin(\theta)$. We introduce the definition of r_r as in (43).

$$r_r = R\theta \tag{43}$$

Generally radians of θ values are larger or equal to their corresponding $\sin(\theta)$ values. Therefore, $r_r > \delta_r$. If we consider the real scenario, from (36) we are able to confirm that $r_w > \delta_w$. The UAV travels in constant v_u in the real scenario. As a result, the UAV will travel the δ_w distance in a much shorter time compared to the r_w distance. As per Remark 2, the time taken to travel δ_r will be much lesser than the time taken to travel r_r. In essence, the proof made till this point certifies that the UAV's relative movement through UA edge reaches point A much faster than the other method reaching via point C.

Let B_{w1} be the final point of the intermediate stage in the real scenario corresponding to B_1. According to the Remark 2, the UAV enters B_1 in the relative scenario and B_{w1} in the real scenario at the same time. Similarly, let B_{w2} be the final point of the intermediate stage in the real scenario which corresponds to B_2 in the relative scenario. The UAV will arrive at B_{w1} at first due to the early entrance to the intermediate stage. Thereafter, the UAV will move directly to the goal in a straight line. Point B_{w1} always will have a lesser distance to the goal compared to B_{w2} (Figure 6b). Therefore, the UAV will be reaching the goal in a much shorter time through $B_{w1}G$. In correspondence, the UAV will relatively move towards the goal through B_1 in a shorter time and reach the goal at $G(t1)$ if the goal reaching time is $t1$. In conclusion, $UAB_1G(t1)$ will be the time-efficient path.

This completes the proof of Theorem 1. We also lay our investigation towards time efficiency comparison between a straight line trajectory (described prior) and non-linear/piecewise linear paths during the hazard stage in this section. Let a UAV travel to a point on the periphery of the obstacle in a non-straight trajectory in t_1 time by being at a constant speed. The UAV can reach the same point in a straight trajectory in the same constant speed in t_2 time, where $t_2 < t_1$. Therefore, if the UAV travels in a constant speed, there will always be a time-efficient linear path to reach a point in the periphery compared to any other non-linear path traveled in the same speed.

It can be confirmed that the PHSO method gives the most time-efficient solution in comparison to PHSI method. Therefore, the PHSO method will be referred as PHS (Purely Heading Solution) through out the rest of the script.

4.4. Comparison of Phs with Pss Method

Theorem 2. *Let T_1 be the completion time of PHS and T_2 be the completion time of PSS. $T_1 < T_2$ at any given collision cone scenario.*

Proof of Theorem 2. A collision cone scenario is presented in Figure 7a and the PHS and PSS corrections are shown in Figure 7b. The vector triangle UA_sB_s represents the correction of PSS and UA_hB_h represents the correction of PHS. Let, $\angle UA_sB_s$ be θ_s and $\angle UA_hB_h$ be θ_h. By applying the law cosines, the magnitude of the relative velocity vector of PSS could be shown as in (44) and the magnitude of the relative velocity vector of PHS could be given by (45). Let, v_{u1} be the speed correction of PSS. The UAV travels in v_{max} prior to the collision detection . □

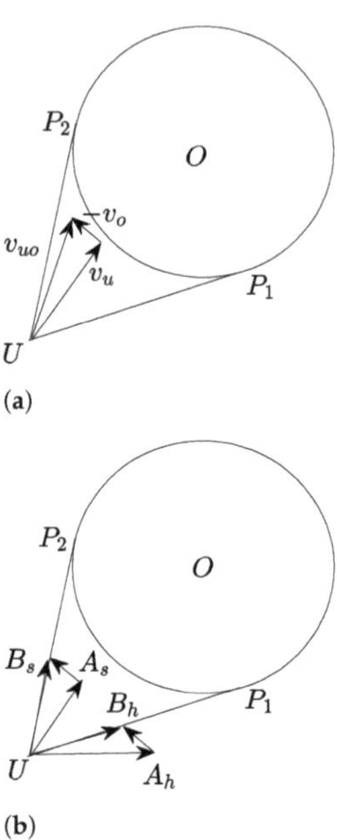

Figure 7. The figures represent a collision scenario and the corresponding *PHS* and *PSS* corrections in form of relative velocity vector triangles: (**a**) Presents the scenario prior to the correction. (**b**) Shows *PHS* and *PSS* corrections.

$$|v_s| = \sqrt{|v_{u1}|^2 + |v_o|^2 - 2|v_{u1}||v_o|\cos(\theta_s)}, \tag{44}$$

$$|v_h| = \sqrt{|v_{max}|^2 + |v_o|^2 - 2|v_{max}||v_o|\cos(\theta_h)}. \tag{45}$$

Due to $|v_h|^2$, $|v_s|^2 \in \mathbb{R}_{>0}$, $|v_{u1}|^2 + |v_o|^2 > 2|v_{u1}||v_o|\cos(\theta_s)$ and $|v_{max}|^2 + |v_o|^2 > 2|v_{max}||v_o|\cos(\theta_h)$. Furthermore, $v_{max} > v_{u1}$ condition will satisfy $v_h > v_s$ at all times. It is clear that with $v_h > v_s$ condition being satisfied, PHS algorithm will make the UAV reach the obstacle's safety boundary in the minimum time. Similar to the other theorems, we will consider the real scenario to investigate the intermediate and the post-hazard stages. Due to the lead of PHS at the hazard stage and as the speed of PSS during the intermediate stage is lesser than v_{max}, the UAV will be made to arrive at the final point of the intermediate stage in a shorter time by the PHS algorithm. As a result, PHS makes the UAV enter the post-hazard stage before PSS. During the post-hazard stage, both PSS and PHS will make the UAV navigate in v_{max}. However, due to the early entrance, PHS navigates the UAV towards the goal in much lesser time when compared with PSS.

It is important to note that in some instances UAV's heading correction of PHS will make the relative vector coincide with UP_1 edge. In these situations, moving towards P_1 becomes more time-efficient than moving towards P_2. Therefore, the theorem still becomes valid. This completes the proof of Theorem 2.

4.5. Comparison of Phs with Speed and Heading Hybrid Method

Hitherto, we have discussed PHS and PSS approaches in detail. Similar to PHS and PSS, a speed variation along with a heading variation becomes feasible when avoiding collisions. Two example hybrid approaches are presented in Figure 8. Let v_{max} and θ_u be the speed and the heading of the UAV prior to the collision. We confirm that $v_u' < v_{max}$. The angle θ_3 is the angle opposite $\vec{v}_{uo3}$ and θ_4 is opposite to $\vec{v}_{uo4}$. It is important to compare the time-efficiency merits of the hybrid method with that of PHS at the outset.

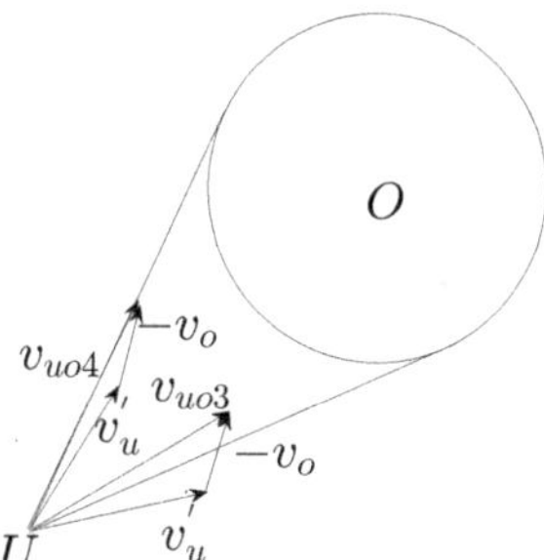

Figure 8. The two relative vectors represent the hazard stages of two hybrid methods. Where, $v_u' < v_u$.

Proposition 1. *There are n (where $n \in \mathbb{R}$) hybrid methods to avoid a collision, given a collision cone scenario. Let T_i be the completion time of a hybrid method, where $i \in \{1 \dots n\}$. Let T_h be the completion time of the PHS of the same collision scenario. Then, $T_i > T_h$.*

Proof of Proposition 1. The resultant velocities are introduced as in (46) and (47).

$$|v_{ou3}| = \sqrt{(|v_u'|)^2 + |v_o|^2 - 2|v_u'||v_o|\cos(\theta_3)}, \tag{46}$$

$$|v_{ou4}| = \sqrt{(|v_u'|)^2 + |v_o|^2 - 2|v_u'||v_o|\cos(\theta_4)}. \tag{47}$$

From Theorem 1 we can establish that v_{uo4} will make the UAV reach the obstacle's safety boundary at the earliest. If v_u' is increased to v_{max} by maintaining the heading, we can assure that the resultant relative velocity will be greater than v_{uo4}. This will eventually be the PHS solution. Any hybrid correction that happens inside the collision cone could be stated in the form of (46). The resultant relative velocity could be given by (47) if the relative velocity vector of any hybrid correction coincides with the edge of the collision cone. Therefore, the PHS method will be more time-efficient than any of the hybrid solutions. According to Theorems 1 and 2, the method that makes the UAV reach the safety margin of the obstacle at first will surely reach the goal ahead of the other methods. $\square$

This completes the proof of Proposition 1. It is clear that PHS becomes the most time-efficient method to avoid a collision in a single obstacle collision cone scenario.

5. Multiple Collision Handling

A multiple collision situation is where a UAV senses multiple obstacles in a given time instance, which would lead to future collisions. In this section, we introduce an algorithm for a general multiple collision situation initially. Thereafter, an enhanced version of the introduced algorithm will be presented to handle complex multiple obstacle scenarios.

5.1. Typical Multiple Collision Handling

A typical multiple collision example is shown in Figure 9a. As per the conclusions made in the previous section, the time-efficiency of the PHS approach out performs all the remaining avoidance methods that could be used in collision cone scenarios. Navigating on full speed on a straight line during the hazard stage stands the crux of the PHS approach. A multiple collision scenario is a combination of several single collision cones. As per the merits discussed, we consider the application of PHS in multiple collision situations. The aim will be to navigate the UAV in full speed on a straight line.

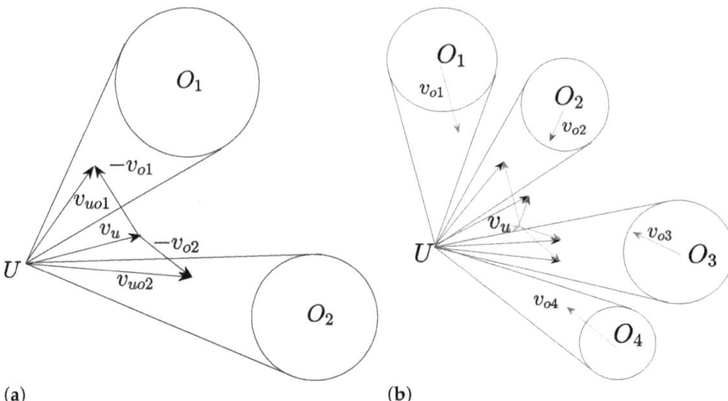

Figure 9. Multiple collision avoidance: (**a**) A typical collision situation with two obstacles. Since, v_{uo1} and v_{uo2} are inside the collision cones the UAV will be colliding with both O_1 and O_2 obstacles. (**b**) presents a complex multiple obstacle scenario. Where, the UAV has a risk colliding with O_2 and O_3 initially. With a clockwise correction, the UAV has a risk of colliding with O_1. With an anti-clockwise correction, the UAV has a risk of colliding with O_4.

Figure 9a depicts a multiple collision scenario with two obstacles. Let R_{O1} be the radius of obstacle O_1 and R_{O2} be the radius of obstacle O_2. As the initial step, collision cones will be constructed with all the available obstacles. Due to the multiple obstacle nature, unlike in the single obstacle problem, both clockwise and anti-clockwise heading corrections should be considered in each collision cone. As an example, v_{uo1} could be made to coincide with UP_1 or UP_2. As some of the relative vectors are much closer to the bottom edge of the collision cone and some others are closer to the top edge of the collision cone, calculating the heading ($\theta_c(t)$) in both directions becomes crucial.

Let the number of sensed obstacles be n. Where, $n \in \mathbb{N}$. With the aid of the PHS algorithm we calculate the clockwise heading correction angle Ω_i for each collision cone $\{i = 1, \ldots, n\}$. Similarly, we calculate the anti-clockwise heading correction angle η_i for each collision cone. From all the Ω_i values, we select the maximum as in (48).

$$\Omega_{max} = max\{\Omega_1, \ldots, \Omega_n\}. \tag{48}$$

Similarly, we repeat the selection for η_i as in (49).

$$\eta_{max} = max\{\eta_1, \ldots, \eta_n\}. \tag{49}$$

Finally, the minimum value from Ω_{max} and η_{max} is selected as in (50).

$$\theta_c(t) = min\{\eta_{max}, \Omega_{max}\}. \tag{50}$$

Equations (48)–(50) should be calculated at every sample time. The calculated $\theta_c(t)$ values from (50) are used throughout the hazard stage. At the last part of the hazard stage the UAV arrives at the safety boundary of one obstacle from the lot. Therefore, the intermediate stage and the post-hazard stage will be similar to the single obstacle scenario. If the UAV encounters another obstacle during the intermediate stage, the collision is handled with a new collision cone.

5.2. Complex Multiple Collision Handling

Figure 9b illustrates a different collision scenario not described before. The UAV in the image has encountered four unattached obstacles. As per the collision cones, UAV has a risk of colliding with O_2 and O_3 initially. If the UAV makes an anticlockwise heading correction, the risk of colliding with O_1 increases. On the other hand, a clockwise correction escalates the risk of impact with O_4. Therefore, we enhance the PHS method made for multiple collisions introduced in Section 5.1. Algorithm 4 introduces the enhancement. Let there be n number of obstacles sensed. Out of the obstacles, the UAV has a risk of immediately colliding with m number of obstacles. Where, $m < n$. χ_i and Y_i (where, $\{i = 1, \ldots, n\}$) refers to χ and Y angles of the ith obstacle (Definition 2).

With the introduction of Algorithm 2, we can confirm that there is always a feasible way of avoiding a collision in a time-efficient manner irrespective of the complexity of the collision.

Algorithm 2 Calculation of $\theta_c(t)$ in a complex situation.

    Input: $\eta_1 \ldots \eta_n, \Omega_1 \ldots \Omega_n$
    Output: $\theta_c(t)$
    while 1 **do**
      $k \leftarrow m$
      $\eta_{max} \leftarrow max\{\eta_1, \ldots, \eta_m\};$
      $\Omega_{max} \leftarrow max\{\Omega_1, \ldots, \Omega_m\};$
      $\theta_c(t) \leftarrow min\{\eta_{max}, \Omega_{max}\};$
      if $\chi_{k+1} > Y_{k+1}$ **then**
        break;
      end if
      $m \leftarrow m + 1;$
    end while

---

6. Results

6.1. Simulation Results

The section assesses the performance of the PSS and PHS algorithms in different collision scenarios. The simulations were conducted in a Matlab based testing platform.

6.1.1. Simulations Related to Pss Algorithm

The UAV in Figure 10 has traveled to a goal located at (300, 300) from the initial starting location. However, it has been obstructed by a dynamic obstacle as shown in Figure 10a. The UAV has maneuvered at a close proximity to the obstacle's surface during the intermediate stage with the aid of (11). As per the velocity profile shown in Figure 10d, the velocity has been around 1.4 ms^{-1} throughout the hazard stage as a result of (5). The UAV has passed the intermediate stage swiftly with the aid of (11) which was introduced in Section 3.1.2.

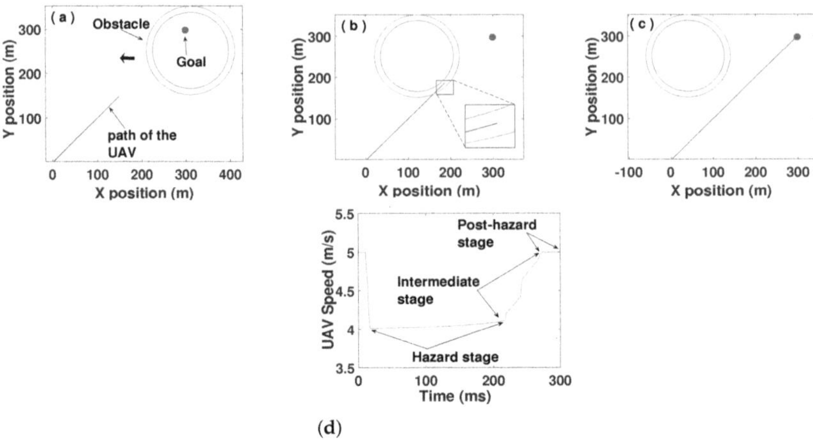

Figure 10. Simulation conducted on the purely speed scenario: (**a**) Shows the hazard stage. Figures (**b**) illustrates the intermediate stage. (**c**) Shows the post hazard stage. (**d**) Shows the speed profile of UAV.

6.1.2. Simulations Related to Phs Algorithm

A typical PHS scenario is presented in Figure 11. The UAV has navigated in a speed of $5\,\text{ms}^{-1}$. As mentioned in Section 4.2.2, an additional safety boundary has been introduced for the intermediate stage. Figure 11a shows the full trajectories of the UAV and obstacle. The operation of the sliding mode controller is quite observable in the hazard stage depicted in Figure 11b. The last point of the hazard stage (which is zoomed in and presented) is clearly positioned in between the obstacle's surface and the exterior safety boundary. Algorithm 1 hasn't allowed the UAV to collide with the obstacle or to reach out of the exterior safety boundary. Figure 11c delineates the post-hazard stage with a straight trajectory directly towards the goal.

6.1.3. Multiple Collision Avoidance Handing

Figure 12 shows a collision avoidance situation with three obstacles with different radii. The UAV has been navigated to the safety margin of obstacle 1 by Equation (50) as presented in Figure 12b. It is important to note that Equation (50) hasn't maneuvered the UAV to the safety boundaries of obstacle 2 and obstacle 3 as a result of potential collisions. The last point of the hazard stage has been zoomed in and presented. In Figure 12, the UAV has avoided the obstacles satisfying the equations introduced in Section 5.1 and have successfully maneuvered to the goal location.

6.1.4. Complex Collision Avoidance

A complex multiple collision avoidance situation is illustrated in Figure 13. The trajectory profiles are presented in Figure 13a. The UAV, which traveled in a $5\,\text{ms}^{-1}$ speed, has initially encountered obstacle 2 and obstacle 3 around (70,70) as potential threats while traveling towards the goal location. The UAV will not be able to make an anti-clockwise heading correction to move towards the safety margin of obstacle 2, as it would enter the collision cone of obstacle 1. Similarly, it does not have the ability to reach the safety margin of obstacle 3, as it would enter the collision cone of obstacle 4. Therefore, with the aid of Algorithm 3 introduced in Section 5.2, the UAV has navigated to the safety margin of obstacle 4 as shown in Figure 13b. After avoiding the collision, the UAV has successfully reached the goal according to Figure 13c.

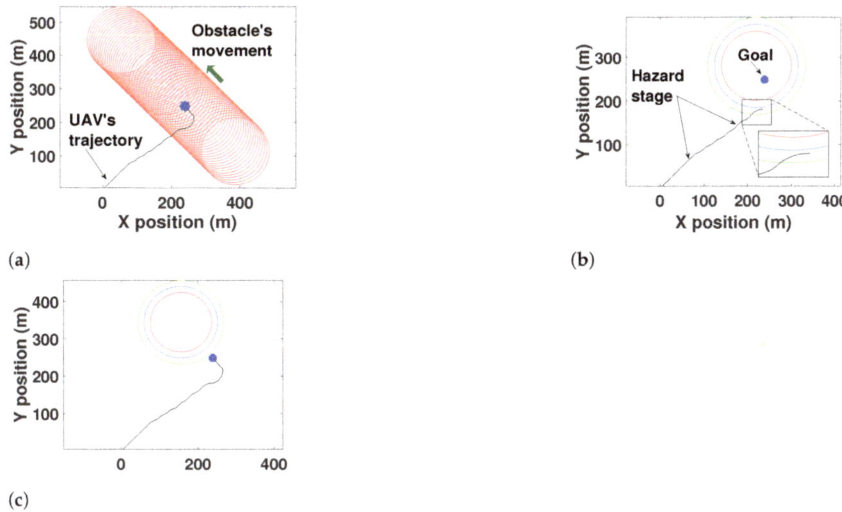

Figure 11. Simulation conducted on the purely heading scenario. The obstacle has traveled at a speed of 4.9 ms^{-1} and moved at a direction of $3\pi/2$ rad from the X-axis. The UAV has traveled towards a goal located at (250,250): (**a**) Shows the full trajectory of the UAV and the obstacle. (**b**) Presents the hazard stage. The point at which the UAV enters the safety margin of the obstacle has been zoomed in and presented. (**c**) Depicts the goal reaching moment (post-hazard stage).

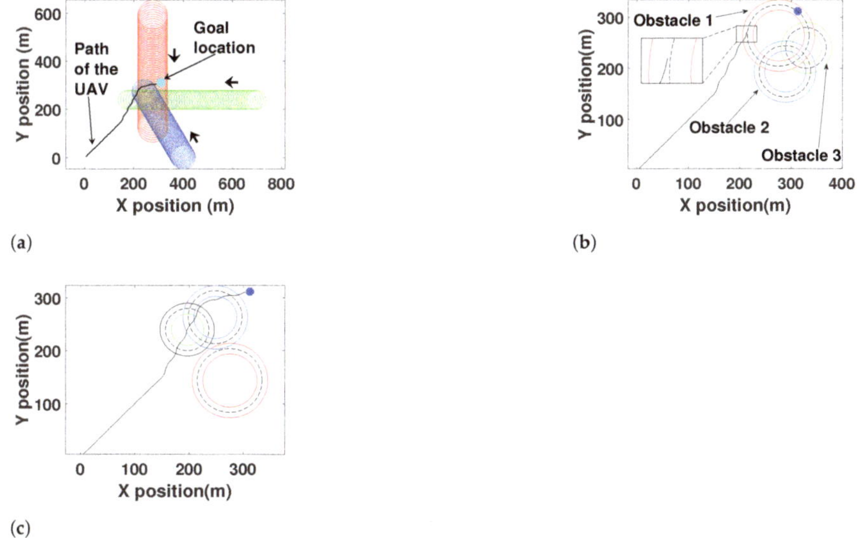

Figure 12. Multiple collision avoidance with three obstacles. Obstacle 1 moves in a speed of 4.8 ms^{-1} in a direction of 1.5708 rad, obstacle 2 moves in a direction of 2.0944 rad at a speed of 3.2 ms^{-1} and obstacle 3 moves at an angle -3.1416 rad in a speed 4.3 ms^{-1}: (**a**) Shows the full trajectory profiles. (**b**) Presents the hazard stage. (**c**) Shows the successful completion of the task during the post-hazard stage.

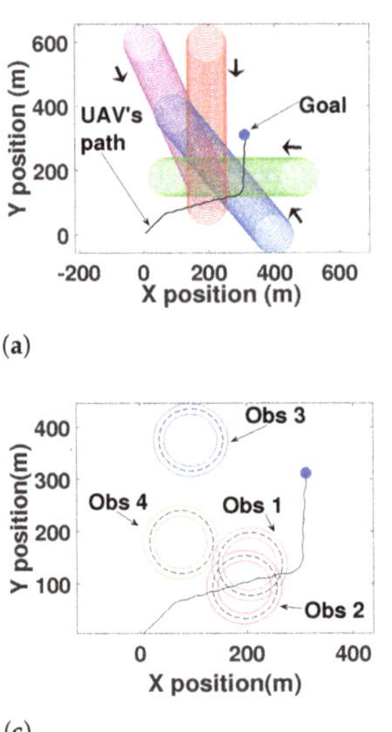

(a)

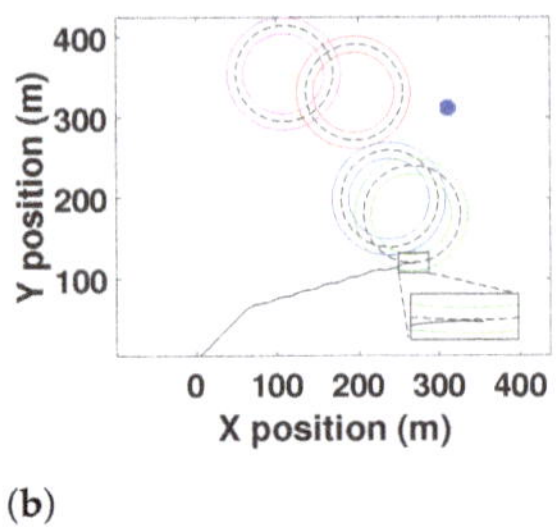

(b)

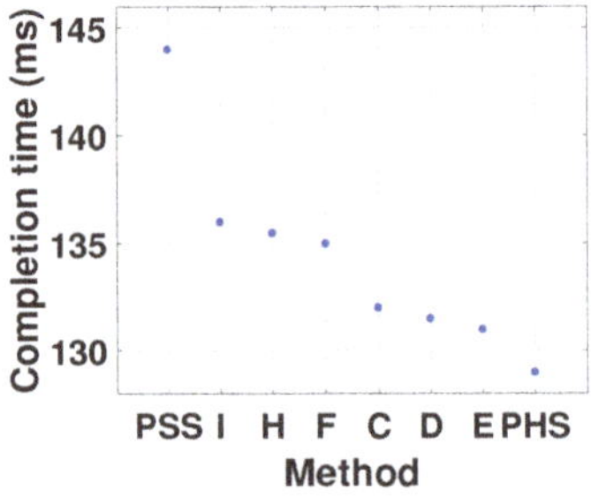

(c)

Figure 13. Complex multiple collision scenario. Obstacle 1 has traveled at a speed of 4.4 ms^{-1} in a heading of 1.15 rad, obstacle 2 has moved with a heading of 1.57 rad with a speed of 4.4 ms^{-1}, obstacle 3 has traveled in a heading direction of 2.25 rad with a speed of 4.2 ms^{-1} and obstacle 4 has traveled at a speed of 3.5 ms^{-1} in a direction of -3.14 rad.: (**a**) Shows the full trajectory profiles. (**b**) Shows the hazard stage and the last point of the hazard stage is zoomed. (**c**) Presents post-hazard stage.

6.1.5. A Simulation to Justify Theorems 1 and 2

The purpose of this simulation is to justify Theorems 1 and 2. In order to test the validity of Theorem 1, we make the UAV travel through the points labeled in Figure 6a. In other words, $\widehat{UEA}$, $\widehat{UDA}$, $\widehat{UCA}$, $\widehat{UFA}$, $\widehat{UHA}$, $\widehat{UIA}$ segments separately and to reach the goal according to PHSI algorithm. Thereafter, the UAV is sent through the UA edge to reach the goal as stated in the PHS algorithm. In order to examine Theorem 2, the collision situation is once more handled using the PSS method. More importantly, the completion time of each and every method is recorded for comparison. Figure 14 presents the completion time results of each method, which justifies Theorems 1 and 2.

Figure 14. Shows the completion times taken by different methods. Labels I, H, F, C, D, E represents the path through each point.

6.1.6. Comparison with Tscc (Time Scaled Collision Cone) Method

Authors in [17] use nonlinear time scaling [18] to avoid collisions by acceleration or retardation. It is the sole approach or benchmark method that proposes a time-efficient collision avoidance paradigm for collision cones as mentioned in the literature review. We have compared our PHS approach with the TSCC method proposed in [17] in Figure 15. Both the algorithms were applied in the same environment, and the task completion time has been recorded. Figure 15a,b illustrates the behavior of the algorithms in a five obstacle environment. To normalize both scenarios and to present in real-time, we have set the max speed of the UAV as 5 ms^{-1}. A completion time of 196.63 ms was recorded by the PHS algorithm and the TSCC has recorded 276 ms to complete the five obstacle avoidance task. By having the same goal position as in Figure 15a,b, we have increased the number of obstacles and recorded the completion time in Figure 15c in order to draw a proper conclusion. In Figure 15c, the number of obstacles has been increased to 10, 15 and 20 (due to the high clutteredness, these situations will not be illustrated). It is more than clear that the PHS method's time-efficiency outperforms that of TSCC method.

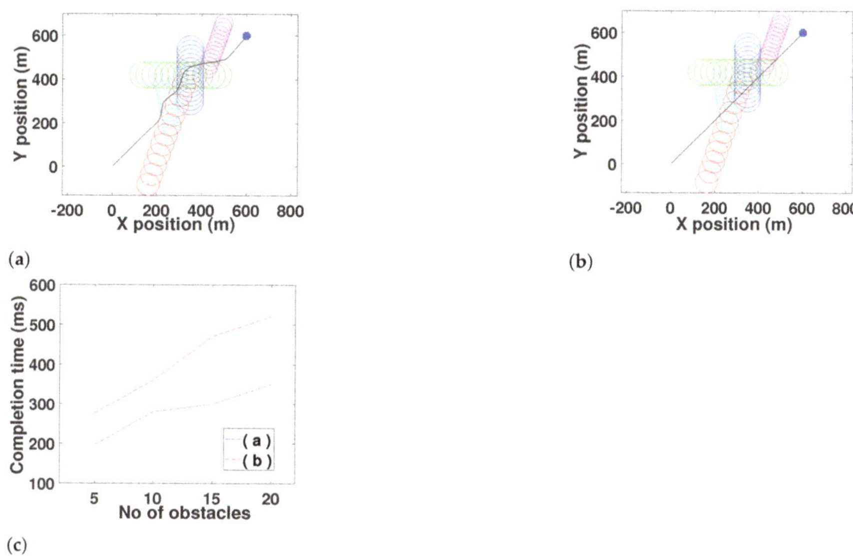

Figure 15. Comparison between PHS and TSCC: (**a**) Shows the trajectory profiles of the PHS method and five different obstacles. (**b**) Shows the trajectory profiles of the same five obstacles and the TSCC method. (**c**) Shows the time taken by PHS and TSCC to complete a task when the number of obstacles are increased.

It is fair to say, that the simulation results presented in this section solidly ratify the theories discussed in Sections 3–5.

6.2. Experimental Results

A set of physical experiments were conducted to ensure the practicality of the PHS method. The practical set-up and equipment used are shown in Figure 16. A Matrice 600 PRO hexacopter is used as the UAV. An Intel NUC mini PC is used to automate the hexacopter. The mini PC is connected to the hexacopter via a USB cable and powered by the battery power of the hexacopter. Since obstacles are UAVs in reality, the available other hexacopters in the laboratory can be used as obstacles for the experiments with a similar automation set-up. However, as the wind condition is not addressed in this research fictitious obstacles are used as a safety measure. In addition, we have built up a wireless local area network (WLAN) with the aid of an Ubiquiti Edge router, an Ubiquiti access point and a laptop computer. All computers are connected to the network so that data such as the location of a hexacopter could be shared among the computers. Ubuntu operating systems are installed in all the computers. In addition, we have installed DJI SDK 3.6 to access the flight controller (A3 pro) of the hexacopter. The algorithm is coded in C++. The experiments were conducted in an open field in Menangle, New South Wales, Australia.

With the aid of the above set-up, we have conducted single obstacle and multiple obstacle related experiments of the PHS algorithm. The τ value of (16) is replaced by a heading angle of 1 rad. As a result, the UAV's controller will be switching between -1 rad, 0 rad and $+1$ rad. The starting location of the UAV will be considered the origin, and all the graph co-ordinates in this section are with respect to the origin. With a reasonable safety boundary, the radius of a hexacopter could be stated as 1.5 m.

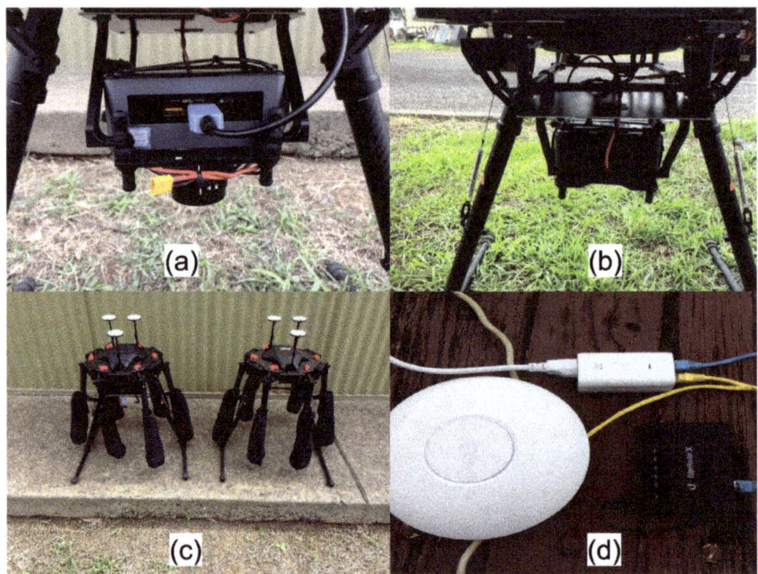

Figure 16. Equipment used for the experiments: (**a**) Shows the NUC mini PC connected to the flight controller via a USB cable. (**b**) Shows the NUC mini PC and the power connection which is provided from the hexacopter batteries. (**c**) Hexacopters available in the laboratory. The radius of a hexacopter with a reasonable safety boundary is around 1.5 m. (**d**) Shows the Edge router and the access point.

Figure 17 presents the single obstacle experiment based on speed. The hexacopter has traveled in a 0.26 ms^{-1} speed in a heading direction of $\pi/3$ rad. The obstacle has traveled in a direction of $-\pi/2$ rad and in a speed of 0.15 ms^{-1}. Figure 17a shows the hazard stage. A sample of the intermediate stage is presented in Figure 17b and it is more than clear that the hexacopter hasn't been collided with the obstacle. Figure 17c presents the goal reaching moment. The complete speed profile is shown in Figure 17d. Some spikes in the speed profile can be observed. This is primarily due to external resistances. Figure 17e has shown the complete trajectory of the obstacle and the hexacopter.

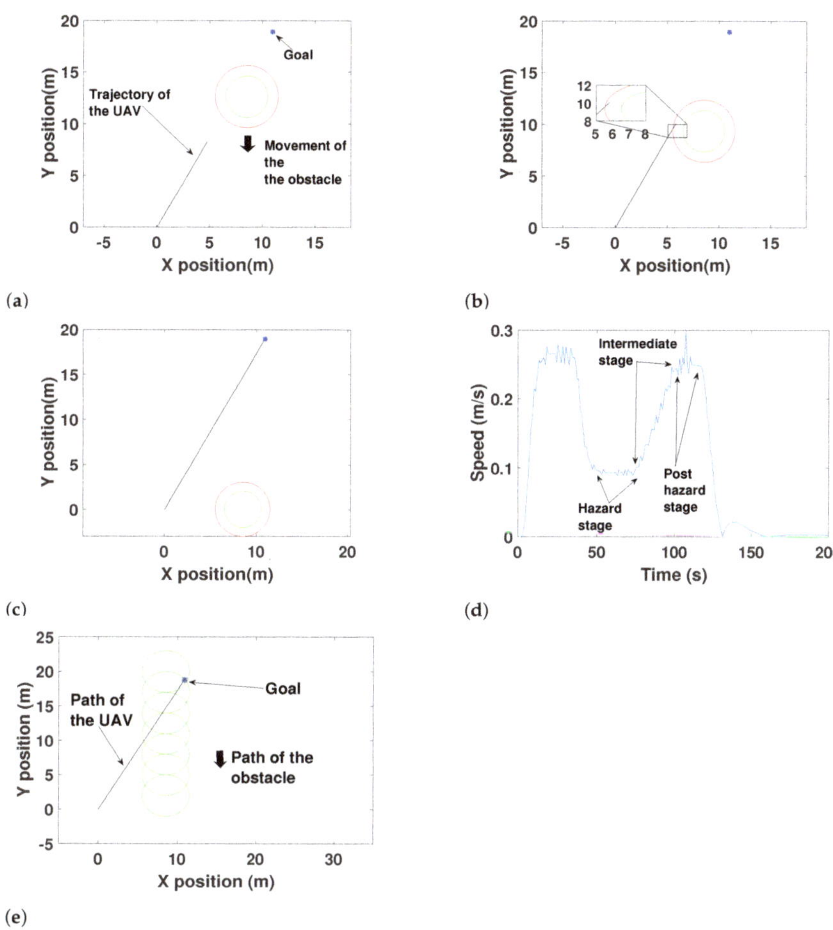

Figure 17. Speed experiment: (**a**) shows the hazard stage. (**b**) illustrates a samples of the intermediate stage. (**c**) shows the post hazard stage. (**d**) shows the speed profile of UAV. (**e**) Complete trajectories of the UAV and the obstacle.

The full trajectory profiles of the UAV and the obstacle is shown in Figure 18a. As the controller is switching between −1 rad and 1 rad, a trajectory oscillation (switching due to the sliding mode controller) is not visible as in the simulations during the hazard stage (Figure 18b). The UAV has successfully reached the safety boundary of the obstacle. Figure 18c shows a sample of the intermediate stage. Even though Algorithm 1 has navigated the UAV successfully, the travel trajectory during the intermediate stage hasn't been smooth when compared to the simulations. The main reason behind this difference is the dynamics of the UAV (hexacopter) and the external influences such as wind. Figure 18d presents the goal reaching moment of the UAV. It is clear that the dynamics haven't allowed the UAV to make a swift heading change to reach the goal. Hence, it has made a 3.14 rad sluggish heading change to reach the goal.

Figure 19 shows a potential multiple collision situation with two obstacles. Figure 19b depicts the successful completion of hazard stage by the use of (50). In Figure 19c, the UAV has reached the goal successfully. However, due to the dynamics and external resistances, it has reached the goal with moderate heading changes. Hence, it hasn't been able to navigate directly towards the goal.

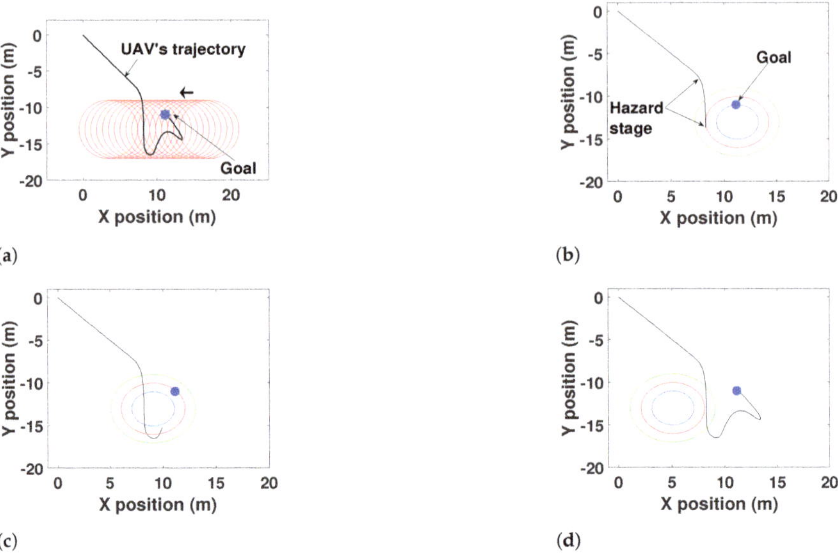

(a)

(b)

(c)

(d)

Figure 18. First experiment conducted on the purely heading scenario with a single obstacle. The hexacopter has navigated in a speed of $0.3\ \mathrm{ms}^{-1}$ towards a goal located at $(10.69, -10.69)$ and has encountered as potential threat by an obstacle traveling at a speed of $0.1\ \mathrm{ms}^{-1}$ in a direction of 3.14 rad.: (**a**) shows the full trajectory of the hexacopter and the obstacle. (**b**) presents the hazard stage. (**c**) shows a sample from the intermediate stage. (**d**) depicts the goal reaching moment (post-hazard stage).

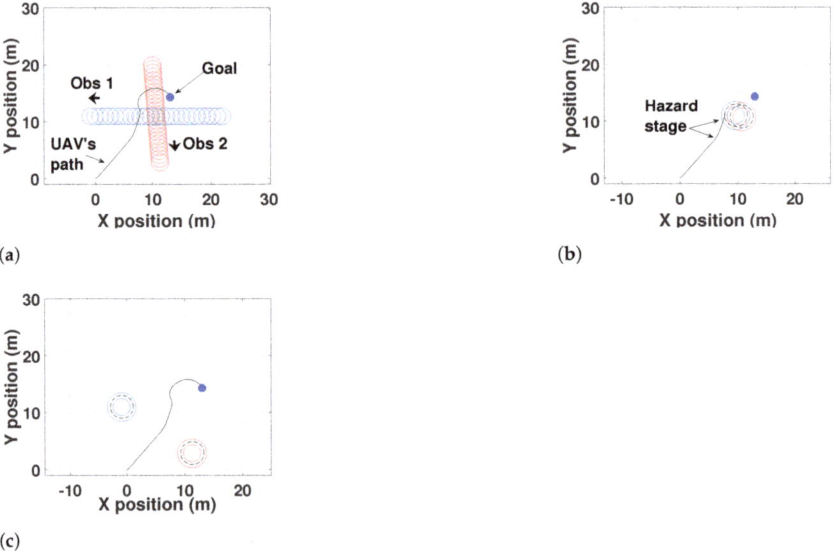

(a)

(b)

(c)

Figure 19. Experiment on avoiding two obstacles. The hexacaopter has traveled at a speed of $0.31\ \mathrm{ms}^{-1}$ towards a goal located around $(13, 14)$. Obstacle 1 has moved in a speed of $0.165\ \mathrm{ms}^{-1}$ in direction of a -3.14 rad, and obstacle 2 has moved in a speed of $0.22\ \mathrm{ms}^{-1}$ in a heading angle of -1.4960 rad: (**a**) shows the full trajectories of the hexacopter and the obstacles. (**b**) shows the hazard stage. (**c**) presents the goal reaching moment during the post-hazard stage.

In summary, the main differences between the experimental results and the simulation results are the dynamics of the UAV and external resistances such as the wind. However, the experimental results firmly validate the PHS algorithm.

7. Conclusions

A UAV could avoid a potential collision with dynamic obstacles either by changing the speed or changing its heading or by changing both speed and heading concurrently. This paper has thoroughly investigated all three possibilities and has shown that the heading based method is time-efficient than the other two. The paper has proposed PHS (Purely Heading Solution) method and has confirmed the time-efficiency with rigorous mathematical proofs. Initially, the PHS method was first implemented for a single obstacle and thereafter enhanced and implemented to avoid multiple and complex collision scenarios. Furthermore, the time-efficiency of PHS was shown to be better when the results were compared with other work in literature. The validity of the proposed method was demonstrated both in simulation and real flight experiments.

As future work, the proposed PHS method will be enhanced to perform in 3D environments. Furthermore, the impact of the wind resistance will be addressed with the aid of an additional controller.

Author Contributions: Conceptualization, M.G. and J.K.; methodology, M.G.; software, M.G.; validation, M.G. and J.K.; formal analysis, M.G.; investigation, M.G.; resources, J.K.; data curation, M.G.; writing—original draft preparation, M.G.; writing—review and editing, M.G.; visualization, M.G.; supervision, J.K.; project administration, J.K.; funding acquisition, J.K. All authors have read and agreed to the published version of the manuscript.

Funding: This research received no external funding.

Institutional Review Board Statement: Not applicable.

Informed Consent Statement: Not applicable.

Data Availability Statement: Not applicable.

Conflicts of Interest: The authors declare no conflict of interest.

References

1. Kavraki, L.E.; Svestka, P.; Latombe, J.C.; Overmars, M.H. Probabilistic roadmaps for path planning in high-dimensional configuration spaces. *IEEE Trans. Robot. Autom.* **1996**, *12*, 566–580. [CrossRef]
2. Hart, P.E.; Nilsson, N.J.; Raphael, B. A formal basis for the heuristic determination of minimum cost paths. *IEEE Trans. Syst. Sci. Cybern.* **1968**, *4*, 100–107. [CrossRef]
3. Stentz, A. Optimal and efficient path planning for partially known environments. In *Intelligent Unmanned Ground Vehicles*; Springer: Berlin/Heidelberg, Germany, 1997; pp. 203–220.
4. LaValle, S.M.; Cheng, P; Resolution complete rapidly-exploring random trees. In Proceedings of the 2002 IEEE International Conference on Robotics and Automation (Cat. No. 02CH37292), Washington, DC, USA, 11–15 May 2002; Volume 1, pp. 267–272.
5. Borenstein, J.; Koren, Y.; The vector field histogram-fast obstacle avoidance for mobile robots. *IEEE Trans. Robot. Autom.* **1991**, *7*, 278–288. [CrossRef]
6. Fox, D.; Burgard, W.; Thrun, S. The dynamic window approach to collision avoidance. *IEEE Robot. Autom. Mag.* **1997**, *4*, 23–33. [CrossRef]
7. Khatib, O. Real-time obstacle avoidance for manipulators and mobile robots. In *Autonomous Robot Vehicles*; Springer: Berlin/Heidelberg, Germany, 1986; pp. 396–404.
8. Chakravarthy, A.; Ghose, D. Obstacle avoidance in a dynamic environment: A collision cone approach. *IEEE Trans. Syst. Man Cybern. Part A Syst. Hum.* **1998**, *28*, 562–574. [CrossRef]
9. Fiorini, P.; Shiller, Z. Motion planning in dynamic environments using velocity obstacles. *Int. J. Robot. Res.* **1998**, *17*, 760–772. [CrossRef]
10. Large, F.; Sckhavat, S.; Shiller, Z.; Laugier, C. Using non-linear velocity obstacles to plan motions in a dynamic environment. In Proceedings of the 7th International Conference on Control, Automation, Robotics and Vision, (ICARCV 2002), Singapore, 2–5 December 2002; IEEE: Piscataway, NJ, USA, 2002; Volume 2, pp. 734–739.
11. Fulgenzi, C.; Spalanzani, A.; Laugier, C. Dynamic obstacle avoidance in uncertain environment combining PVOs and occupancy grid. In Proceedings of the 2007 IEEE International Conference on Robotics and Automation, Rome, Italy, 10–14 April 2007; IEEE: Piscataway, NJ, USA, 2007; pp. 1610–1616.

12. Kluge, B.; Prassler, E. Reflective navigation: Individual behaviors and group behaviors. In Proceedings of the IEEE International Conference on Robotics and Automation, 2004. Proceedings. ICRA'04. 2004, New Orleans, LA, USA, 26 April–1 May 2004; IEEE: Piscataway, NJ, USA, 2004; Volume 4, pp. 4172–4177.

13. Van den Berg, J.; Lin, M.; Manocha, D. Reciprocal velocity obstacles for real-time multi-agent navigation. In Proceedings of the 2008 IEEE International Conference on Robotics and Automation, Pasadena, CA, USA, 19–23 May 2008; IEEE: Piscataway, NJ, USA, 2008; pp. 1928–1935.

14. Snape, J.; Van Den Berg, J.; Guy, S.J.; Manocha, D. Independent navigation of multiple mobile robots with hybrid reciprocal velocity obstacles. In Proceedings of the 2009 IEEE/RSJ International Conference on Intelligent Robots and Systems, St. Louis, MO, USA, 10–15 October 2009; IEEE: Piscataway, NJ, USA, 2009; pp. 5917–5922.

15. Guy, S.J.; Chhugani, J.; Kim, C.; Satish, N.; Lin, M.; Manocha, D.; Dubey, P. Clearpath: Highly parallel collision avoidance for multi-agent simulation. In Proceedings of the 2009 ACM SIGGRAPH/Eurographics Symposium on Computer Animation, New Orleans, LA, USA, 1–2 August 2009; pp. 177–187.

16. Singh, A.K.; Krishna, K.M. Reactive collision avoidance for multiple robots by non linear time scaling. In Proceedings of the 52nd IEEE Conference on Decision and Control, Florence, Italy, 10–13 December 2013; IEEE: Piscataway, NJ, USA, 2013; pp. 952–958.

17. Gopalakrishnan, B.; Singh, A.K.; Krishna, K.M. Time scaled collision cone based trajectory optimization approach for reactive planning in dynamic environments. In Proceedings of the 2014 IEEE/RSJ International Conference on Intelligent Robots and Systems, Chicago, IL, USA, 14–18 September 2014; IEEE: Piscataway, NJ, USA, 2014; pp. 4169–4176.

18. Gnanasekera, M.; Katupitiya, J. A Time Optimal Reactive Collision Avoidance Method for UAVs Based on a Modified Collision Cone Approach. In Proceedings of the 2020 IEEE/RSJ International Conference on Intelligent Robots and Systems, Las Vegas, NV, USA, 25–29 October 2020; IEEE: Piscataway, NJ, USA, 2020; pp. 5685–5692.

 *drones*

Article

Structure-from-Motion 3D Reconstruction of the Historical Overpass Ponte della Cerra: A Comparison between MicMac® Open Source Software and Metashape®

Matteo Cutugno [1,*] , **Umberto Robustelli** [2] and **Giovanni Pugliano** [1]

[1] DICEA, University of Naples Federico II, 80125 Naples, Italy
[2] Department of Engineering, University of Naples Parthenope, 80143 Naples, Italy
* Correspondence: matteo.cutugno@unina.it

Abstract: In recent years, the performance of free-and-open-source software (FOSS) for image processing has significantly increased. This trend, as well as technological advancements in the unmanned aerial vehicle (UAV) industry, have opened blue skies for both researchers and surveyors. In this study, we aimed to assess the quality of the sparse point cloud obtained with a consumer UAV and a FOSS. To achieve this goal, we also process the same image dataset with a commercial software package using its results as a term of comparison. Various analyses were conducted, such as the image residuals analysis, the statistical analysis of GCPs and CPs errors, the relative accuracy assessment, and the Cloud-to-Cloud distance comparison. A support survey was conducted to measure 16 markers identified on the object. In particular, 12 of these were used as ground control points to scale the 3D model, while the remaining 4 were used as check points to assess the quality of the scaling procedure by examining the residuals. Results indicate that the sparse clouds obtained are comparable. MicMac® has mean image residuals equal to 0.770 pixels while for Metashape® is 0.735 pixels. In addition, the 3D errors on control points are similar: the mean 3D error for MicMac® is equal to 0.037 m with a standard deviation of 0.017 m, whereas for Metashape®, it is 0.031 m with a standard deviation equal to 0.015 m. The present work represents a preliminary study: a comparison between software packages is something hard to achieve, given the secrecy of the commercial software and the theoretical differences between the approaches. This case study analyzes an object with extremely complex geometry; it is placed in an urban canyon where the GNSS support can not be exploited. In addition, the scenario changes continuously due to the vehicular traffic.

Keywords: photogrammetry; unmanned aerial vehicle (UAV); free-and-open-source software (FOSS); MicMac; Metashape; 3D model; accuracy; sparse point cloud

Citation: Cutugno, M.; Robustelli, U.; Pugliano, G. Structure-from-Motion 3D Reconstruction of the Historical Overpass Ponte della Cerra: A Comparison between MicMac® Open Source Software and Metashape®. *Drones* **2022**, *6*, 242. https://doi.org/10.3390/drones6090242

Academic Editors: Arianna Pesci, Giordano Teza and Massimo Fabris

Received: 3 August 2022
Accepted: 4 September 2022
Published: 6 September 2022

Publisher's Note: MDPI stays neutral with regard to jurisdictional claims in published maps and institutional affiliations.

1. Introduction

In recent years, technological advancements in the unmanned aerial vehicle (UAV) industry have drastically transformed survey techniques for 3D model reconstruction. These improvements exploit the evolution of algorithms from computer vision, which once was considered the bottleneck of such techniques. Now, they require less time and are mostly automated.

Moreover, the performance of free-and-open-source software (FOSS) for image processing is increasing, allowing users to conduct the entire photogrammetric process without being obliged to purchase an expensive software license. These two crucial trends present opportunities for both researchers and surveyors. Indeed, with consumer UAVs and FOSS, one can potentially survey an object at a considerably lower cost compared with the past, when UAVs were available to only a few, and for professional results, one had to purchase expensive software licenses. The current tendency in the photogrammetric community is to employ an increased number of images to ensure high overlap values. Today, several photogrammetric software solutions are available, both commercial and FOSS. Table 1 presents an overview of the available photogrammetric software packages. It was demonstrated that Structure-from-Motion (SfM) algorithms perform well in many applications. At this time, SfM 3D surveys became a viable cost-effective alternative surveying method to aerial LiDAR [1]. In addition, 3D photogrammetry has improved in terms of point density and geometric accuracy through increased overlap between images, improved radiometry, and significant progress in multi-view matching; moreover, the graphics processing unit (GPU) computation power has increased [2] continuously; regardless, the advantages and disadvantages of photogrammetry and LiDAR compensate for each other. Horizontal errors in photogrammetry are usually smaller than those in LiDAR, whereas LiDAR can obtain higher vertical than horizontal accuracy [3].

Table 1. Overview of available photogrammetric software packages.

Name	OS	Pricing
3DFlow Zephyr [4]	Windows	360 EUR/month
Autodesk Recap [5]	Windows	55 EUR/month
Agisoft Metashape [6]	Windows, macOS, Linux	4075 EUR
BAE Systems SOCET GXP [7]	Windows	On request
Bentley ContextCapture [8]	Windows	from 211 EUR/month
ColMap [9]	Windows, macOS, Linux	Free
Drone Deploy [10]	Windows, macOS, Android, iOS	299 EUR/month
Planetek IMAGINE [11]	Windows	On request
Meshroom [12]	Windows, Linux	Free
MicMac [13]	Windows, macOS, Linux	Free
Multi-view environment [14]	Windows, macOS	Free
Photometrix IWitness Pro [15]	Windows	986 EUR
PhotoModeler [16]	Windows	from 50 EUR/month
Pix4D Mapper [17]	Windows, macOS, Android, iOS	from 185 EUR/month
PMS AG Elcovision 10 [18]	Windows	On request
OpenDroneMap WebODM [19]	Windows, macOS	from 50 EUR
OpenMVG [20]	Windows, macOS, Linux	Free
RealityCapture [21]	Windows	3220 EUR
SimActive Correlator 3D [22]	Windows	from 250 EUR/month
Regard3D [23]	Windows, macOS, Linux	Free
Trimble InPho [24]	Windows	On request
VisualSFM [25]	Windows, macOS, Linux	Free

In the past decade, several photogrammetric applications employing UAVs have been documented. UAVs have been efficiently employed used in various fields, including geomorphological analyses [26–28], hydraulics modelling [29], agriculture and forest analyses [30–33], emergency management support [34–36], infrastructure monitoring [37–39], and cultural heritage monitoring and 3D reconstruction [40–43]. The literature also provides numerous examples of combined photogrammetric and LiDAR surveys [44–46].

This research aims to determine if MicMac® can be employed for 3D complex object reconstruction in non-optimal survey conditions, obtaining results comparable with a commercial software solution. MicMac® has been employed by various researchers both for aerial and terrestrial photogrammetry. Griffiths and Burningham [47] compared PhotoScan® and MicMac® concluding that the latter can generate significantly more accurate Digital Surface Models (DSMs) because a more complex lens distortion model is included. Altman et al. [48] tested MicMac® for terrestrial photogrammetry highlighting that feature extraction results from MicMac® are comparable with those from Photoscan®. It is possible to produce complete models in MicMac® as well, but it likely requires a more refined image set. Jaud et al. [49] compared the DSM obtained by MicMac® and Photoscan® concluding that both software can provide satisfying results; nonetheless Photoscan® is more straightforward to use but its source code is not open, whereas MicMac® is recommended for experimented users as it is more flexible. In the present work, we focused our attention on the capabilities of MicMac® to reconstruct a complex object with aerial imagery obtained with consumer UAV. For comparison, the same dataset was processed by either the FOSS and the commercial software package. The qualities of the generated tie point (TP) clouds were analyzed by evaluating the results of the bundle block adjustment (BBA) examining the image residuals and the 3D errors on ground control points (GCPs). Subsequently, the 3D errors on check points (CPs), which are not included in the BBA, are investigated. Additionally, the quality of the results was assessed based on the statistical analysis of the Cloud-to-Cloud distances and the relative accuracy.

The remainder of this paper is organized as follows: Section 2 reports a brief introduction to photogrammetry highlighting the differences and similarities of approach by the two software packages investigated. Section 3 presents the case study of the Ponte della Cerra overpass, and Section 4 provides the comparison of the results of the reconstructed models obtained with and a Agisoft Metashape®. Section 5 draws conclusions, including a discussion of the results as well as future developments. Lastly, Appendixes A and B define the workflow of the FOSS and the commercial software package, respectively.

2. Material and Methods

Nowadays, SfM photogrammetry has mainly replaced traditional photogrammetry. The principles of Sfm photogrammetry can be found in [50]. Figure 1 depicts the main stages of this process along with the identification of the roles in the investigated software packages, namely Agisoft Metashape® Pro and MicMac®. Regarding the acquisition scheme, it is known that image acquisition depends primarily on the type of object. For approximately planar objects the parallel method is useful. In this acquisition scheme, none of the images will cover the entire scene; instead, every image is a tile of the scene. The present work is about the 3D reconstruction of a historical overpass; given the complexity of the object, it was split into three separate parts, namely the north facade, the south facade, and the extrados. When surveying each facade also images capturing the corresponding side of the intrados were collected for a total number of collected images equal to 222. The facades, as well as the extrados, were treated as planar objects taking horizontal and vertical images, respectively. Then, to improve object reconstruction, to reduce shadow areas, and to include some common elements for projects merging, 52 oblique images were collected. Lastly, one of the most important parameters to consider when planning a survey using the parallel method is the overlap between images. Two different overlaps must be considered: the along-track overlap, and the cross-track overlap. In the present survey, the mean along-overlap was equal to 80%, while the cross-track was equal to 55%.

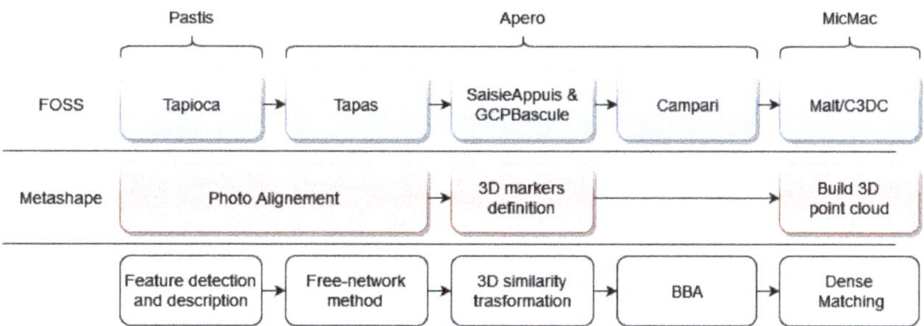

Figure 1. Comparison between the commands in the two software packages investigated. The top row indicates commands for the FOSS, the middle row refers to commercial software, and the bottom row reports the relative photogrammetric processing stages.

2.1. MicMac® Photogrammetric Processing

The photogrammetric process (TP search, estimation of camera poses, densification) is conducted using MicMac® [13,51,52]. MicMac® is a FOSS (Cecill-B license) photogrammetric suite developed by IGN® (French National Geographic Institute) and ENSG® (French

National School for Geographic Sciences). It can be used in a variety of 3D reconstruction scenarios [13]. MicMac® allows the creation of both 3D models and ortho-imagery. The software, which aims to be a cross-platform project, can run on all main operating systems (Windows, Mac OS, and Linux), although, in our experience, has revealed more stable and complete under Linux environment. MicMac® processing chain is completely under user control and most of the parameters can be fine-tuned. MicMac® comprises several tools, each of which is described in the dedicated wiki page [53]. For TP extraction MicMac® uses the Pastis algorithm that is no more than an interface to the well-known SIFT++, a lightweight distribution of the Scale-invariant feature transform (SIFT) [54]. Based on advances in image feature recognition, characteristic image objects can be automatically detected, described, and matched between images. After that, Apero® starts from tie points generated by Pastis and computes external and internal orientations compatible with these measurements [55]. The knowledge of the interior orientation of a camera used for image acquisition is a fundamental requisite for precise photogrammetric object reconstruction. Parameters such as principal distance, principal point coordinates regarding the image coordinate system, and some correction terms for lens distortion, etc., are determined by camera calibration [56]. Nowadays, photogrammetric camera calibration is usually carried out along with the calculation of object coordinates within a self-calibrating bundle adjustment. In MicMac® several distortion models are implemented, including Radial, Fraser, and Fisheye. All theoretical and practical aspects concerning bundle block adjustment with MicMac® are described in MicMac® official documentation [13,55]. The main processing steps of MicMac® can be summarized as follows:

- Tie point computation: the Pastis tool uses the SIFT++ algorithm [57] for the tie point pair generation. This algorithm creates an invariant descriptor that can be used to identify the points of interest matching them even under a variety of perturbing conditions (scale changes, rotation, changes in illumination, viewpoints, or image noise). In this work, this was achieved with Tapioca, a tool interface of SIFT++,
- External orientation: in this step external orientations of the cameras are computed. The relative orientations were computed with the Tapas tool following the free-network approach; this approach involves a calculation of the exterior parameters in an arbitrary coordinate system [58],
- Bundle Block Adjustment: this step includes also the internal parameters, and, for this reason is know as "Self-Calibration"; this is conducted by introducing at least three control points and integrate them within the computation matrix. MicMac® solves the BBA with the Levenberg–Marquardt (L-M) method [59]. The L-M method is in essence the Gauss–Newton method enriched with a damping factor to handle rank-deficient Jacobian matrices [60]. This stage was achieved by exploiting GCPBascule and Campari tools.

2.2. Agisoft Metashape Photogrammetric Processing

To compare the quality of the generated point cloud, the same dataset was also processed in commercial software, namely Metashape® software package by Agisoft (ver. 1.7.2 build 12070, 2021) [6]. Metashape® is a stand-alone software product that performs the photogrammetric processing of digital images and generates 3D spatial data to be used in GIS applications, cultural heritage documentation, and visual effects production as well as for indirect measurements of objects of various scales [6]. The standard photogrammetric pipeline of Metashape is reported below:

- The first step of the photogrammetric processing starts with feature matching across the images: Metashape detects points in the source images which are stable under viewpoint and lighting conditions and generates a descriptor for each point based on its local neighborhood; then, these descriptors are used later to detect correspondences across the photos. This is similar to the well-known SIFT approach but uses different algorithms for a slightly higher alignment quality [61],

- The second stage comprehends the computation of camera intrinsic and extrinsic orientation parameters: Metashape uses a proprietary algorithm to find approximate camera locations and refines them later using a bundle-adjustment algorithm. This should be similar to Bundler algorithm by Snavely et al. (see [62,63]).

The specific algorithms implemented in the software package are not detailed in the manual for business reasons; nevertheless, a description of the Structure-from-Motion (SfM) procedure in Metashape and commonly used parameters are described in [64,65]. It can be noted that during the alignment step, the accuracy was set to high and not to highest; this was to provide the same conditions with respect to MicMac® processing (TP search on full resolution images); in fact, according to Metashape manual [6], at high accuracy setting the software works with the images at the original size, while highest accuracy setting provides an upscaling of the images by a factor of 4. The software is user-friendly, but the adjustment of parameters is limited to pre-defined values. It can be noted that the user manual describes mainly the general workflow and gives only very limited details regarding the theoretical basis of the underlying calculations and the associated parameters as confirmed in [49,58]. On the contrary, at each step of MicMac® processing, the user can choose any numerical value, whereas Metashape® only offers preset values; for example, during the alignment step one can choose among ultra-low, low, medium, high, and ultra-high, whereas MicMac® allows a multitude of choices. According to [66], the alignment time depends mainly on the number of images while densification step depends also on their resolution and overlap. Those characteristics of the dataset determine the RAM capacity needed, which often is the hardware bottleneck when dealing with photogrammetric commercial software packages. In contrast to commercial solutions, MicMac® is generally less demanding in RAM given that it writes all temporary files on the physical memory instead of storing them in the cache memory. In the present study, both photogrammetric processes were computed on the same workstation, namely an Asus computer with 16 Gb Ram, 1 Tb SSD storage, Intel Core i5-3330 CPU at 3.00GHz processor, and Nvidia Quadro 600 GPU. Metashape® does not support the graphic processing unit of the workstation employed in this work [66]; on the other hand, according to [53], MicMac® provides support for GPU processing starting from dense matching. Hence, to assure the same processing conditions, we do not use GPU support at all.

3. Case Study

3.1. Ponte della Cerra Overpass (Italy)

The object surveyed is an overpass in the Vomero neighborhood of Naples, Italy, as shown in Figure 2; it was built in the seventeenth century and is functional to realize the overpass of Suarez street over Conte della Cerra street. The overpass structure comprises Neapolitan yellow tuff blocks; the geometric model can be assimilated to a barrel vault whose generating curve is a lowered arch with a span light of 16.50 m with a belt deflection of 3.20 m. The whole width, comprising the left and right abutments, is equal to 21.40 m. The overpassing roadway comprises a single carriageway, with one lane in each direction and a total width of 13.45 m, including parking areas, bus stops, and two sidewalks each of which 3.27 m wide. The structure is subject to degradation phenomena and, for this reason, was secured by applying a tessellated mesh to prevent the detachment of the damaged parts. It also has some prestigious coats of arms on both facades. The survey presented here is finalized to the geometric measurements and 3D modeling useful for the structural studies conducted in [67].

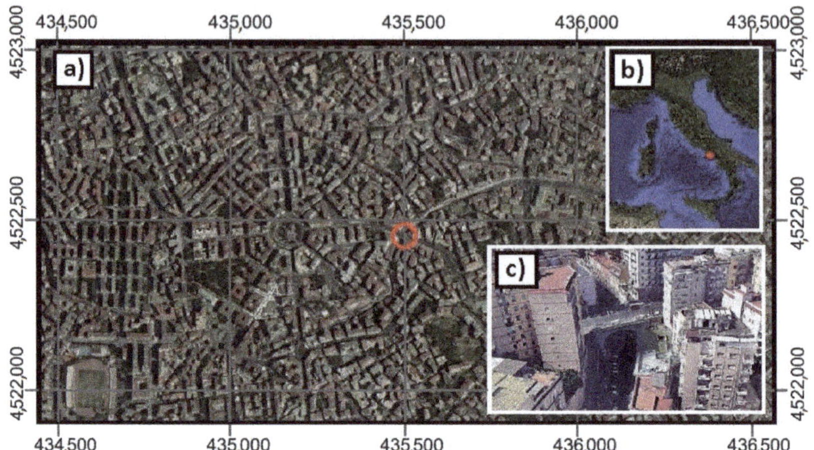

Figure 2. Test area: (**a**) location map; (**b**) localization in Southern Italy; (**c**) Conte della Cerra overpass.

3.2. Unmanned Aerial Vehicle Photogrammetric Flight

For the field test, as shown in Figure 3, a consumer UAV from DJI model Mavic 2 Pro was used to survey the test area. It weighs about 1 kg and is equipped with a $1''$ CMOS Hasselblad optic sensor. The characteristics of the sensor are reported in Table 2. For exhaustive technical documentation, one can refer to [68]. The UAV is equipped with a dual-frequency multi-constellation (GPS and GLONASS) receiver which in normal scenarios delivers geotagged images with an accuracy of about 10 m; in this particular application, since the location of the survey is a highly degraded urban canyon where the UAV was surrounded by tall buildings, the geotagging procedure failed for several images; moreover, part of the survey was conducted under the overpass removing all chance to fix the GNSS position at all; therefore, after some failures in the alignment step, the coordinates stored in the EXIF file attached to each image was erased. This procedure allows the success of the photogrammetric process but drastically increases computing times, since the software, ignoring the position of each picture taken must search homologous points for every pair of images. In normal conditions, the availability of the position of the takes speeds up the processing since the software searches for homologous points only in overlapped pairs of images.

Figure 3. DJI Mavic 2 pro.

Table 2. Camera features and image settings.

Camera Model	Hasselblad L1D-20c
Focal length	10.3 mm
Image format	jpeg
Image width	5472 pixel
Image height	3648 pixel
Exposure time	1/80 s
ISO sensitivity	400
Pixel size	2.41 μm × 2.41 μm

3.3. Dataset Description

The acquired dataset comprises both horizontal, nadiral, and oblique images. According to [69], to improve detail reconstruction and minimize holes in the 3D model, it is convenient to collect also oblique images. Indeed, oblique images are useful to reduce the shadow areas in which data can not be acquired, especially when dealing with complex objects, as in this case. In particular, the following types of images were acquired, depending on the part of the overpass being reconstructed:

- For the south facade, 111 images (12 oblique and 99 horizontal) of which there are:
  - 21 images at 10 m from the object;
  - 90 images at 4 m from the object.

- For the north facade, 70 images (29 oblique and 41 horizontal) of which there are:
  - 32 images at 10 m from the object;
  - 38 images at 4 m from the object.

- For the extrados, 41 images at a flying altitude of 40 m of which there are:
  - 29 nadiral;
  - 12 oblique.

The photogrammetric survey was conducted according to Comité International de Photogrammétrie (CIPA) 3 × 3 Rules (see [70]). These simple rules, written, tested, and published at the CIPA Symposium in Sofia in 1988, should be observed for photography with non-metric cameras. The rules are divided into geometric rules (control, wide-area stereo photo cover, and detail stereo photo cover), camera rules (camera properties, camera calibration, and image exposure), and procedural rules (record photo layout, log the metadata, and archive). To reach a proper level of detail on the facades presenting some details such as coats-of-arms, images were taken at two different distances, as suggested in [70]; at 10 m of distance the wide-area stereo photo cover while at 4 m close-up images for detail stereo photo cover.

A critical aspect when dealing with the 3D reconstruction of such complex objects where images of facades were taken at very low height as if they were captured like a terrestrial acquisition scheme is to ensure that the algorithm can identify the same key points as tie points between images. To achieve this, the survey strategy focused on having a noteworthy image overlap, both cross- and along-track. Therefore the mean along-overlap was equal to 80% while the cross-track to 55%.

3.4. Ground Control Points

The scaling procedure of the 3D model was created based on a support survey conducted with a professional TOPCON total station. This choice was influenced by two main factors: the impracticability of employing a GNSS receiver due to the highly degraded urban canyon, as well as the type of object we were attempting to model. Thus, an open polygonal was created, consisting of five points on the ground, as shown in Figure 4; then, making station on those points, a detailed survey was carried out to measure several support points, from which 12 of them were selected to model scaling: 4 points on the south facade, 4 points on the north facade, and 4 points on the extrados. The GCPs taken were chosen with a preference for well-identifiable features such as tuff edges and net-holding anchors as shown in Figures 5–7. All the points surveyed were taken using a local coordinate system. Attention was paid to be sure that points that were gathered were also visible in the model. Given the historicity of the overpass, we were not allowed to place dedicated targets during the survey; thus, the accuracy related to the natural points chosen was set to 0.005 m, to take into account the uncertainty associated with the type of feature.

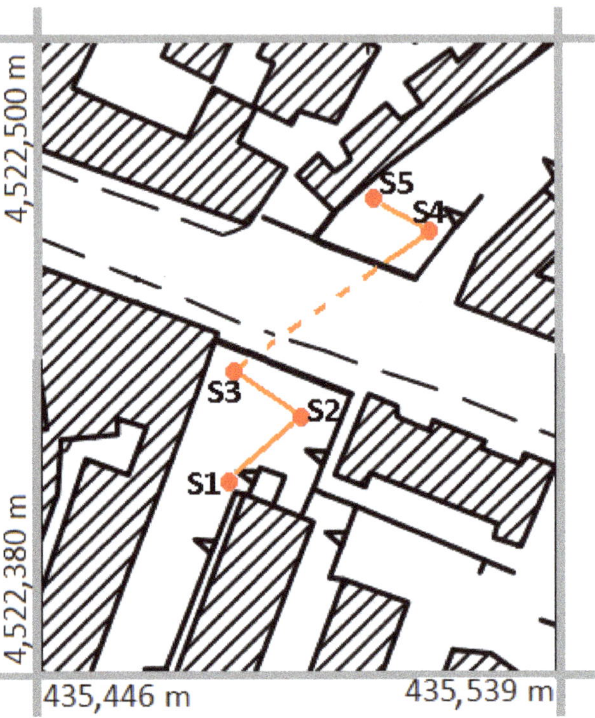

Figure 4. Open polygonal created for the topographic support survey projected on cartography.

Figure 5. View of the south facade with markers locations. GCPs are represented in red while CP is represented in green.

Figure 6. View of the north facade with markers locations. GCPs are represented in red while CP is represented in green.

Figure 7. View of the north side part of the extrados with markers locations. The left image refers to the left part of the extrados while the right image to the right part. GCPs are represented in red while CPs are represented in green.

4. Results

4.1. Internal and External Orientation Results

In Figure 8a, visual representation of the orientation results of BBA by Metashape® and MicMac® is shown. Panel (a) reports the image poses represented with blue rectangles oriented based on Metashape® external orientation; panel (b) depicts the camera poses, represented with pyramid-shaped 3D objects, computed by MicMac®. The images reported here have a different visualization given that the two software use different strategies to display camera poses information.

To choose the best camera calibration model for the specific case study, various relative orientation processes were carried out: Radial, Fraser, Four15x4, Brown, and Ebner. Some of them did not converge at all while others exhibit high image residuals. For this specific dataset, it was found that the best trade-off between convergence achievement and residual minimization was the Fraser model [71]. It is a radial model for camera self-calibration, with decentric and affine parameters. In particular, 12 degrees of freedom are taken into account: 1 for focal length (F), 2 for principal point (PP_1 and PP_2), 2 for distortion center ($Cdist_1$ and $Cdist_2$), 3 for coefficients of radial distortion (r_3, r_5, and r_7), 2 for decentric parameters (P_1 and P_2), and 2 for affine (in-plane distortion) parameters (b_1 and b_2). The parameters computed by the self-calibration algorithm implemented in MicMac® and then imposed in Metashape® are reported in Table 3.

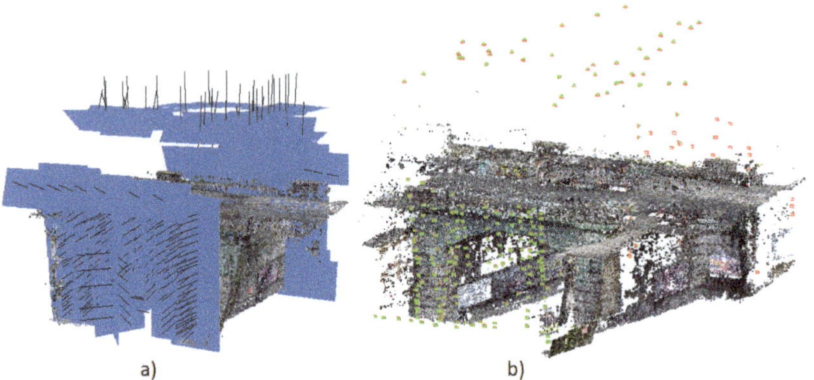

Figure 8. Results of the external orientation process showing the positions and attitudes of each camera station, together with the TP clouds generated from the respective feature matching process: (**a**) Metashape®; (**b**) Apero®.

Table 3. Fraser camera model self-calibration parameters.

Parameter	Symbol	Value (pix)
Focal length	F	4276.067
Principal Point coordinates	PP_1 PP_2	2702.974 1836.010
Distortion center coordinates	$Cdist_1$ $Cdist_2$	2686.749 1809.302
Radial distortion coefficients	r_3 r_5 r_7	-8.259×10^{-10} 7.756×10^{-17} 4.838×10^{-24}
Decentric parameters	P_1 P_2	2.469×10^{-7} 2.251×10^{-7}
Affine parameters	b_1 b_2	1.017×10^{-4} 1.802×10^{-4}

Parameters computed with the self-calibration method were then imposed in Metashape® camera calibration to set the same conditions for results comparison. The Apero® tool of MicMac® generates external and internal orientations of the camera. The relative orientations were computed with the Tapas tool in two steps: first on a small set of images and then by using the calibration obtained as an initial value for the global orientation of all images. Then, the Campari command is used to compensate for heterogeneous measurements [13]. In Metashape®, camera alignment by bundle adjustment is achieved by detecting common tie points and match them on images to compute the external camera orientation parameters for each picture. Then, it proceeds to solve for camera internal and external orientation parameters using an algorithm to find approximate camera locations refining them later using the BBA. Figure 9 depicts the probability density estimation of image residuals expressed in pixels. It was obtained by calculating the normalized histogram of image residuals. The height of each bar is the number of samples of images residual falling in that bin divided by the width of the bin and the total number of elements in the sample. In this way, the area of each bar is the relative number of observations. This type of normalization is commonly used to estimate the PDF. Both histograms were calculated considering a bin width of 0.03 pixels. By observing the figure it can be noticed that MicMac® image residual distribution is long-tailed, whereas Metashape® is short-tailed. Indeed, the image residuals standard deviation obtained with MicMac® is higher. In particular, the image residual standard deviation for MicMac® is equal to 0.277 pixels, whereas for Metashape®, it is equal to 0.173 pixels. Regarding the mean image residuals, the two software show similar values: MicMac® reaches the value of 0.770 pixels and Metashape® attests to 0.735 pixels.

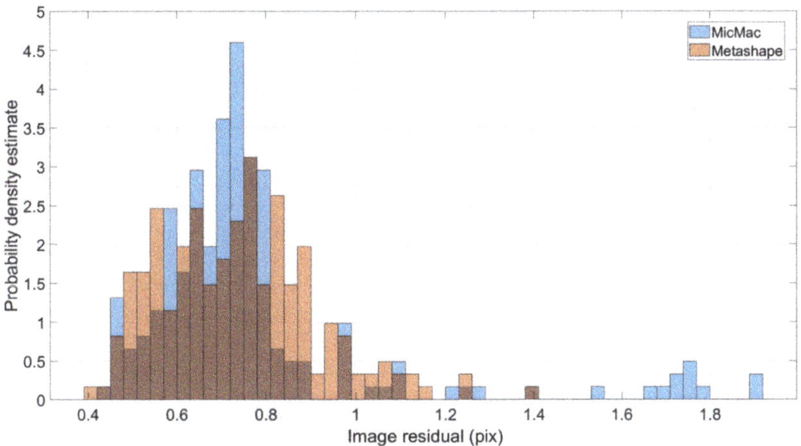

Figure 9. Image residuals probability density function estimates. On *x*-axis are reported images residuals expressed in pixels while on *y*-axis is the probability density estimation. Blu and red bins represent MicMac® and Metashape® respectively.

4.2. TP Clouds Results

The point cloud obtained with the commercial software was used as a reference for the quality assessment of the TP cloud generated by MicMac®. The analysis presented was conducted with the FOSS CloudCompare ver. 2.11.3 (Anoia 64-bits) and Matlab R2021b Statistic toolbox. Various properties of the clouds were compared: number of points, point densities, GCP errors, and CP errors. Table 4 depicts the main properties of the resulting TP clouds.

Table 4. Summary of the main properties of the TP clouds.

		Number of Points (Points)	Mean Surface Density (Points/m^2)	Std Surface Density (Points/m^2)
MicMac®	extrados	648,197	7503	8276
	north facade	660,720	8863	7537
	south facade	2,265,025	32,192	25,250
Metashape®	extrados	243,213	1410	1004
	north facade	263,958	1947	984
	south facade	430,178	2428	1489

Table 4 demonstrates that the MicMac® TP cloud has significantly more points than the Metashape® cloud; this is addressed to the way the points are managed in the two software packages. Indeed, Metashape® merges TPs to obtain a single point via the *gradual selection* algorithm, whereas MicMac® does not realize this merging; in fact, if one zooms in a MicMac® TP and/or dense cloud, a cluster of points will be found. A post-processing decimation can be conducted in MicMac® via the tool Schnaps to clean and reduce tie points. This reflects in the other metrics reported here (computed taking into account a sphere with a radius of 0.1 m): the surface density of the TP cloud generated by MicMac® has a mean of 7503 points/m^2, 8863 points/m^2, and 32,192 points/m^2 for the extrados, the north, and the south facades, respectively. The corresponding values for the Metashape® TP cloud are 1410 points/m^2, 1947 points/m^2, and 2428 points/m^2. Concerning the standard deviations of the surface's densities, the MicMac® TP cloud shows values equal to 8276 points/m^2, 7537 points/m^2, and 25,250 points/m^2, for extrados, north, and south facades, respectively. Those values for the Metashape® TP cloud are 1004 points/m^2, 984 points/m^2, and 1489 points/m^2, respectively. Lastly, another key aspect to take into account is the computation time. According to the Metashape® manual [66], it does not support the graphic processing unit of the workstation employed in this work; on the other hand, according to [53], MicMac® provides support for GPU processing starting from dense matching. Hence, to assure the same conditions, we do not use GPU support at all. MicMac® took 23.48 h to complete the photogrammetric pipeline, while Metashape® took 45% less time, namely 12.34 h.

In Figure 10 are shown the TP clouds obtained with both software. As confirmed by graphical visualization, the MicMac® cloud is denser, especially on the facades. Moreover, Metashape® reveals a more homogeneous reconstruction of the whole model, whereas MicMac® reveals some holes in intrados reconstruction.

Figure 10. The 3D view of the RGB TP clouds obtained with MicMac® and Metashape®. Panel (**a**) refers to MicMac® TP cloud. Panel (**b**) refers to Metashape® TP cloud.

Tables 5 and 6 depict the 3D errors in GCPs and CPs for both software packages. Regarding MicMac® GCPs errors, except for marker P60, all 3D errors are less than 0.018 m; on the other hand, regarding Metashape®, markers P07, P09, P18, and P60 exhibit a 3D error of about 0.025 m. Three of these belong to the south facade suggesting that this was the most difficult part to reconstruct. This is confirmed by analyzing the MicMac® errors on the same markers: except for these and for marker P68, they are the only errors greater than 0.010 m. Table 6 shows that, except for marker P50, 3D errors on CPs are comparable between software packages. Table 7 contains the corresponding error statistics. The errors are of the same order of magnitude. Regarding the GCPs, MicMac® shows smaller error statistics; in fact, the mean 3D error and the standard deviation are 0.010 m and 0.007 m, respectively; the same metrics for Metashape® are 0.015 m and 0.008 m. Conversely, regarding the CPs, the mean 3D error for MicMac® is 0.037 m with a standard deviation of 0.017 m while the mean 3D error on CPs obtained with Metashape® is 0.031 m with a standard deviation equal to 0.015 m.

Table 5. Summary of MicMac® and Metashape® residuals on GCPs.

Marker Label	MicMac® 3D Err. (m)	Metashape® 3D Err. (m)
P07	0.017	0.026
P09	0.014	0.023
P18	0.018	0.026
P33	0.005	0.013
P35	0.004	0.004
P54	0.005	0.006
P55	0.005	0.009
P60	0.023	0.025
P63	0.006	0.009
P66	0.007	0.012
P67	0.003	0.013
P68	0.013	0.009

Table 6. Summary of MicMac® and Metashape® residuals on CPs.

Marker Label	MicMac® 3D Err. (m)	Metashape® 3D Err. (m)
P20	0.020	0.023
P50	0.048	0.017
P64	0.024	0.034
P69	0.055	0.052

Table 7. Statistics on ground control points and check points errors.

	GCP 3D Error		CP 3D Error	
	Mean (m)	Std (m)	Mean (m)	Std (m)
MicMac®	0.010	0.007	0.037	0.017
Metashape®	0.015	0.008	0.031	0.015

4.3. Relative Accuracy

To assess the relative accuracy of both reconstructed models, four measurements were computed between various GCPs and CPs. The true distances were calculated on the basis of the total station support survey; afterward, the same measurements were realized on the 3D models exploiting CloudCompare software. In Figure 11 is reported an example for the measurement P7-P9 while Table 8 summarizes all the measurements carried out. The errors obtained are comparable; however, it can be noted that Metashape® shows always positive values, whereas for MicMac®, three out of four values are negative. Moreover, the errors for Metashape® ranges from 0.007 m to 0.018 m while for MicMac® from 0.004 m to 0.018 m.

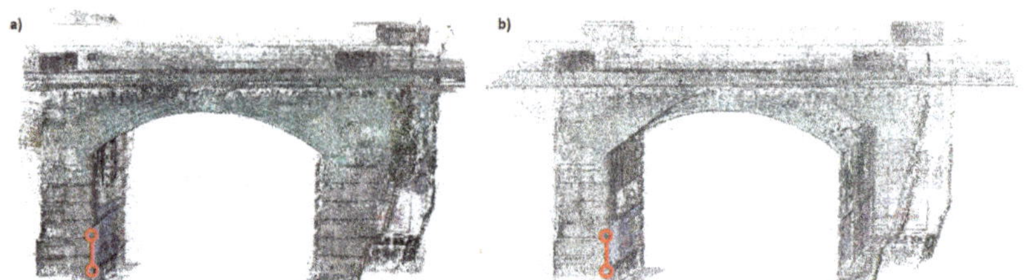

Figure 11. Relative model accuracy measurement between markers P7 and P9 displayed with model screenshots: (**a**) MicMac®; (**b**) Metashape®.

Table 8. Relative accuracy measurements summary.

Control Line	True Dist. (m)	MicMac®		Metashape®	
		Meas. (m)	3D Err. (m)	Meas. (m)	3D Err. (m)
P7–P9	2.389	2.394	0.005	2.396	0.007
P11–P18	15.396	15.390	−0.006	15.405	0.009
P9–P35	5.442	5.438	−0.004	5.460	0.018
P56–P68	6.959	6.941	−0.018	6.968	0.009

4.4. Cloud-to-Cloud Distance

In this section, the Cloud-to-Cloud distance, computed exploiting the M3C2 plugin of CloudCompare software package [72], is shown. The principle of nearest neighbour distance is used to compute distances between two points: for each point in the compared cloud, the nearest point in the reference cloud is searched and their Euclidean distance is computed [73]. Figure 12 shows the results of Cloud-to-Cloud distance computation. In the present case, the radius of the sphere where the algorithms search for the nearest neighbors was set to 0.100 m, the reference cloud chosen is the Metashape® one, and the graphic results are visualized in color scale. The figure provides a visual representation of the Cloud-to-Cloud distance; by analyzing the figure, it emerges that the Cloud-to-Cloud distances are homogeneous mainly on the facades while some misalignment between TP clouds is present under the intrados; nonetheless, this misalignment is under 0.002 m. Figure 13 shows the Cloud-to-Cloud distance probability density function (PDF) estimate between MicMac® and Metashape® TP clouds; the mean distance value between the two clouds is equal to 0 mm while the standard deviation is equal to 0.254 mm. Observing the figure, it can be noted that more than 95% of distance values are included in the interval [−1, 1] mm. These results confirm the high-level performance reachable by MicMac®, given that the two TP clouds are averagely superimposable.

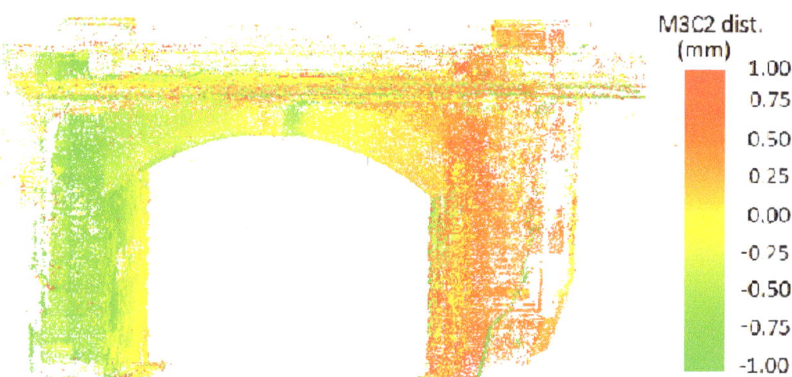

Figure 12. Cloud-to-Cloud distance of MicMac® and Metashape® TP clouds computed with M3C2 plugin in CloudCompare.

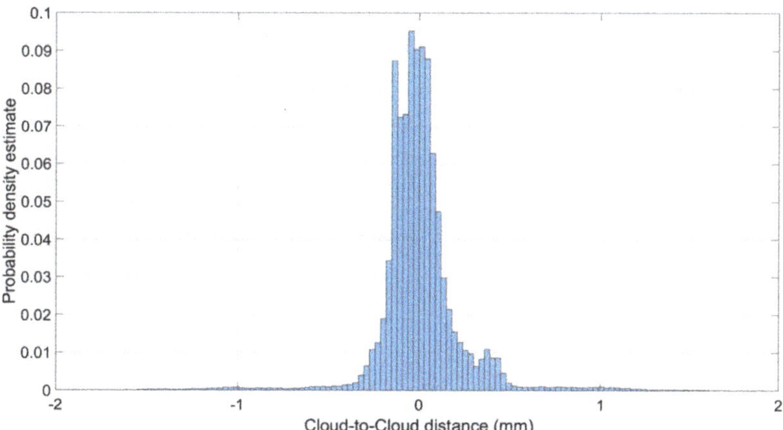

Figure 13. MicMac® and Metashape® Cloud-to-Cloud distance probability density estimate. On the *x*-axis are reported Cloud-to-Cloud distances expressed in pixels while on the *y*-axis the probability density estimation.

4.5. Other Products

Both software can produce several other products derived from the TP cloud. First, the densification algorithms can generate the dense cloud. After that, textured and tiled models can be obtained. In addition, raster products are obtainable, such as Digital Elevation Model (DEM) and Orthophotos. In this section, for example, the dense clouds generated by both software are shown and their main features are compared. In Figures 14 and 15 are reported the 3D views of the dense clouds obtained with MicMac® and Metashape®, respectively.

Table 9 shows the number of points and the surface densities after densification. According to the Metashape® manual [66] and the MicMac® wiki [53], to ensure proper comparability, the same parameters were employed: in Metashape®, the accuracy was set to high, whereas in MicMac®, when launching the Malt tool, the parameter ZoomF was set to 2. As can be seen in Table 9, for the extrados, the dense cloud obtained with MicMac® has about 20% more points than the Metashape® dense cloud. Regarding the north facade the FOSS-derived dense cloud has two times the points than the commercial one. Lastly, the MicMac® dense cloud of the south facade has about 35% more points than the corresponding from Metashape®. However, a comparison between dense clouds from different software packages is something hard to achieve for the different algorithms and strategies employed; hence, for this product, we limit to show the resulting values. To deepen this specific aspect a dedicated study should be conducted.

Figure 14. 3D view of the RGB dense cloud obtained with MicMac®.

Figure 15. 3D view of the RGB dense cloud obtained with Metashape®.

Table 9. Summary of the main properties of the dense clouds.

		Number of Points (Points)	Mean Surface Density (Points/m^2)	Std Surface Density (Points/m^2)
MicMac®	extrados	12,445,984	170,382	30,687
	north facade	19,490,308	268,875	175,817
	south facade	51,604,250	1,000,355	480,666
Metashape®	extrados	10,543,654	133,463	101,654
	north facade	9,341,169	118,795	65,411
	south facade	38,479,447	534,579	224,562

5. Conclusions and Future Works

Unmanned aerial vehicles are playing a major role in the acquisition of geospatial information. With the significantly lower costs associated with UAV surveys and the high quality of the derived products, this is a technique that proves to be useful for several applications and most certainly a valid alternative to traditional techniques. Due to the payload limitation of prosumer UAVs, the reliability of results may not be as high as other techniques, e.g., classical aerial photogrammetry and LiDAR. For this reason, an accuracy assessment of different software packages can reveal truly interesting. The present work represents a case study of the reconstruction of a historical overpass in Italy carried out with MicMac® open source software and then compared to Metashape® commercial photogrammetry solution. Particular attention was paid to the image acquisition procedure since, as well-known in literature, it has a relevant impact on the precision of the results.

The quality of the photogrammetric results were assessed by analyzing image residuals, the statistics of GCPs and CPs errors, the relative accuracy assessment, and the Cloud-to-Cloud distance. Regarding the mean image residuals, the two software show similar values: MicMac® reaches the value of 0.770 pixels and Metashape® attests to 0.735 pixels. The standard deviations are 0.277 pixels and 0.173 pixels for MicMac® and Metashape®, respectively. Analyzing the 3D errors on CPs, MicMac® revealed slightly worse statistics; in fact, the mean 3D error is equal to 0.037 m with a standard deviation of 0.017 m while the mean 3D error on CPs obtained with Metashape® is 0.031 m with a standard deviation equal to 0.015 m. For what it concerns the relative accuracy assessment, four linear distances were measured on the 3D models revealing mean errors of 0.008 m and 0.011 m for MicMac® and Metashape®, respectively. Lastly, the Cloud-to-Cloud distance analysis was carried out between the TP clouds: it can be noted that more than 95% of distance values are included in the interval $[-1, 1]$ mm. The probability density estimate has a zero mean. Besides this numerical analysis that demonstrates similar results, a visual investigation reveals that the Metashape® TP cloud presents fewer holes than MicMac® TP cloud, especially under the overpass. A relevant advantage of MicMac® resides in the accessibility to intermediate results and in the capability of finely tuning practically all parameters. The trade-off of this extreme manageability lies in the difficulty of usage. Conversely, the nontrivial license cost of the commercial software is justified by its ease of use and its user-friendly interface, which allows anyone with minimal experience to perform the photogrammetric 3D model reconstruction but has few intermediate output capabilities and limited parameters control. It should be highlighted that a comparison between software is something hard to achieve given the secrecy of the commercial software, the long computation times, and the theoretical difference underlying the approaches. Moreover, the object reconstructed is particularly complex being a historical overpass with several details. Lastly, the overpass is placed in an urban canyon where the GNSS support can not be exploited and the scenario changes continuously due to vehicular traffic. The analyses conducted show that the two software packages have comparable products qualities even if the software approaches are different: an open-source project fully customizable versus a commercial black-box software; hence the experiment confirms the potentiality of FOSS for photogrammetric applications. We can state that MicMac® can produce professional results comparable with products generated by a market-leading commercial solution. The future work will consider other types of objects to reconstruct via photogrammetry. This will define more completely the behaviour of the FOSS investigated when dealing with different geometries. Moreover, in-depth analyses of other products, e.g., DEM and Orthomosaic, will be carried out. Another interesting future goal will be to further improve the cost-effectiveness of the equipment; this can be achieved by employing a compact ultra low-cost UAV. Indeed, in the literature, there are few relevant studies concerning the utilization of a cost-effective platform for surveying and 3D reconstruction, so far. This can represent a promising avenue of research to define a correct procedure for 3D model generation and quality assessment employing ultra-low-cost equipment.

Author Contributions: Conceptualization, M.C., U.R. and G.P.; methodology, M.C., U.R. and G.P.; software, M.C.; validation, M.C. and U.R.; formal analysis, M.C., U.R. and G.P.; investigation, M.C., U.R. and G.P.; data curation, M.C. and G.P.; writing—original draft preparation, M.C. and U.R.; writing—review and editing, M.C., U.R. and G.P.; visualization, M.C. and U.R.; supervision, U.R. and G.P. All authors have read and agreed to the published version of the manuscript.

Funding: This research received no external funding.

Institutional Review Board Statement: Not applicable.

Informed Consent Statement: Not applicable.

Data Availability Statement: Not applicable.

Conflicts of Interest: The authors declare no conflict of interest. The funders had no role in the design of the study; in the collection, analyses, or interpretation of data; in the writing of the manuscript, or in the decision to publish the results.

Abbreviations

The following abbreviations are used in this manuscript:

CIPA	Comité International de Photogrammétrie Architecturale
CMOS	Complimentary Metal-Oxide-Semiconductor
CP	Check Point
DEM	Digital Elevation Model
ENSG	French National School for Geographic Sciences
EXIF	Exchangeable Image File
FOSS	Free and Open-Source Software
GCP	Ground Control Point
GIS	Geographical Information System
GLONASS	Global'naja Navigacionnaja Sputnikovaja Sistema
GPS	Global Positioning System
GPU	Graphics Processing Unit
GNSS	Global Navigation Satellite System
IGN	French National Geographic Institute
LiDAR	Laser Detection and Ranging
SIFT	Scale-Invariant Feature Transform
TP	Tie Point
UAV	Unmanned Aerial Vehicle

Appendix A. MicMac® Processing Pipeline

Once the image dataset is ready for processing (one must manually remove inappropriate images), the first command to run is *Tapioca* where the software computes the TPs:

> *mm3d Tapioca MulScale "DJI.*.JPG" 500 -1*

The argument MulScale means that MicMac® computes TPs for images in low resolution (500) and then for the highest resolution (-1). This implementation should speed up the process. The next command is *Tapas* for internal and external orientation parameter computation:

> *mm3d Tapas Fraser "DJI.*.JPG" Out=All-Rel*

When dealing with large datasets a convenient procedure is to run the *Tapas* command over a sub-dataset or another dataset with the same camera setting; in this manner, when running the command for the whole dataset, the process will be less time-consuming. *Tapas* creates a named directory that contains the camera calibration file, with camera parameters (focal length, PPP, distortion parameters), and the orientation file for each image, with camera orientation, TPs used for orientation, and rotation parameters. Further important information is stored in the file named "Residuals.xml" (image residual, number of TPs used per image, etc.). Then, the command *Apericloud* is needed to generate a visualization of the 3D TP cloud previously generated by *Tapas* along with cameras position:

mm3d Apericloud "DJI.*.JPG" All-Rel

At this point, the measurement process must be conducted and it can be conducted with the graphic tool *SaisieAppuisInitQT* as follows:

mm3d SaisieAppuisInitQT "DJI_[8||9].JPG" All-Rel 0001 GCP.xml

where "All-Rel" is the previously generated oriented model folder, "0001" is the name of the marker and "GCP.xml" is a properly-formatted XML file containing the information about the markers (name, coordinate X, coordinate Y, coordinate Z, uncertainties). A minimum number of three support points must be defined, each of which on at least two images. The computation of the absolute orientation follows:

mm3d GCPBascule ".*JPG" Ori-All Ori-All-Basc GCP.xml GCP-S2D.xml

where "Ori-All-Basc" is the output name of the GCPs orientation folder. The previously defined GCPs must be validated on the other image that composes the dataset; this is conducted with the command *SaisieAppuisPredictQT*, as follows:

mm3d SaisieAppuisPredicQT "DJI*.*JPG" Ori-All-Basc GCP.xml GCP-Final.xml

The computation of the absolute orientation is updated on the basis of the GCPs validated on the whole dataset with the following command:

mm3d GCPBascule ".*JPG" Ori-All Ori-All-Basc2 GCP.xml GCP-Final-S2D.xml

The last step of the measurement process is computing the final adjustment with *Campari*:

mm3d Campari ".*JPG" Ori-All-Basc2 Ori-Terrain GCP=[GCP.xml,0.02,GCP-Final-S2D.xml,0.5]

where 0.02 is the support point accuracy, and 0.5 is the pixel accuracy of the linking points. Once the measurement process is accomplished, the *AperiCloud* tool is launched to create the 3D TP cloud on which the user can define the 3D mask to limit the densification area, as follows:

mm3d AperiCloud ".*JPG" Ori-Terrain

mm3d SaisieMasqQT AperiCloud_Ori-Terrain.ply

Once the mask is created, the 3D reconstruction (densification) can be conducted via the following:

mm3d C3DC MicMac "DJI_*.*JPG" Ori-Terrain Masq3D=AperiCloud_Ori-Terrain.ply Out=C3DC_MicMac_ponte.ply

Once the photogrammetric process is finished, products other than TP and dense clouds can be built, primarily the following:

- Tawny, which creates the orthorectified depth maps image;
- GrShade, which creates a faded relief image; and,
- to8Bits, which creates a hypsometric color image.

In Figure A1 is depicted the pipeline of MicMac® processing limited to the TP cloud generation and scaling. Since the generation of further products are not of interest of the present work, the flowchart does not take into account the subsequent steps, e.g., dense cloud and orthomosaic generation.

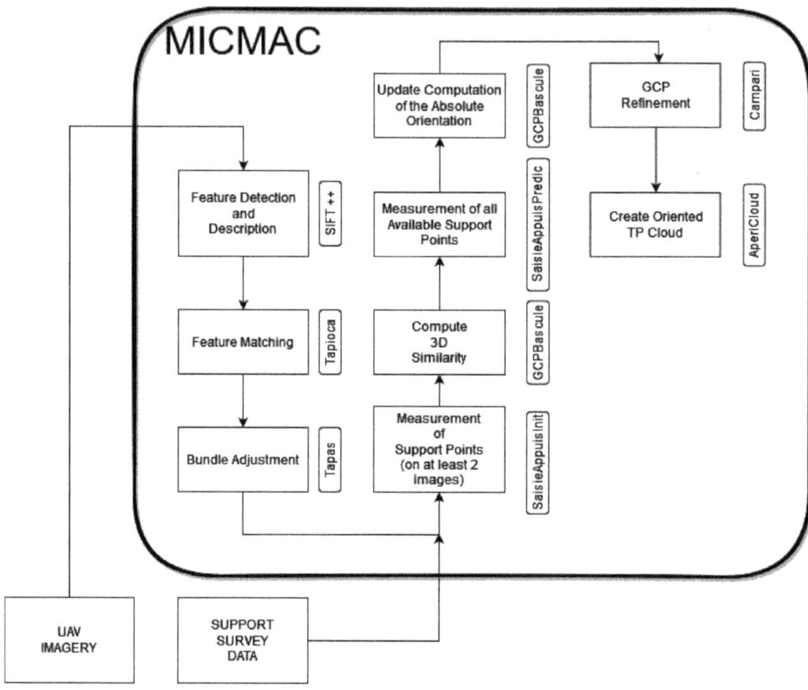

Figure A1. Flowchart of MicMac® processing pipeline. Please note that the pipeline is limited to the TP cloud generation and scaling.

Appendix B. Agisoft Metashape® Processing Pipeline

The described standard workflow can be summarized as follows:

- Add photos to the project;
- Align photos to create the TP cloud (accuracy = high, key point limit = none, TP limit = none);
- Build dense cloud to densify the TP cloud;
- Build mesh to create a triangular mesh on the point cloud and to obtain a surface model;
- Build texture to create a texture and wrap it on the model; and,
- Insert markers and define which are GCPs and which CPs.

References

1. Berrett, B.E.; Vernon, C.A.; Beckstrand, H.; Pollei, M.; Markert, K.; Franke, K.W.; Hedengren, J.D. Large-Scale Reality Modeling of a University Campus Using Combined UAV and Terrestrial Photogrammetry for Historical Preservation and Practical Use. *Drones* **2021**, *5*, 136. [CrossRef]
2. Leberl, F.; Bischof, H.; Pock, T.; Irschara, A.; Kluckner, S. Aerial Computer Vision for a 3D Virtual Habitat. *IEEE Comput.* **2010**, *43*, 24–31. [CrossRef]

3. Lichti, D.; Gordon, S.; Stewart, M.; Franke, J.; Tsakiri, M. Comparison of digital photogrammetry and laser scanning, laser scanner behaviour and accuracy, close-range imaging, long-range vision. In Proceedings of the ISPRS Commission V, Symposium, Corfu, Greece, 2–6 September 2002; pp. 39–44.

4. 3DFlow SRL Website. Available online: https://www.3dflow.net/it/software-di-fotogrammetria-3df-zephyr (accessed on 10 March 2022).

5. Autodesk Inc. Available online: https://www.autodesk.it/products/recap/overview (accessed on 10 March 2022).

6. Agisoft Website. Available online: https://www.agisoft.com/ (accessed on 11 March 2022).

7. BAE Systems. Available online: https://www.geospatialexploitationproducts.com/content/ (accessed on 31 March 2022).

8. Bentley Systems Inc. Available online: https://www.bentley.com/it/products/brands/contextcapture (accessed on 10 March 2022).

9. ColMap Documentation. Available online: https://colmap.github.io/index.html (accessed on 10 March 2022).

10. DroneDeploy. Available online: https://www.dronedeploy.com/ (accessed on 10 March 2022).

11. Planetek Italia s.r.l. Available online: https://www.planetek.it/prodotti/tutti_i_prodotti/imagine_photogrammetry_lps (accessed on 1 August 2022).

12. AliceVision. Available online: https://alicevision.org/#meshroom (accessed on 10 March 2022).

13. MicMac Wiki. Available online: https://micmac.ensg.eu/index.php/Accueil (accessed on 10 March 2022).

14. University of Darmstadt. Available online: https://www.gcc.tu-darmstadt.de/home/proj/mve/ (accessed on 10 March 2022).

15. Photometrix Software. Available online: https://www.photometrix.com.au/it/iwitness/ (accessed on 31 March 2022).

16. Photomodeler Technologies. Available online: https://www.photomodeler.com/ (accessed on 10 March 2022).

17. Pix4d. Available online: https://www.pix4d.com/product/pix4dmapper-photogrammetry-software (accessed on 10 March 2022).

18. PMS AG. Available online: https://en.elcovision.com/ (accessed on 31 March 2022).

19. OpenDroneMap. Available online: https://www.opendronemap.org/webodm/ (accessed on 10 March 2022).

20. OpenMVG Github Page. Available online: https://github.com/openMVG/openMVG/wiki (accessed on 10 March 2022).

21. Capturing Reality Website. Available online: https://https://www.capturingreality.com/ (accessed on 10 March 2022).

22. SimActive Inc. Available online: https://www.simactive.com/correlator3d-mapping-software-features.html (accessed on 10 March 2022).

23. Regard3D. Available online: https://www.regard3d.org/ (accessed on 10 March 2022).

24. Trimble Inc. Available online: https://geospatial.trimble.com/products-and-solutions/inpho (accessed on 10 March 2022).

25. Wu, C. Available online: http://ccwu.me/vsfm/ (accessed on 10 March 2022).

26. Śledź, S.; Ewertowski, M.; Piekarczyk, J. Applications of unmanned aerial vehicle (UAV) surveys and Structure from Motion photogrammetry in glacial and periglacial geomorphology. *Geomorphology* **2021**, *378*, 107620. 2021.107620. [CrossRef]

27. Tomczyk, A.M.; Ewertowski, M.W. UAV-based remote sensing of immediate changes in geomorphology following a glacial lake outburst flood at the Zackenberg river, northeast Greenland. *J. Maps* **2020**, *16*, 86–100. 9146. [CrossRef]

28. Fabbri, S.; Grottoli, E.; Armaroli, C.; Ciavola, P. Using High-Spatial Resolution UAV-Derived Data to Evaluate Vegetation and Geomorphological Changes on a Dune Field Involved in a Restoration Endeavour. *Remote Sens.* **2021**, *13*, 1987. [CrossRef]

29. Lama, G.F.C.; Crimaldi, M.; Pasquino, V.; Padulano, R.; Chirico, G.B. Bulk Drag Predictions of Riparian Arundo donax Stands through UAV-Acquired Multispectral Images. *Water* **2021**, *13*, 1333. [CrossRef]

30. Goodbody, T.R.; Coops, N.C.; Marshall, P.L.; Tompalski, P.; Crawford, P. Unmanned aerial systems for precision forest inventory purposes: A review and case study. *For. Chron.* **2017**, *93*, 71–81. [CrossRef]

31. Feng, Q.; Liu, J.; Gong, J. UAV Remote Sensing for Urban Vegetation Mapping Using Random Forest and Texture Analysis. *Remote Sens.* **2015**, *7*, 1074–1094. [CrossRef]

32. Liang, X.; Wang, Y.; Pyörälä, J.; Lehtomäki, M.; Yu, X.; Kaartinen, H.; Kukko, A.; Honkavaara, E.; Issaoui, A.E.I.; Nevalainen, O.; et al. Forest in situ observations using unmanned aerial vehicle as an alternative of terrestrial measurements. *For. Ecosyst.* **2019**, *6*, 20. [CrossRef]

33. Belcore, E.; Pittarello, M.; Lingua, A.; Lonati, M. Mapping Riparian Habitats of Natura 2000 Network (91E0*, 3240) at Individual Tree Level Using UAV Multi-Temporal and Multi-Spectral Data. *Remote Sens.* **2021**, *13*, 1756. [CrossRef]

34. Rajan, J.; Shriwastav, S.; Kashyap, A.; Ratnoo, A.; Ghose, D. Chapter 6—Disaster management using unmanned aerial vehicles. In *Unmanned Aerial Systems*; Advances in Nonlinear Dynamics and Chaos (ANDC); Koubaa, A., Azar, A.T., Eds.; Academic Press: Cambridge, MA, USA, 2021; pp. 129–155. [CrossRef]

35. Luo, C.; Miao, W.; Ullah, H.; McClean, S.; Parr, G.; Min, G. Geological Disaster Monitoring Based on Sensor Networks. In *Unmanned Aerial Vehicles for Disaster Management*; Geological Disaster Monitoring Based on Sensor Networks: Singapore, 2019; pp. 83–107. [CrossRef]

36. Erdelj, M.; Natalizio, E. UAV-assisted disaster management: Applications and open issues. In Proceedings of the 2016 International Conference on Computing, Networking and Communications (ICNC), Kauai, HI, USA, 15–18 February 2016; pp. 1–5. [CrossRef]

37. Congress, S.S.; Puppala, A.J.; Lundberg, C.L. Total system error analysis of UAV-CRP technology for monitoring transportation infrastructure assets. *Eng. Geol.* **2018**, *247*, 104–116. [CrossRef]

38. Maltezos, E.; Skitsas, M.; Charalambous, E.; Koutras, N.; Bliziotis, D.; Themistocleous, K. Critical infrastructure monitoring using UAV imagery. In Proceedings of the RSCy 2016 Fourth International Conference on Remote Sensing and Geoinformation of the Environment, Paphos, Cyprus, 4–8 April 2016. [CrossRef]
39. Ham, Y.; Han, K.; Lin, J.; Golparvar-Fard, M. Visual monitoring of civil infrastructure systems via camera-equipped Unmanned Aerial Vehicles (UAVs): A review of related works. *Vis. Eng.* **2016**, *4*, 1. [CrossRef]
40. Lo Brutto, M.; Garraffa, A.; Meli, P. UAV Platforms for Cultural Heritage Survey: First Results. *ISPRS Ann. Photogramm. Remote Sens. Spat. Inf. Sci.* **2014**, *II-5*. [CrossRef]
41. Themistocleous, K. The Use of UAVs for Cultural Heritage and Archaeology. In *Remote Sensing for Archaeology and Cultural Landscapes: Best Practices and Perspectives Across Europe and the Middle East*; Springer International Publishing: Cham, Switzerland, 2020; pp. 241–269. [CrossRef]
42. Bakirman, T.; Bayram, B.; Akpinar, B.; Karabulut, M.F.; Bayrak, O.C.; Yigitoglu, A.; Seker, D.Z. Implementation of ultra-light UAV systems for cultural heritage documentation. *J. Cult. Herit.* **2020**, *44*, 174–184. [CrossRef]
43. Barazzetti, L.; Binda, L.; Cucchi, M.; Scaioni, M.; Taranto, P. Photogrammetric reconstruction of the My Son G1 temple in Vietnam. *Int. Arch. Photogramm. Remote Sens. Spat. Inf. Sci.* **2009**, *38*, 8.
44. Righetti, G.; Serafini, S.; Brondi, F.; Church, W.; Garnero, G., Survey of a Peruvian Archaeological Site Using LiDAR and Photogrammetry: A Contribution to the Study of the Chachapoya. In *Computational Science and Its Applications*, 21st ed.; Springer: Cham, Switzerland, 2021; pp. 613–628.
45. Liu, Q.; Li, S.; Tian, X.; Fu, L. Dominant Trees Analysis Using UAV LiDAR and Photogrammetry. In Proceedings of the IGARSS 2020—2020 IEEE International Geoscience and Remote Sensing Symposium, Waikoloa, HI, USA, 26 September–2 October 2020; pp. 4649–4652. [CrossRef]
46. Marques Freguete, L.; Chu, T.; Starek, M. Mapping with LIDAR and structure-from-motion photogrammetry: Accuracy assessment of point cloud over multiple platforms. In *Proceedings of Remote Sensing Technologies and Applications in Urban Environments VI*; SPIE Remote Sensing; SPIE: Bellingham, WA, USA, 2021; p. 11. [CrossRef]
47. Griffiths, D.; Burningham, H. Comparison of pre-and self-calibrated camera calibration models for UAS-derived nadir imagery for a SfM application. *Prog. Phys. Geogr. Earth Environ.* **2019**, *43*, 215–235. [CrossRef]
48. Altman, S.; Xiao, W.; Grayson, B. Evaluation of low-cost terrestrial photogrammetry for 3d reconstruction of complex buildings. *ISPRS Ann. Photogramm. Remote Sens. Spat. Inf. Sci.* **2017**, *4*, 199–206. [CrossRef]
49. Jaud, M.; Passot, S.; Le Bivic, R.; Delacourt, C.; Grandjean, P.; Le Dantec, N. Assessing the Accuracy of High Resolution Digital Surface Models Computed by PhotoScan® and MicMac® in Sub-Optimal Survey Conditions. *Remote Sens.* **2016**, *8*, 465. [CrossRef]
50. Smith, M.; Carrivick, J.; Quincey, D. Structure from motion photogrammetry in physical geography. *Prog. Phys. Geogr. Earth Environ.* **2016**, *40*, 247–275. [CrossRef]
51. Deseilligny, M.; Clery, I. APERO, an open source bundle adjusment software for automatic calibration and orientation of set of images. *Int. Arch. Photogramm. Remote Sens. Spatial Inf. Sci.* **2011**, *XXXVIII-5/W16*, 269–276. [CrossRef]
52. Pierrot-Deseilligny, M. MicMac, Apero, Pastis and Other Beverages in a Nutshell! Available online: http://logiciels.ign.fr/IMG/pdf/docmicmac-2.pdf (accessed on 11 April 2022).
53. MicMac Wiki. Available online: https://micmac.ensg.eu/index.php/MicMac_tools (accessed on 1 March 2022).
54. Lowe, D. Distinctive Image Features from Scale-Invariant Keypoints. *Int. J. Comput. Vis.* **2004**, *60*, 91. [CrossRef]
55. Deseilligny, M.; Bosser, P.; Pichard, F.; Thom, C. UAV onboard photogrammetru and GPS positioning for earthworks. *Int. Arch. Photogramm. Remote Sens. Spatial Inf. Sci.* **2015**, *XL-3/W3*, 293–298. [CrossRef]
56. Peipe, A.J.; Tecklenburg, B.W. *Photogrammetric Camera Calibration Software—A Comparison*; ISPRS Commission V, WG V/1; Elsevier: Hannover, Germany, 2006.
57. Vedaldi, A. *An Open Implementation of the SIFT Detector and Descriptor*; UCLA CSD Technical Report 070012; UCLA CSD: Los Angeles, CA, USA, 2007. [CrossRef]
58. Murtiyoso, A.; Grussenmeyer, P.; Börlin, N.; Vandermeerschen, J.; Freville, T. Open Source and Independent Methods for Bundle Adjustment Assessment in Close-Range UAV Photogrammetry. *Drones* **2018**, *2*, 3. [CrossRef]
59. Rupnik, E.; Daakir, M.; Deseilligny, M.P. MicMac—A free, open-source solution for photogrammetry. *Open Geospat. Data Softw. Stand.* **2017**, *2*, 1–9. [CrossRef]
60. Nocedal, J.; Wright, S.J. Conjugate gradient methods. In*Numerical Optimization*; Springer: Berlin/Heidelberg, Germany, 2006; pp. 101–134.
61. Chiabrando, F.; Donadio, E.; Rinaudo, F. SfM for orthophoto to generation: A winning approach for cultural heritage knowledge. *Int. Arch. Photogramm. Remote Sens. Spat. Inf. Sci.* **2015**, *40*, 91. [CrossRef]
62. Snavely, N.; Seitz, S.; Szeliski, R. Photo tourism: Exploring photo collections in 3D. *ACM Trans. Graph.* **2006**, *25*, 835–846. [CrossRef]
63. Snavely, N.; Seitz, S.; Szeliski, R. TI—Modeling the World from Internet Photo Collections. *Int. J. Comput. Vis.* **2008**, *80*, 189–210. [CrossRef]
64. Verhoeven, G. Taking computer vision aloft: Archaeological three-dimensional reconstructions from aerial photographs with PhotoScan. *Archeol. Prospect.* **2011**, *18*, 67–73. [CrossRef]

65. Doneus, M.; Verhoeven, G.; Fera, M.; Briese, C.; Kucera, M.; Neubauer, W.. From deposit to point cloud: A study of low-cost computer vision approaches for the straightforward documentation of archaeological excavations. *Geoinformatics* **2011**, *6*, 81–88. [CrossRef]
66. Agisoft LLC. Available online: https://www.agisoft.com/pdf/metashape-pro_1_7_en.pdf (accessed on 15 September 2021).
67. Colarullo, M. Rilievo Topografico e Valutazione della Sicurezza del "Ponte della Cerra" in Napoli. Bachelor's Thesis, Federico II University of Naples, Napoli, Italy, 2021.
68. DJI. Available online: https://www.dji.com/it/mavic-2/info#specs (accessed on 13 March 2022) .
69. Pádua, L.; Adão, T.; Hruška, J.; Marques, P.; Sousa, A.; Morais, R.; Lourenço, M.; Sousa, J.; Peres, E. UAS-based photogrammetry of cultural heritage sites: A case study addressing Chapel of Espírito Santo and photogrammetric software comparison. In Proceedings of the GARSS 2020 IEEE International Geoscience and Remote Sensing Symposium, Waikoloa, HI, USA, 26 September–2 October 2020.
70. Atkinson, K.B. Close Range techniques and machine vision. *Photogramm. Rec.* **1994**, *14*, 1001–1003. [CrossRef]
71. Fraser, C.S. Digital camera self-calibration. *ISPRS J. Photogramm. Remote Sens.* **1997**, *52*, 149–159. [CrossRef]
72. Lague, D.; Brodu, N.; Leroux, J. Accurate 3D comparison of complex topography with terrestrial laser scanner: Application to the Rangitikei canyon (N-Z). *ISPRS J. Photogramm. Remote Sens.* **2013**, *82*, 10–26. [CrossRef]
73. Shen, Y.; Lindenbergh, R.; Wang, J. Change Analysis in Structural Laser Scanning Point Clouds: The Baseline Method. *Sensors* **2016**, *17*, 26. [CrossRef] [PubMed]

 *drones*

Article

The Bathy-Drone: An Autonomous Uncrewed Drone-Tethered Sonar System

Antonio L. Diaz [1,*], Andrew E. Ortega [2], Henry Tingle [1], Andres Pulido [1], Orlando Cordero [2], Marisa Nelson [1], Nicholas E. Cocoves [1], Jaejeong Shin [1], Raymond R. Carthy [3], Benjamin E. Wilkinson [2] and Peter G. Ifju [1]

[1] Mechanical & Aerospace Engineering, University of Florida, Gainesville, FL 32611, USA
[2] Geomatics Program, University of Florida, Gainesville, FL 32611, USA
[3] U.S. Geological Survey, Florida Cooperative Fish and Wildlife Research Unit, University of Florida, Gainesville, FL 32611, USA
[*] Correspondence: tony52892@ufl.edu; Tel.: +1-352-294-2829

Abstract: A unique drone-based system for underwater mapping (bathymetry) was developed at the University of Florida. The system, called the "Bathy-drone", comprises a drone that drags, via a tether, a small vessel on the water surface in a raster pattern. The vessel is equipped with a recreational commercial off-the-shelf (COTS) sonar unit that has down-scan, side-scan, and chirp capabilities and logs GPS-referenced sonar data onboard or transmitted in real time with a telemetry link. Data can then be retrieved post mission and plotted in various ways. The system provides both isobaths and contours of bottom hardness. Extensive testing of the system was conducted on a 5 acre pond located at the University of Florida Plant Science and Education Unit in Citra, FL. Prior to performing scans of the pond, ground-truth data were acquired with an RTK GNSS unit on a pole to precisely measure the location of the bottom at over 300 locations. An assessment of the accuracy and resolution of the system was performed by comparison to the ground-truth data. The pond ground truth had an average depth of 2.30 m while the Bathy-drone measured an average 21.6 cm deeper than the ground truth, repeatable to within 2.6 cm. The results justify integration of RTK and IMU corrections. During testing, it was found that there are numerous advantages of the Bathy-drone system compared to conventional methods including ease of implementation and the ability to initiate surveys from the land by flying the system to the water or placing the platform in the water. The system is also inexpensive, lightweight, and low-volume, thus making transport convenient. The Bathy-drone can collect data at speeds of 0–24 km/h (0–15 mph) and, thus, can be used in waters with swift currents. Additionally, there are no propellers or control surfaces underwater; hence, the vessel does not tend to snag on floating vegetation and can be dragged over sandbars. An area of more than 10 acres was surveyed using the Bathy-drone in one battery charge and in less than 25 min.

Keywords: bathymetry; uncrewed aircraft; sonar; tethered; retention pond; hydrology; survey; recreational sonar; drone

Citation: Diaz, A.L.; Ortega, A.E.; Tingle, H.; Pulido, A.; Cordero, O.; Nelson, M.; Cocoves, N.E.; Shin, J.; Carthy, R.R.; Wilkinson, B.E.; et al. The Bathy-Drone: An Autonomous Uncrewed Drone-Tethered Sonar System. *Drones* **2022**, *6*, 294. https://doi.org/10.3390/drones6100294

Academic Editors: Arianna Pesci, Giordano Teza and Massimo Fabris

Received: 29 July 2022
Accepted: 12 August 2022
Published: 10 October 2022

Publisher's Note: MDPI stays neutral with regard to jurisdictional claims in published maps and institutional affiliations.

1. Introduction

This paper introduces a unique drone-based method to perform bathymetry on relatively small waterways with high spatial and temporal resolution. There is a wide variety of methods employed to perform bathymetry for an innumerable variety of scales and applications. Satellites can provide large-scale surveys of large bodies of water [1–3], such as lakes, bays, gulfs, and oceans, while drones and small uncrewed watercraft are increasingly used for smaller waterbodies such as rivers, inlets, retention ponds, boat basins, shipping channels, and nearshore applications [4–6]. The benefit of increased field operator safety, reduced fatigue and environmental exposure, and more accurate raster patterns are primary motivations behind uncrewed systems. The literature on drone-based and uncrewed watercraft-based bathymetry shows that a variety of sensors have been utilized, each having their advantages and limitations. They can be grouped into airborne methods, such as photogrammetry, lidar, radar, and fluid lensing, and immersed methods such as sonar and underwater photogrammetry. Uncrewed surface vehicles (USVs) and uncrewed underwater vehicles (UUVs) allow for sensors, such as acoustic sensors, since these vehicles are in contact with the water, while aerial drones can only utilize sensors that operate with a standoff from the water surface. Typically, USVs and UUVs are slower-moving with a smaller sensor swath than aerial drones and, thus, limit the area covered during operation. In the following paragraphs, we review typical sensors, platforms, and applications associated with small UAS and uncrewed watercraft, providing the motivation and background for the development of the Bathy-drone.

Remote sensing instruments used for uncrewed bathymetry are grouped into optical or acoustic sensing which can each be accomplished with passive or active methods. Optical sensors can actively measure reflected energy, such as with immersed range-gated camera systems [7] or water-penetrating green lidar, which can reach depths of 40–50 m in clear waters [8,9]. Passively sensing reflected or scattered light is accomplished with hyper- or multispectral imaging [2,10,11] or strictly in the visible light spectrum [12], and the imagery may be georeferenced with structure from motion (SfM) [13–18] and photogrammetry [19].

Acoustic sensors typically rely on active sensing by emitting acoustic waves and measuring the reflected, scattered, and absorbed energy. Acoustic methods used in uncrewed bathymetry include sonar technologies such as multibeam and single-beam echo sounders (MBES or SBES respectively), side-scan sonars (SSS), and phase-measuring side-scan sonar (PMSS) [20]. Water-penetrating radar (WPR) [21–24] and Doppler velocity logger (DVL) or acoustic Doppler current profilers (ADCP) [25] are also active acoustic sensors that have been used in uncrewed bathymetry. In the oil and gas industry, seismographic or sub-bottom profiler (SBP) [26] sensing, either active or passive, is also a popular method of mapping.

Ancillary sensors are crucial to both optical and acoustic techniques and typically correct for positioning, heading, attitude, tide, and sound velocity. Respectively, these corrections can be accomplished with Global Navigation Satellite Systems (GNSS) antenna, gyrocompass, Inertial Measurement Unit (IMU) or Inertial Navigation System (INS), marigraph, and sound velocity profiler.

The desired data characteristics depend on the application, and uncrewed solutions are still developing. For example, deep-sea exploration can involve crewed expeditions focused on extensive coverage to capture geological features within a resolution of 50 m. Navigable coastal and inland waterways, which typically are shallow waters, have been surveyed by multiple uncrewed bathymetric platforms to meet the required rigorous navigational mapping safety standards set by agencies such as the International Hydrographic Organization (IHO) or the US Army Corps of Engineers, which can require a resolution to detect features of 0.5 m and accuracy of below 1 m uncertainty depending on under-keel clearance [27,28]. Categorizing habitats and infrastructure is another application of uncrewed bathymetric platforms and demands understanding the bottom composition of the observed area. Under any of a vast variety of scenarios, the specific application requirements guide the selection of appropriate platforms and sensors.

Bathymetry systems can be categorized as sensing-immersed such as USV and UUV or above-the-water such as satellites and UAS. Satellite altimetry provides entire models of the Earth based on radar readings of water height and slope, induced by local gravity of subsea geological features, and it is independent of water clarity with spatial resolution varying between 1 and 12 km [2,3]. Satellites such as the Hyperion hyperspectral sensor onboard NASA's EO-1 platform have collected bathymetry above coastal waters in large swaths of ~7 km to a spatial resolution of 30 m and 1–20 m of water depth depending on clarity [2]. The large swath coverage of satellites is excellent for capturing expansive geological features, but the resolution is not useful for safe maritime navigation. Satellites have also used camera sensors for depth estimation methods through wave celerity inversion, leading to sub-meter resolution, in less than 35 m clear water [29]. Fluid lensing photogrammetry from UAS also uses camera sensors and can result in up to centimeter-scale resolution, as the technique filters for advantageous images with magnifying wave conditions [30]. The trend seen in photogrammetry is also observed with UAS radar altimetry [18] compared to satellite altimetry: a shorter sensor to area of interest distance typically leads to refined accuracy, precision, and resolution while sacrificing swath coverage. Lidar bathymetry must strike a balance, and, when paired with UAS such as Reigl's bathycopter [31], an observation height of 500–600 m can lead to accuracy of 3–5 cm at depths of 0–4 m in small to medium rivers of width 5–25 m. This tradeoff provides each optical sensor and platform pairing with unique utility, but they are all challenged by obstructions such as grassy bottom composition or flotsam, suspended particles impacting water clarity, water refraction (when not immersed) [32], and overhanging structures such as foliage or cliffs.

Acoustic sensors have been paired with USV, UUV, and UAS to great success in research and from commercial suppliers. USV with SBES, MBES, SSS, PMBS, and ADCP have all led to successful bathymetric surveys. Most acoustic sensors are deployed on UUV or USV and do not need to account for water surface reflection or refraction. Depending on the sensor and frequencies used, acoustic sensors on USV can typically record depths of 0.2 m to thousands of meters regardless of turbidity. Centimeter resolution can be achieved with sonar on USV but varies due to factors such as sensor frequency and depth. There are a variety of USV hull shapes [5] to accommodate different sensors and marine environments, such as small-waterplane-area twin hull (SWATH) [33,34], catamaran [35–37], and V hull. Commercial and research USV such as the Seafloor Systems Hydrone [38], Jetyak [39], and Searobotics Hycat [40] have successfully conducted bathymetric surveys for navigational purposes and characterization of the bottom environment in coastal and inland bodies of water. Observations from USV platforms are slower than airborne platforms due to the closer proximity, reducing the swath coverage area of the sensor, and the increased drag on the platform in water compared to air. USV and ROV deployment is also challenging as it typically requires a crewed vessel to arrive at the study site or a boat ramp for field operators to lift the vehicle into the water. USV and ROV are also typically electrically self-propelled and can struggle to steer in an autonomous mission safely and accurately against strong currents and winds [41].

Hybridization of platforms and sensor packages has led to new combinations of sensor fusion and enabled novel deployment opportunities and results. For example, combining aquatic and aerial sensors such as camera photogrammetry and sonar can improve the results of the independent sensors, especially in transitionary areas where one sensor has advantages over the other [42,43]. A unique and emerging platform hybrid is that of the tethered acoustic sensor to a UAS. Historically, this configuration can be seen on crewed helicopters with dipping sonar for submarine detection [44], and several towable sonar packages exist for crewed marine vessels such as for analyzing coral reefs [45]. Most UAS with tethered sonar uses small lightweight SBES with limited ancillary sensors [17,43,46–48]. Bobber-shaped casting sonar from Lowrance is typically used, which lacks ancillary sensors such as IMU but is excellent for extending the UAS range to take sparse point measurements by minimizing payload weight. The Bathy-drone proposed in this paper (Figure 1) is a novel approach to uncrewed bathymetry, configured as a tethered, hull-enclosed sonar with IMU and Real Time Kinematics Global Navigation Satellite System (RTK-GNSS) that can autonomously gather bathymetric data.

As evidenced by the literature, there is a trend of developing efficient uncrewed systems that are practical, inexpensive, and easy to deploy, while providing high spatial and temporal resolution for bathymetry on small bodies of water, such as ponds, rivers, boat basins, shipping lanes, and pre-construction and nearshore applications. The University of Florida has developed a system that incorporates a drone that drags a vessel/platform via a tether that can be equipped with a variety of sensors such as sonar or underwater cameras. It has advantages over USVs since the system can be flown to the survey location; thus, many surveys can be initiated from a land-based ground station, and no boats or boat ramps are needed if the location is within the FAA-required visual line of sight. Since the vessel has no propulsion system (propellers), floating vegetation does not hamper operation. In the remainder of this manuscript, we describe the system design, the experimental procedure to assess the design, the results from our assessment, and a description of future work.

Figure 1. Bathymetry system during autonomous flight for ground-truthing at University of Florida Plant Science Research Citra, Florida. Photo credit: author's UF UASRP lab.

2. System Design

The design of the vessel for the Bathy-drone system is an important consideration for the system to perform over a wide range of speeds without capsizing in turns while maintaining a level attitude and housing the sonar components. The vessel must track in a straight line behind/below the drone to ensure that the passive vessel path is similar to the commanded drone path. Unlike most powered vessels, the driving force produced by the drone through the tether does not act in a direction parallel to the water surface. Instead, it acts upward at an angle that is determined by the length of the tether and the height of the drone above the water. Thus, the upward component of the force vector must be considered in the vessel design.

During the process of designing the hull, three basic shapes were tested, a soft-edge V-hull shape, a trimaran shape, and a skiff-like planing (which planes or skims on the water surface versus a hull that parts the water) hull. Each of these was satisfactory for straight-line portions of the mission patterns, but the lateral resistance, especially in the bow of the V-hull and trimaran, led to capsizing issues in the turns. This was documented by repeated experiments that were conducted over a wide velocity and corner radius range. The soft-edge skiff-like planing hull shape proved robust in both straight/level tracking and corner turns once a trim plate and fins were added.

The location of the tether attachment point on the vessel proved to be important to balance the forces on the hull through the entire speed range. The first attempt placed the tether attachment point on the nose of the vessel, but this led to longitudinal pitch-up as a function of speed. After analysis of the free body diagram (Figure 2) at a constant velocity, it was determined that, when attaching the tether so that the force passes through the center of gravity (CG), trim and level conditions were achieved through the entire speed range. Figure 2 shows the forces acting on the hull at a constant velocity. The tether provides a constant pulling force that is dependent on the drag force, which is horizontal, while the pulling force is angled upward. Both are dependent on the speed of the vessel/drone.

Other forces on the hull include an upward buoyancy force and a hydrodynamic force on the forward portion of the rocker line. The latter is speed-dependent and tends to produce a longitudinal pitching moment about the CG that results in the bow pitching upward as the speed increases. To compensate for this pitching moment, a trim plate was added behind the CG, angled downward, to produce a speed-dependent pitching moment in the opposite direction of the speed-dependent bow pitching moment. The trim plate is like a horizontal stabilizer on an aircraft. By placing the tether attachment location so that the force acts through the CG, and with the angle of the trim-tab empirically optimized, the vessel tracked the level through the speed range of 0 through 24 km/h. By incorporating two fins on the trim-tab, tracking of the vessel improved, in addition to allowing turns without capsizing. The fins act as a pivot point where horizontal forces through the tether swing the hull's bow around, as shown in Figure 3. The soft edges of the planing hull bow rocker provide little lateral resistance; thus, the bow swings around smoothly in the turns. This was not the case for the V-hull and trimaran shapes.

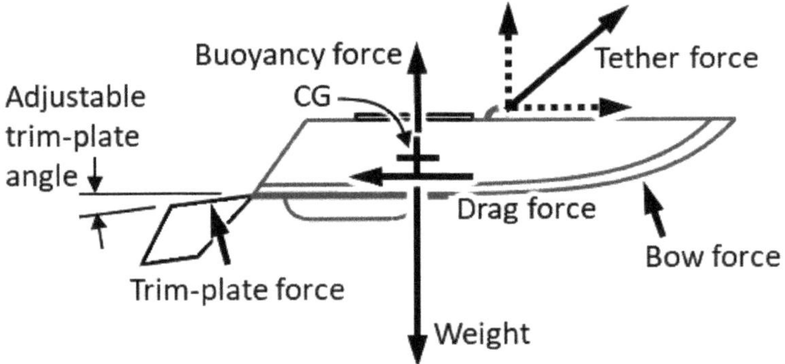

Figure 2. Forces at a constant speed on the hull. The center of gravity is labeled CG. Figure credit: author's UF UASRP lab.

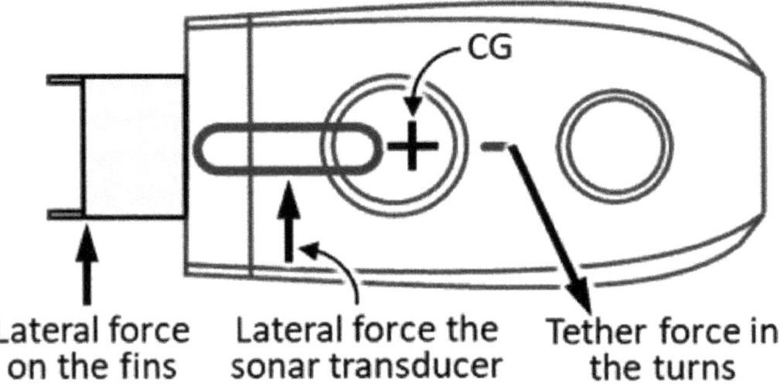

Figure 3. In a turn, the tether force acts laterally with respect to the hull. The center of gravity is labeled CG. Figure credit: author's UF UASRP lab.

Additional considerations in the design included providing adequate volume in the hull to house the sonar console and ports to provide easy access to the microSD cards, batteries, and console keypad. The CG was also factored into the design so that, at 0 km/h, the hull floated level. This was achieved by balancing the buoyancy force with the center of gravity of the vessel. Small adjustments were made to achieve this by slight shifts to the CG and small weight plates mounted on the top of the trim plate. The aluminum trim-plate also acts as an electrical ground for the sonar transducer, mounted on the bottom in the aft portion of the hull.

The sonar unit is a low-cost commercial off-the-shelf (COTS) recreational fish-finder by Lowrance®, a company from Tulsa, OK, USA, model Elite ti7, with an active scan transducer. The Lowrance sonar unit features a NMEA serial data port which allows for communication between other marine electronics. By using this port, the Lowrance can output serial data packets to the ground station using an onboard telemetry link. The telemetry link is an RFD900+ long-range 900 MHz transmitter RFDesign® a company from Brisbane, Australia, that can send data up to 30 km in optimal conditions. The NMEA protocol uses a variety of 'sentences' to transmit individual data streams. The Lowrance menu allows for the selection of NMEA sentences and at what frequency they are output. The Bathy-drone currently outputs the GPS position and the sonar depth values using the following packet types: DBT—depth below transducer, DPT—depth of water, RMC—recommended minimum navigation information, GGA—global positioning system fix data, HDG—magnetic heading, deviation, variation, GLL—geographic position, latitude/longitude, and VTG—track made good and ground speed. These sentences are broadcast using the RFD900 telemetry link and are received by the ground station computer to interpret the data using the Reefmaster software package. Details of Reefmaster use are found in the ground station and data processing discussion. The sonar imagery down- and side-scan data are saved directly to the onboard microSD card to conserve telemetry bandwidth. A Pixhawk Cube orange with a Here+ RTK system was included in this design (Figure 4) for future work with RTK + IMU integration.

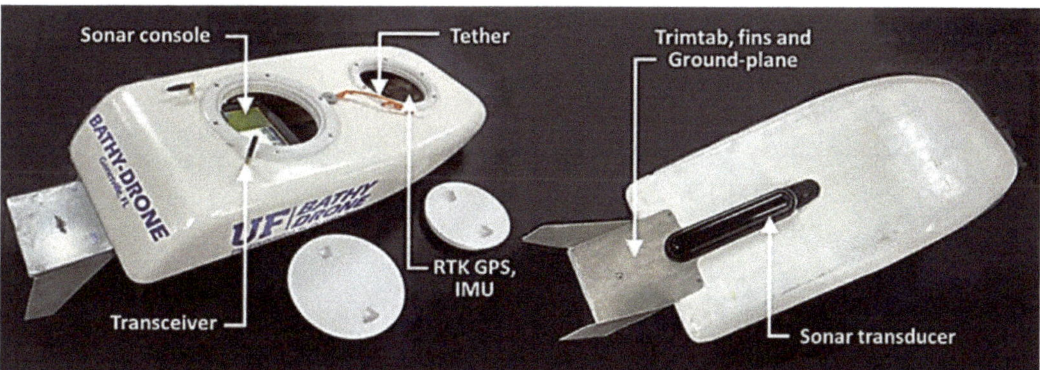

Figure 4. Isometric view of the Bathy-drone from the top showing sonar screen through the open hatches (**left**) and from the bottom with fins and transducer (**right**). Where RTK GPS is Real Time Kinematics corrected Global Positioning System and IMU is Inertial Measurement Unit. Photo credit: author's UF UASRP lab.

The Bathy-drone vessel is designed to be multirotor UAS agnostic if the drone can carry the weight of the vessel portion. The interface between the UAS and the vessel can be as simple as a knot. Any drone can, therefore, tow the vessel without any prior modification. The DJI® (a company from Nanshan, Shenzhen, China) Matrice 600 drone was utilized to test the Bathy-drone vessel, featuring 30 min of battery life and 6 kg payload capacity [49]. An alternate platform, the Alta X drone from Freefly Systems®, a company from Woodinville, WA, USA [50], has similar performance metrics. The current iteration of the Bathy-drone weighs just under 14 lbs and is easily airlifted by the Matrice 600 to be shuttled to the water surface.

3. Experimental Procedures

Traditional methods of pond survey included a field crew operating a small watercraft to measure the pond bottom in a systematic grid pattern. Measurements were taken via a level pole, range pole, or sounding wire and were interpolated to create an estimated surface of the pond bottom (Figures 5 and 6) [42,51].

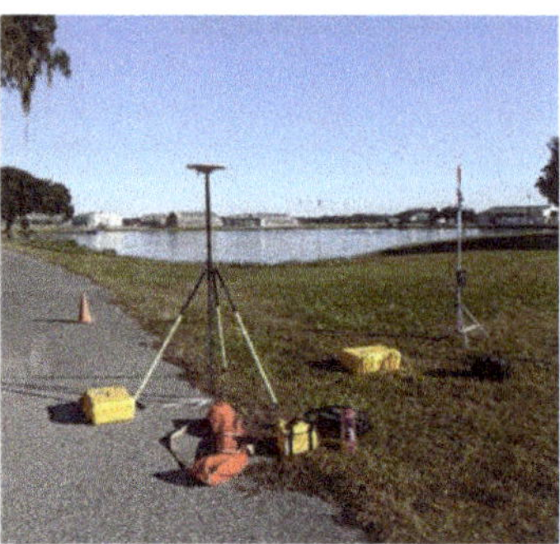

Figure 5. The moving rover and the data collector taking topographic points throughout the pond from a barge (**left** image). Custom 3D printed silt foot for level rod (**center**). The base station set up on a nail driven into the asphalt (**left** tripod) and the radio (**right** tripod) (**right** image) transmitting the RTK GPS corrections. The base nail was measured for 8 h as static observations and submitted to NOAA OPUS. Topographic points collected across different days were translated so that the base station points aligned with the NOAA OPUS solution. Photo credit: UF UASRP; student researchers pictured on the left image.

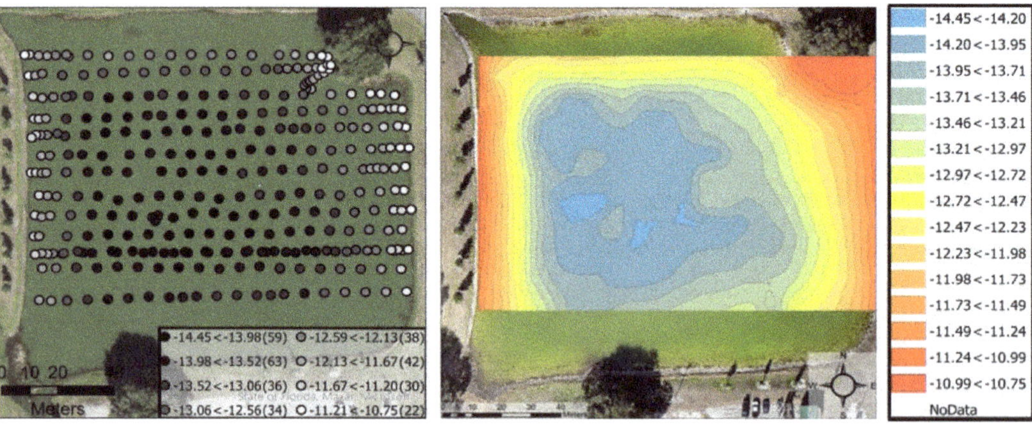

Figure 6. Ground-truthing data consisting of 324 RTK-corrected points with a minimum depth of 0.17 m, maximum of 3.87 m, mean of 2.30 m, and standard deviation of 1.03 m (**left**). Local polynomial interpolation of the ground-truth data (**right**). Horizontal coordinates are in NAD 1983 (2011) State Plane Florida West FIPS 0902 (meters) and vertical coordinates are in height above ellipsoid (meters).

The University of Florida Plant Science Research and Education Unit (UF PSREU) is a 1086 acre facility which conducts specialty crop research. The UF Unmanned Aircraft Systems Research Program (UASRP) uses this expansive facility to test uncrewed aircraft systems in an unpopulated area. On this property, there is a 5 acre retention pond in which we tested the bathy-drone system. Measurements from a graduated level rod were used to measure the bottom surface of the pond and compare the obtained sonar data to traditional survey methodology (Figure 5, center image). To take direct measurements, a Trimble® (a company from Sunnyvale, CA, USA) RTK GNSS survey system was attached to the top of the level rod. RTK GNSS uses a stationary base station to transmit corrections to a moving rover with a radio or internet data link (right image). These computed corrections allow the rover to obtain centimeter-level accuracy point measurements. Data from the base station were entered into NOAA OPUS [52] to translate the observations gathered into the national spatial reference system (Figure 6). The ground-truthing experiments used the Trimble SPS855 as the base station receiver and the SPS986 as the rover platform. Corrections were transmitted using a high-power 35 W TDL450 radio. The level rod and observation barge were tethered to a rope that was strung across the length of the pond (Figure 5, left image). The rope was graduated with markings every 7.6 m (25 feet). Each of the points surveyed for the ground-truthing had a report of associated observation notes and statistics such as dilution of precision, number of satellites, horizontal precision (DRMS), vertical precision (1 sigma), and tilt distance. As a summary of the measurement uncertainty associated with the ground-truthing points gathered, the reported tilt distance average was 6 cm with a standard deviation of 5 cm, horizontal precision average was 1 cm with a standard deviation of 1 cm, and the vertical precision average was 2 cm with a standard deviation of 1 cm.

The file containing the waypoints of an autonomous flight mission was created in an external software package such as ArcGIS® Pro a company from Redlands, CA, USA or Google Earth® a company from Mountain View, CA, USA. The ground-truthing mission used ArcGIS® Pro point creation tools to create equally spaced waypoints and the line creation tool for the flight lines. The "export to KMZ" geoprocessing tool was then used to create a file compatible with DJI®, China, ground station pro software and with the DJI® Matrice 600 multirotor drone. Ground Station Pro allows operators to set mission parameters such as speed and elevation and transmits the preplanned mission to the drone. The pilot can see the mission status, as well as battery and GPS signal levels, live throughout the flight mission. The autonomous turn settings are changed from stop and turn to continuous radius turns. Radius turns prevent capsizing from abrupt slackening and tugging of the towline. The radius is programmed to be as large as possible between transects which can vary in spacing depending on the resulting resolution desired from the sonar data. Alternatively, autonomous missions can be planned on the open-source software Mission Planner.

The current flight field procedures were developed for ground-truthing of the system and gathering data for isobaths that tie to geodetic datums. The vessel electronics were turned on and set to record before takeoff of the UAS. Standard flight operating procedures and safety were followed. The pilot navigated the drone to the first waypoint lifting the vessel above any obstructions. The drone was lowered to the flight altitude of 6 m (20 feet), placing the vessel on the water. With the drone and vessel at the starting waypoint, the autonomous mission was enabled. The autopilot flew the drone and vessel to the recovery point while the pilot maintained the ability to interfere if necessary. A top-of-water surface reading was taken with RTK GNSS to offset the sonar measurements and translate the measurements to the coordinate system of the geodetic datums. Figure 7 shows a typical path for both the multirotor drone (solid red line) and the vessel (white points).

ReefMaster is a commercial software package from Australia used for the interpretation and analysis of data from recreational fish finders and sonar units. ReefMaster supports data import from Lowrance and Hummingbird brand sonar units and can export generic point data such as shapefiles and CSV files. The combination of USV with fish finder sonar data processed through Reefmaster has proven results in bathymetry, ecological classification, and bottom characterization in multiple projects [15,53].

The software acts as part of the ground station during the flight in the field by receiving data over a telemetry link. Live data visualization on the ground station is achieved by processing depth and position updates using the NMEA protocol. Live telemetry only supports the transmission of depth value data; raw sonar data must be saved to and read from the microSD card. The Bathy-drone sonar package from Lowrance produces an sl2 file format to store the raw data from the three-in-one sonar transducer onboard. Reefmaster software can import the sl2 file from the Lowrance microSD between missions for sonar data visualization on the ground station. Rapid on-site isobaths and bottom hardness plots can be created on the ground station running Reefmaster to inform the next mission. Additionally, Reefmaster can export CSV and shapefiles for further processing on Excel, MATLAB, Python, and ArcGIS Pro.

The live link prevents loss of data during malfunctions such as power loss and adds redundancy to the onboard storage. The on-field review helps ensure that the operators met the mission requirements before returning to the office.

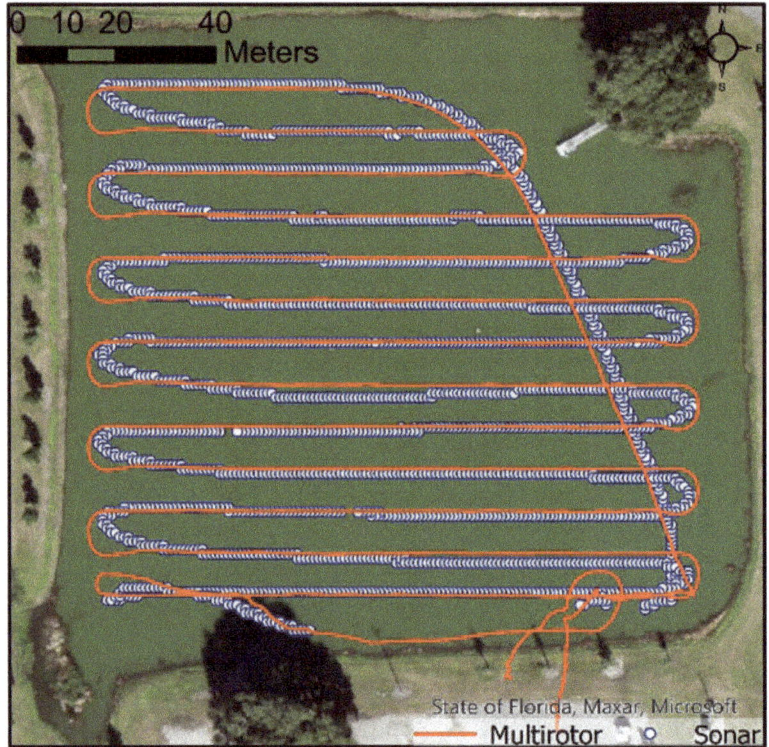

Figure 7. Comparison of GNSS-recorded flight path of multirotor drone and the path of the sonar payload vessel. Photo credit: UF UASRP using satellite imagery.

4. Results

The on-field results generated by preliminary processing of the raw sonar data on Reefmaster with further processing on ArcGIS Pro included isobaths (Figure 8), bottom hardness (Figure 9), and sonar side-scan imagery (Figure 10). According to the experimental procedures in Section 3, the primary desired outputs from the surveys included isobaths and bottom hardness contours. These data are presented here as contour plots in Figures 8 and 9.

To compare the ground-truth survey points to the sonar depth data, both datasets were imported into ArcGIS pro. The sonar data were converted to the same coordinate system as the ground-truth points, such that horizontal coordinates were in State Plane Florida West FIPS 0902 (meters) using the 2011 realization of NAD 1983 and vertical coordinates were in height above ellipsoid (meters). In addition, the sonar data must have the water level subtracted, which is currently measured at the start of the flight using the RTK GNSS level rod. Subtracting the water surface elevation is like accounting for the length of a range pole or, in this case, the level rod, when taking rover measurements with RTK GNSS.

The ArcGIS Pro geostatistical wizard is a toolset that guides users through the process of making interpolation-based surfaces in a step-by-step manner. The geostatistical wizard uses point data to create an interpretive raster surface. To use this tool, the sonar data are imported into ArcGIS from Reefmaster as a point shapefile, with each point representing an individual sonar depth observation. The geostatistical wizard uses the point sonar data with a selected area interpolation method. Local polynomial interpolation was used to create the surface, and the extents were clipped in the processing environments tab (Figure 8). The surface interpolation was restricted to the limits of the sonar data capture to minimize extrapolation. The GA layer to points tool was then used to compare the acquired ground truth survey points to the interpolated sonar data. The GA layer to points tool is used to measure the accuracy of a predicted surface by comparing it to known point observations. In this case, accuracy was assessed by comparing the predicted surface output from the sonar to the points measured using the RTK GNSS level rod (Figures 11 and 12).

The detailed measure of accuracy indirectly captures the horizontal accuracy of the sonar data within the depth measurements. To isolate horizontal accuracy from vertical accuracy, the quoted accuracy of the GNSS devices used can be considered. The Trimble RTK GNSS equipment is positionally accurate with a horizontal accuracy of 1–2 cm, while Pixhawk Cube Orange GNSS RTK internal to the vessel can be as accurate as 3 cm, and the Lowrance Elite Ti2 claims 20 m RMS positioning accuracy [54]. A simple preliminary experiment comparing the three devices observed that the Lowrance GPS can waiver from the RTK GNSS devices by 1.5–3 m. However, the uncorrected GPS is, in principle, mostly translated horizontally and still captures the relative path taken by the vessel [55].

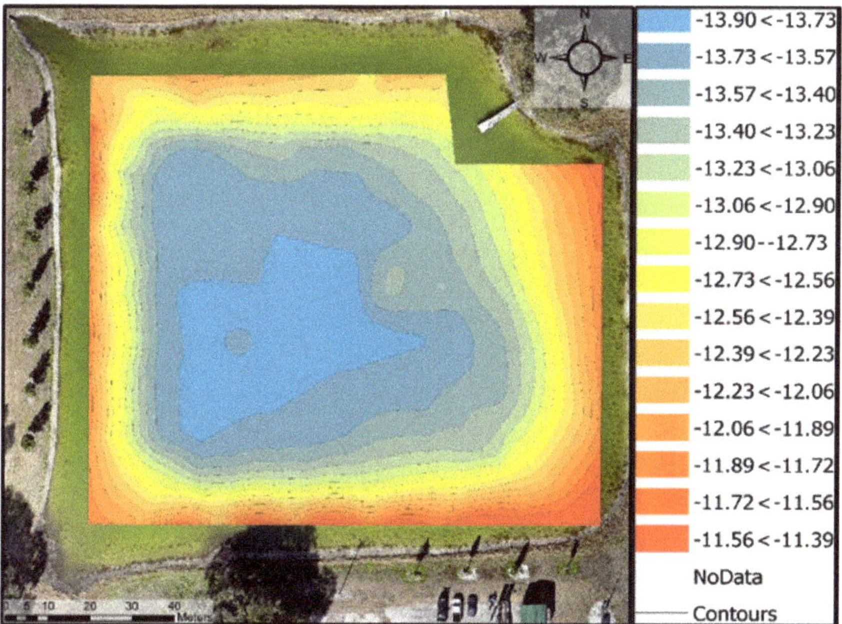

Figure 8. Local polynomial interpolation of sonar data from a cross boustrophedon flight pattern at 4.5 mph. Superimposed on photogrammetry data gathered by author's UF UASRP lab. Photo credit: UF UASRP.

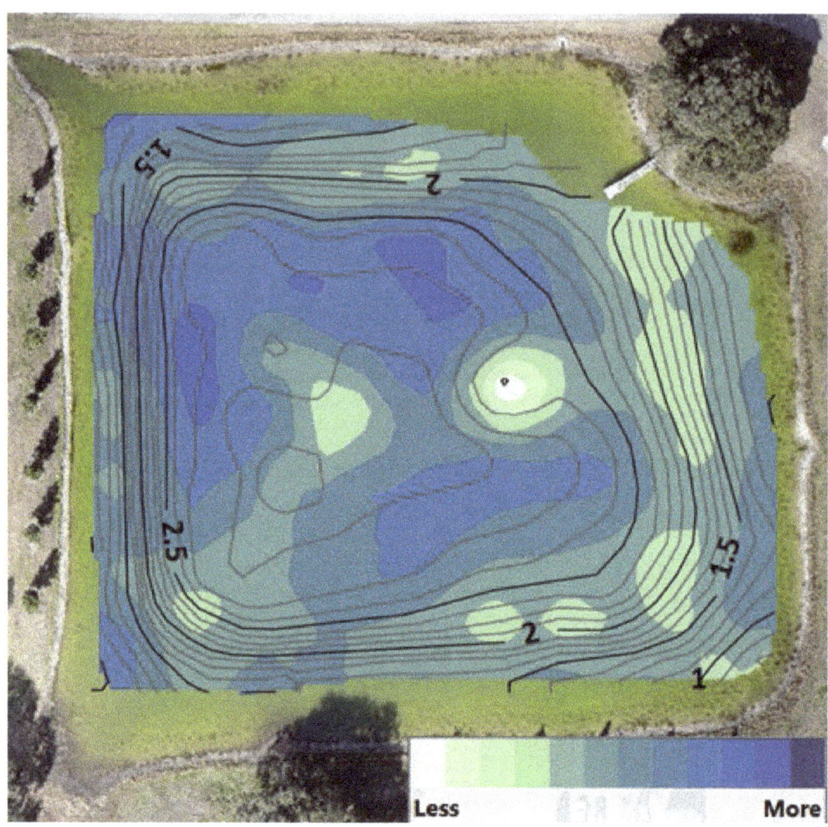

Figure 9. Bottom hardness as a measure of acoustic backscatter where light colors are softer and darker colors are harder. The bottom hardness color plot is overlayed with isobaths in meters. Superimposed on photogrammetry data gathered by author's UF UASRP lab.

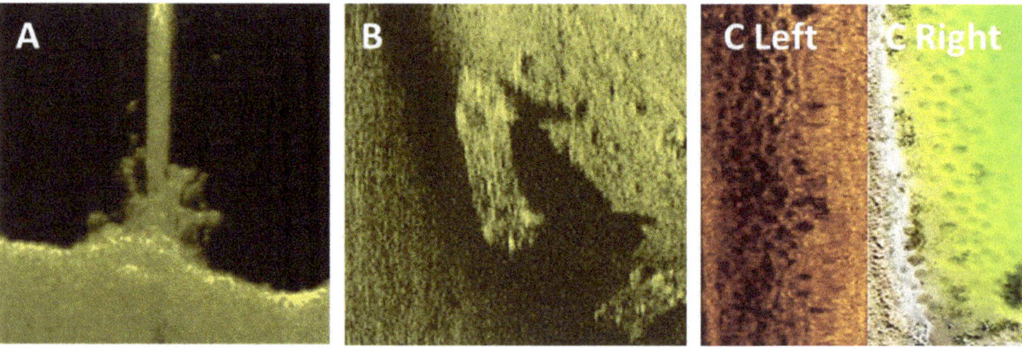

Figure 10. Sonar of Canal C-11, bridge pile at Fort Lauderdale Florida, showing accumulated vegetation and scour undermining the pile (**A**). This shows the potential for the drone bathymetry system to be used for inspection of civil infrastructure. Side-scan sonar of submersed vehicle in a quarry in northern Florida (**B**). ((**C**), **Left**) Sidescan sonar image of tilapia nesting beds captured at the Citra retention pond compared with photogrammetry ((**C**), **Right**). Photo credit: UF UASRP.

To compare one unique flight to another, raster surfaces were created. The GA layer to raster tool was used to create raster surfaces from each of the obtained datasets. The individual raster was compared using the raster calculator, which was used to subtract the two from each other. By subtracting the two rasters, the distribution of the difference between the two surfaces became apparent (Figure 13).

A second approach to analyze the precision of the system depth measurements is to compare the values at each intersection of the north–south (NS) and east–west (EW) raster paths, as illustrated in Figure 13 where two example intersections are highlighted with a star. The difference in the two depth values from NS and EW lawnmower paths was computed using the black intersection line shown in Figure 14. The intersecting points were found by dividing the raster pattern into straight-line segments, excluding the curved part of the trajectories, and then running an algorithm that iterated for all the NS and the EW paths to find the samples with the minimum distance between each other (ideally zero for exact intersections). The 10 closest points from the intersection points were used to fit a 10th-order polynomial of the depth values, and the precision was calculated as the difference in the polynomial depth values at the intersection points. A total of 144 intersecting points were used for this precision analysis. The result is summarized as a histogram in Figure 15, where the mean and median values of all the precision calculations were -1.37 cm and 0.89 cm respectively, and the standard deviation was 19.9 cm.

System analysis of depth reflects horizontal and vertical errors. These errors were characterized statistically from the ground-truth data and measured in multiple approaches for accuracy and precision. The accuracy measures point toward potential improvement if IMU and RTK data are used to correct the depth values. This conclusion is drawn from the sonar consistently reading deeper than the ground-truth data on average by 21.6 cm, with most of the error focused on the deeper area, as highlighted by the scatterplot (Figure 12). The deeper this system is deployed without correction, the more pronounced small-angle changes in attitude will impact accuracy. The major limiting factor of improving accuracy, to be addressed by incorporating IMU and RTK corrections, is also supported by the relative precision between independent surveys.

Ensuring the vessel is properly placed in the water at the first waypoint requires pilot skill and familiarity with the system. Improvements to this issue will be discussed as future work. The Bathy-drone at its current stage has successfully demonstrated rapid deployment in difficult-to-access waters to gather bathymetry, backscatter, and sonar imagery affordably.

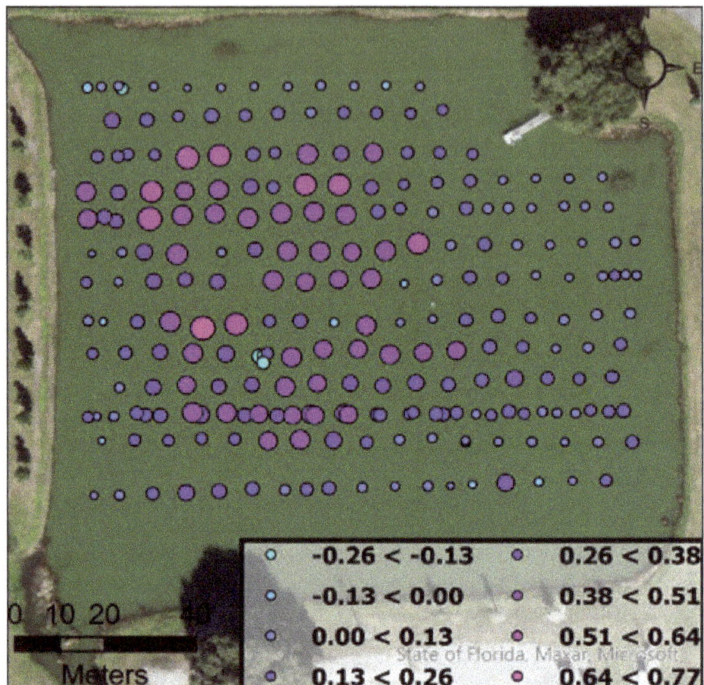

Figure 11. Residuals (m) between local polynomial interpolation of combined sonar transects and ground-truth data obtained from the Real Time Kinematics (RTK) corrected graduated rod measurements. Positive residuals indicate that sonar readings are deeper than ground truth. The size of the points corresponds with the residual magnitude, and the color corresponds with the direction. Photo credit: UF UASRP using satellite imagery.

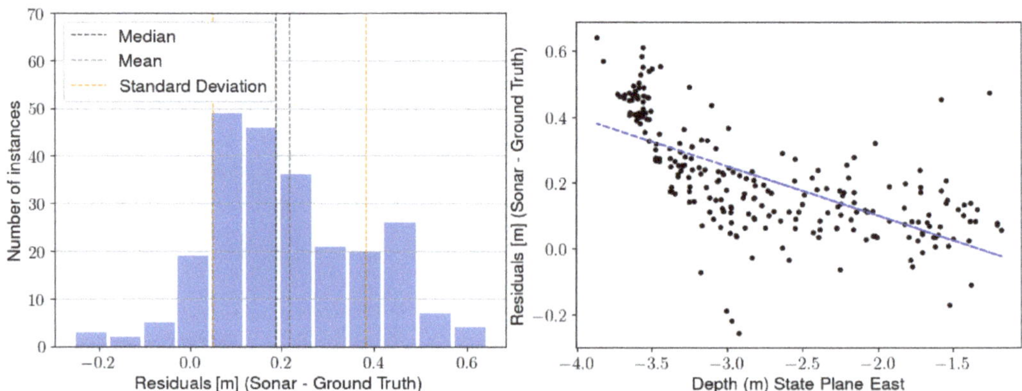

Figure 12. Histogram and summary statistics of residuals where the mean is 21.6 cm, the median is 18.7 cm, and the standard deviation is 16.7 cm (**left**). Scatterplot of residuals and the relationship with depth, where the blue dash linear line of best fit demonstrates decreasing residual with decreasing depth (**right**).

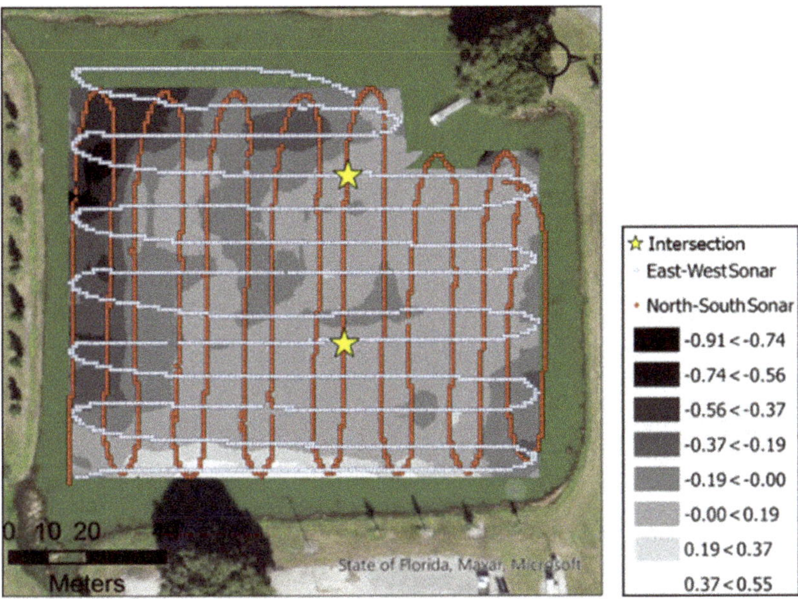

Figure 13. Visualization of east–west interpolated surface versus that of north–south transects. The mean residual is −2.64 cm, the median is 0.95 cm, and the standard deviation is 16.98 cm. The stars labeled "intersection" are example calculation locations for the depth difference between NS and EW. Photo credit: UF UASRP using satellite imagery.

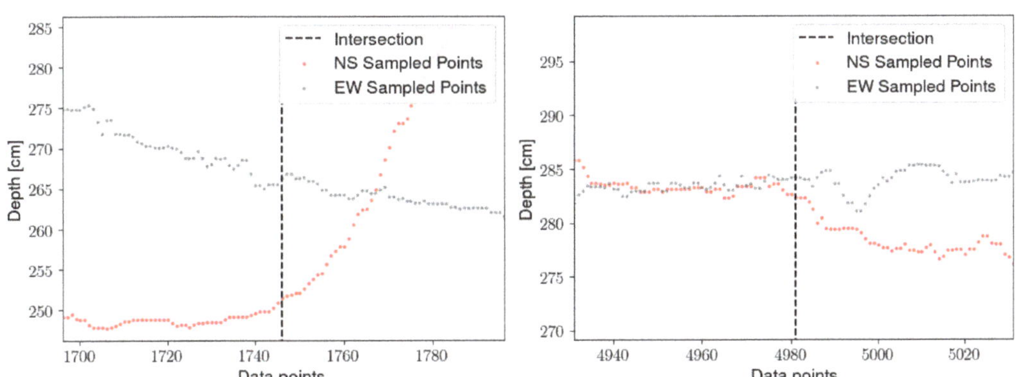

Figure 14. Depth values of the data values shown in red in Figure 13 on the upper star and lower star, respectively. The point where the two intersect is shown (black line). The difference for the left plot is 14.9 cm, and the difference for right plot is −1.69 cm.

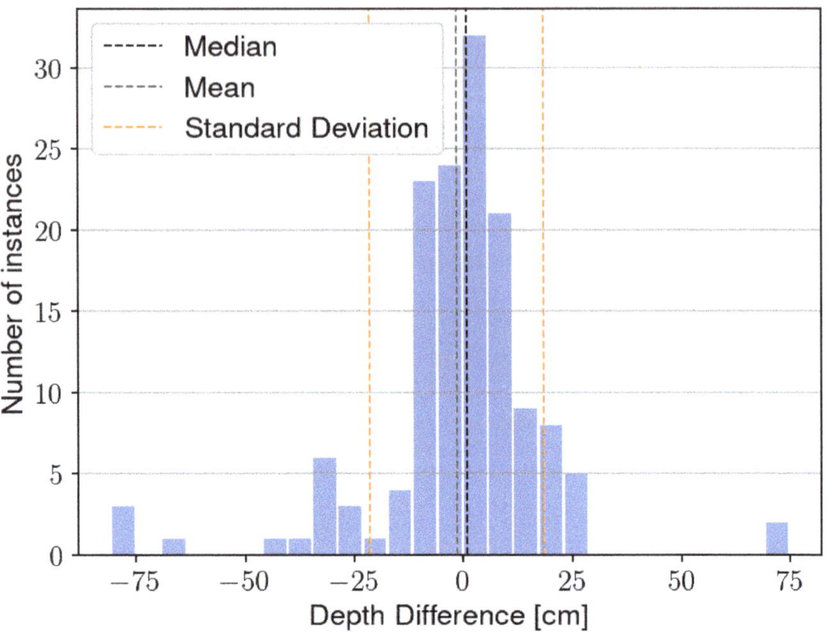

Figure 15. Histogram (blue) of precision analysis: difference of depth measurements at the intersections of NS and EW lawnmower paths.

5. Future Work

The multirotor towable bathymetric system presented in this paper results from the multiple cycles of the iterative design process. Through the design, manufacture, assembly, field testing, and data processing cycle, there were lessons learned and insights into the project's potential future direction. A literature review and collaboration with field operators in bridge infrastructure inspection and marine geomatics guided the current design and will continue to guide future additions. Currently, the bathymetric system is limited to open areas to avoid obstacles that can interfere with the safe operation of the multirotor UAS. Obstacles such as bridges, pilings, and trees are commonly near bodies of water. To avoid these hazards, active vessel control with a servo-actuated rudder would allow for autonomous control semi-independent of the multirotor UAS. Active control of the vessel necessitates that the multirotor flight altitude changes actively, an active winch mechanism, or both to maintain consistent tether tension. In either case, an emergency break-away tether connection can help reduce the chance of losing the UAS into the water if the vessel is snagged.

A new hull design inspired by field testing finished the composite layup manufacturing process. The new design intends to reduce capsizing during aggressive turns and the initial placement on the water, thereby reducing pilot fatigue and extending mission efficiency. Reduced capsizing is achieved with a rounded hull deck and low center of gravity akin to self-righting crewed vessels. A larger hatch allowing for easier access to internal payload will potentially also be tested for structural integrity and environmental intrusion. The stabilizing fins and rudder are reduced in depth to increase operational capability in shallow areas further and avoid groundings, flotsam, and floating vegetation.

The additional step of measuring the area's water level to be scanned so that the data can be tied to a coordinate reference system, which can be simplified with an onboard RTK GNSS receiver. The current bathymetric system had both the RTK GNSS receiver and the IMU installed for experimentation, but the position and attitude data would ultimately be processed onboard to correct the sonar and broadcast the result live to the ground station.

Upgrading the SBES sonar sensor could improve the swath coverage and reduce interpolation between transects. Additionally, underwater geometric features can be submerged in the pond to assess the resolution of locating and identifying structures of interest.

Currently, the sonar imagery generated from side-scan and down-scan is only used for qualitative assessment and cursory characterization of the marine ecology and environment, obstacles, and infrastructure (Figure 10). Ground-truthing the bottom hardness measures will provide more confidence in quantifying the backscatter data. Generating point clouds for three-dimensional reconstruction of the sonar data will assist in bottom visualization. The sonar imagery can also be used for object identification and avoidance or to inform further exploration of areas of interest (AOIs). Integrating communication between the drone and the vessel can allow the vessel to command the drone to explore a new AOI.

6. Conclusions

The uncrewed Bathy-drone is a novel configuration for rapid bathymetry and bottom characterization. The tethered vessel is towed autonomously on a preprogrammed mission by a multirotor drone. The surface vessel has a fixed recreational COTS sonar with down-scan, side-scan, and chirp capabilities gathering sonar imagery, bathymetry, and bottom hardness data. The raw field data are stored on board using a microSD card and transmitted live via telemetry link. This paper discussed state-of-the-art approaches in small, uncrewed bathymetry, the Bathy-drone design, and ground-truthing of the system against traditional RTK GNSS. An assessment of the accuracy and precision of the system based on the ground-truthing results provides insight toward future ancillary sensors. Field operation of the Bathy-drone system highlighted novel advantages unique to the tethered configuration. Surveys are based on land as the multirotor airlifts the vessel to the first waypoint on water allowing for rapid deployment and reduced crew fatigue. Boat docks are unnecessary as the Bathy-drone can fly over obstacles such as mud flats, sandbars, tree lines, and fences. Transport of the system and ground station to the area of interest via crewed boat or vehicle is convenient due to the low weight and volume. The propulsion is provided by the multirotor drone allowing the vessel to glide over swift current, debris, floating vegetation, and the ground. Speeds of 0–24 km/h (0–15 mph) were tested, and an area of more than 40,000 m^2 (10 acres) was surveyed with one battery charge in less than 25 min. Field testing will continue to inform new applications and iterations. Implementation of the future work discussed will continue to expand the capabilities and applications of the system.

Author Contributions: Conceptualization of the multirotor tethered sonar configuration, A.E.O. and P.G.I.; methodology of ground-truthing experiments, A.L.D., A.E.O., P.G.I. and B.E.W. software developed and implemented for data processing, A.L.D., A.E.O., A.P. and J.S.; validation, A.L.D., A.E.O. and A.P.; formal analysis, A.L.D., A.E.O. and A.P.; investigation, A.L.D., A.E.O., A.P., H.T., O.C., M.N. and N.E.C.; resources, P.G.I., B.E.W. and R.R.C.; data curation, A.L.D., A.E.O. and A.P.; writing—original draft preparation, A.L.D., A.E.O., P.G.I. and A.P.; writing—review and editing, A.L.D., A.E.O., A.P., B.E.W. and P.G.I.; visualization, A.L.D., A.E.O. and A.P.; supervision, P.G.I., J.S., B.E.W. and R.R.C.; project administration, P.G.I., J.S., B.E.W. and R.R.C.; funding acquisition, P.G.I. and R.R.C. All authors read and agreed to the published version of the manuscript.

Funding: This work was internally funded by the University of Florida.

References

1. Ashphaq, M.; Srivastava, P.K.; Mitra, D. Review of Near-Shore Satellite Derived Bathymetry: Classification and Account of Five Decades of Coastal Bathymetry Research. *J. Ocean. Eng. Sci.* **2021**, *6*, 340–359. [CrossRef]
2. Lee, Z.P.; Casey, B.; Parsons, R.; Goode, W.; Weidemann, A.; Arnone, R. Bathymetry of Shallow Coastal Regions Derived from Space-Borne Hyperspectral Sensor. In Proceedings of the OCEANS 2005 MTS/IEEE, Washington, DC, USA, 17–23 September 2005; Volume 3, pp. 2160–2170.

3. Wölfl, A.-C.; Snaith, H.; Amirebrahimi, S.; Devey, C.W.; Dorschel, B.; Ferrini, V.; Huvenne, V.A.I.; Jakobsson, M.; Jencks, J.; Johnston, G.; et al. Seafloor Mapping–The Challenge of a Truly Global Ocean Bathymetry. *Front. Mar. Sci.* **2019**, *6*, 283. [CrossRef]

4. Kutser, T.; Hedley, J.; Giardino, C.; Roelfsema, C.; Brando, V.E. Remote Sensing of Shallow Waters–A 50 Year Retrospective and Future Directions. *Remote Sens. Environ.* **2020**, *240*, 111619. [CrossRef]

5. Manley, J.E. Unmanned Surface Vehicles, 15 Years of Development. In Proceedings of the OCEANS 2008, Quebec, QC, Canada, 15–18 September 2008; pp. 1–4. [CrossRef]

6. Rowley, J. Autonomous Unmanned Surface Vehicles (USV): A Paradigm Shift for Harbor Security and Underwater Bathymetric Imaging. In Proceedings of the OCEANS 2018 MTS/IEEE Charleston, Charleston, SC, USA, 22–25 October 2018. [CrossRef]

7. Mariani, P.; Quincoces, I.; Haugholt, K.; Chardard, Y.; Visser, A.; Yates, C.; Piccinno, G.; Reali, G.; Risholm, P.; Thielemann, J. Range-Gated Imaging System for Underwater Monitoring in Ocean Environment. *Sustainability* **2018**, *11*, 162. [CrossRef]

8. Irish, J.L.; White, T.E. Coastal Engineering Applications of High-Resolution Lidar Bathymetry. *Coast. Eng.* **1998**, *35*, 47–71. [CrossRef]

9. Costa, B.M.; Battista, T.A.; Pittman, S.J. Comparative Evaluation of Airborne LiDAR and Ship-Based Multibeam SoNAR Bathymetry and Intensity for Mapping Coral Reef Ecosystems. *Remote Sens. Environ.* **2009**, *113*, 1082–1100. [CrossRef]

10. Sandidge, J.C.; Holyer, R.J. Coastal Bathymetry from Hyperspectral Observations of Water Radiance. *Remote Sens. Environ.* **1998**, *65*, 341–352. [CrossRef]

11. Jérôme, L.; Gentile, V.; Demarchi, L.; Spitoni, M.; Piégay, H.; Mróz, M. Bathymetric Mapping of Shallow Rivers with UAV Hyperspectral Data. In Proceedings of the 5th International Conference on Telecommunications and Remote Sensing, Milan, Italy, 10 October 2016; pp. 43–49.

12. Bergsma, E.W.J.; Almar, R.; Rolland, A.; Binet, R.; Brodie, K.L.; Bak, A.S. Coastal Morphology from Space: A Showcase of Monitoring the Topography-Bathymetry Continuum. *Remote Sens. Environ.* **2021**, *261*, 112469. [CrossRef]

13. Kan, H.; Katagiri, C.; Nakanishi, Y.; Yoshizaki, S.; Nagao, M.; Ono, R. Assessment and Significance of a World War II Battle Site: Recording the USS Emmons Using a High-Resolution DEM Combining Multibeam Bathymetry and SfM Photogrammetry. *Int. J. Naut. Archaeol.* **2018**, *47*, 267–280. [CrossRef]

14. Hatcher, G.A.; Warrick, J.A.; Ritchie, A.C.; Dailey, E.T.; Zawada, D.G.; Kranenburg, C.; Yates, K.K. Accurate Bathymetric Maps From Underwater Digital Imagery Without Ground Control. *Front. Mar. Sci.* **2020**, *7*, 525. [CrossRef]

15. Raber, G.T.; Schill, S.R. A Low-Cost Small Unmanned Surface Vehicle (SUSV) for Very High-Resolution Mapping and Monitoring of Shallow Marine Habitats. In Proceedings of the Remote Sensing of the Ocean, Sea Ice, Coastal Waters, and Large Water Regions 2019, Strasbourg, France, 14 October 2019; Bostater, C.R., Neyt, X., Viallefont-Robinet, F., Eds.; SPIE: Washington, DC, USA, 2019; p. 3.

16. Brodie, K.L.; Bruder, B.L.; Slocum, R.K.; Spore, N.J. Simultaneous Mapping of Coastal Topography and Bathymetry From a Lightweight Multicamera UAS. *IEEE Trans. Geosci. Remote Sens.* **2019**, *57*, 6844–6864. [CrossRef]

17. Alvarez, L.; Moreno, H.; Segales, A.; Pham, T.; Pillar-Little, E.; Chilson, P. Merging Unmanned Aerial Systems (UAS) Imagery and Echo Soundings with an Adaptive Sampling Technique for Bathymetric Surveys. *Remote Sens.* **2018**, *10*, 1362. [CrossRef]

18. Bandini, F.; Sunding, T.P.; Linde, J.; Smith, O.; Jensen, I.K.; Köppl, C.J.; Butts, M.; Bauer-Gottwein, P. Unmanned Aerial System (UAS) Observations of Water Surface Elevation in a Small Stream: Comparison of Radar Altimetry, LIDAR and Photogrammetry Techniques. *Remote Sens. Environ.* **2020**, *237*, 111487. [CrossRef]

19. Lejot, J.; Delacourt, C.; Piégay, H.; Fournier, T.; Trémélo, M.-L.; Allemand, P. Very High Spatial Resolution Imagery for Channel Bathymetry and Topography from an Unmanned Mapping Controlled Platform. *Earth Surf. Processes Landf.* **2007**, *32*, 1705–1725. [CrossRef]

20. Borrelli, M.; Smith, T.L.; Mague, S.T. Vessel-Based, Shallow Water Mapping with a Phase-Measuring Sidescan Sonar. *Estuaries Coasts* **2021**, *45*, 961–979. [CrossRef]

21. Campbell, K.E.J.; Ruffell, A.; Pringle, J.; Hughes, D.; Taylor, S.; Devlin, B. Bridge Foundation River Scour and Infill Characterisation Using Water-Penetrating Radar. *Remote Sens.* **2021**, *13*, 2542. [CrossRef]

22. Bandini, F.; Kooij, L.; Mortensen, B.K.; Caspersen, M.B.; Olesen, D.; Bauer-Gottwein, P. *Mapping Inland Water Bathymetry with Ground Penetrating Radar (GPR) on Board Unmanned Aerial Systems (UASs)*; Research Square: Durham, NC, USA, 2021; In Review. [CrossRef]

23. Ruffell, A.; King, L. Water Penetrating Radar (WPR) in Archaeology: A Crannog Case Study. *J. Archaeol. Sci. Rep.* **2022**, *41*, 103300. [CrossRef]

24. Ruffell, A.; Parker, R. Water Penetrating Radar. *J. Hydrol.* **2021**, *597*, 126300. [CrossRef]

25. Kasvi, E.; Salmela, J.; Lotsari, E.; Kumpula, T.; Lane, S.N. Comparison of Remote Sensing Based Approaches for Mapping Bathymetry of Shallow, Clear Water Rivers. *Geomorphology* **2019**, *333*, 180–197. [CrossRef]

26. Jouve, G.; Caudron, C.; Matte, G. Gas Detection and Quantification Using IXblue Echoes High-Resolution Sub-Bottom Profiler and Seapix 3D Multibeam Echosounder from the Laacher See (Eifel, Germany). In Proceedings of the OCEANS 2021, San Diego–Porto, CA, USA, 20–23 September 2021; IEEE: San Diego, CA, USA, 2021; pp. 1–4.

27. IInternational Hydrographic Organization. *IHO Standards for Hydrographic Surveys, Edition 6.0.0, September 2020*; International Hydrographic Organization: Monaco, 2020; 41p.

28. *Hydrographic Surveying ENGINEER MANUAL*; US Army Corps of Engineers: Washington, DC, USA, 2013.

29. Almar, R.; Bergsma, E.W.J.; Maisongrande, P.; de Almeida, L.P.M. Wave-Derived Coastal Bathymetry from Satellite Video Imagery: A Showcase with Pleiades Persistent Mode. *Remote Sens. Environ.* **2019**, *231*, 111263. [CrossRef]

30. Chirayath, V.; Earle, S.A. Drones That See through Waves–Preliminary Results from Airborne Fluid Lensing for Centimetre-Scale Aquatic Conservation. *Aquat. Conserv. Mar. Freshw. Ecosyst.* **2016**, *26*, 237–250. [CrossRef]

31. Mandlburger, G.; Pfennigbauer, M.; Wieser, M.; Riegl, U.; Pfeifer, N. Evaluation of a Novel Uav-Borne Topo-Bathymetric Laser Profiler. *ISPRS-Int. Arch. Photogramm. Remote Sens. Spat. Inf. Sci.* **2016**, *XLI-B1*, 933–939. [CrossRef]

32. Xu, W.; Guo, K.; Liu, Y.; Tian, Z.; Tang, Q.; Dong, Z.; Li, J. Refraction Error Correction of Airborne LiDAR Bathymetry Data Considering Sea Surface Waves. *Int. J. Appl. Earth Obs. Geoinf.* **2021**, *102*, 102402. [CrossRef]

33. Mahacek, P.; Berk, T.; Casanova, A.; Kitts, C.; Kirkwood, W.; Wheat, G. Development and Initial Testing of a SWATH Boat for Shallow-Water Bathymetry. In Proceedings of the OCEANS 2008, Quebec City, QC, Canada, 15–18 September 2008; IEEE: Quebec City, QC, Canada; pp. 1–6.

34. Beck, E.; Kirkwood, W.; Caress, D.; Berk, T.; Mahacek, P.; Brashem, K.; Acain, J.; Reddy, V.; Kitts, D.C.; Skutnik, J. SeaWASP: A Small Waterplane Area Twin Hull Autonomous Platform for Shallow Water Mapping. *Mar. Technol. Soc. J.* **2009**, *43*, 6–12.

35. Ferreira, H.; Almeida, C.; Martins, A.; Almeida, J.; Dias, N.; Dias, A.; Silva, E. Autonomous Bathymetry for Risk Assessment with ROAZ Robotic Surface Vehicle. In Proceedings of the OCEANS 2009-EUROPE (OCEANS), Bremen, Germany, 11–14 May 2009; IEEE: Bremen, Germany, 2009; pp. 1–6.

36. Wajs, J.; Kasza, D. Development of Low-Cost Unmanned Surface Vehicle System for Bathymetric Measurements. *IOP Conf. Ser. Earth Environ. Sci.* **2021**, *684*, 012033. [CrossRef]

37. Seto, M.L.; Crawford, A. Autonomous Shallow Water Bathymetric Measurements for Environmental Assessment and Safe Navigation Using USVs. 5. In Proceedings of the OCEANS 2015-MTS/IEEE, Washington, DC, USA, 19–22 October 2015; IEEE: Washington, DC, USA, 2015.

38. Specht, M.; Specht, C.; Szafran, M.; Makar, A.; Dąbrowski, P.; Lasota, H.; Cywiński, P. The Use of USV to Develop Navigational and Bathymetric Charts of Yacht Ports on the Example of National Sailing Centre in Gdańsk. *Remote Sens.* **2020**, *12*, 2585. [CrossRef]

39. Kimball, P.; Bailey, J.; Das, S.; Geyer, R.; Harrison, T.; Kunz, C.; Manganini, K.; Mankoff, K.; Samuelson, K.; Sayre-McCord, T.; et al. The WHOI Jetyak: An Autonomous Surface Vehicle for Oceanographic Research in Shallow or Dangerous Waters. In Proceedings of the 2014 IEEE/OES Autonomous Underwater Vehicles (AUV), Oxford, MS, USA, 6–9 October 2014; IEEE: Oxford, MS, USA, 2014; pp. 1–7.

40. SR-Surveyor Class | Autonomous Surface Vehicles | SeaRobotics. Available online: https://www.searobotics.com/products/autonomous-surface-vehicles/sr-surveyor-class (accessed on 25 May 2022).

41. Specht, M.; Specht, C.; Lasota, H.; Cywiński, P. Assessment of the Steering Precision of a Hydrographic Unmanned Surface Vessel (USV) along Sounding Profiles Using a Low-Cost Multi-Global Navigation Satellite System (GNSS) Receiver Supported Autopilot. *Sensors* **2019**, *19*, 3939. [CrossRef]

42. Specht, M.; Stateczny, A.; Specht, C.; Widźgowski, S.; Lewicka, O.; Wiśniewska, M. Concept of an Innovative Autonomous Unmanned System for Bathymetric Monitoring of Shallow Waterbodies (INNOBAT System). *Energies* **2021**, *14*, 5370. [CrossRef]

43. Bandini, F. Hydraulics and Drones: Observations of Water Level, Bathymetry and Water Surface Velocity from Unmanned Aerial Vehicles. 79. Ph.D. Thesis, Technical Univerity of Denmark, Lyngby, Denmark, 2017.

44. Yoash, R.; Atkinson, M.P.; Kress, M. Where to Dip? Search Pattern for an Antisubmarine Helicopter Using a Dipping Sensor. *Mil. Oper. Res.* **2022**, *23*, 3939.

45. Sasano, M.; Imasato, M.; Yamano, H.; Oguma, H. Development of a Regional Coral Observation Method by a Fluorescence Imaging LIDAR Installed in a Towable Buoy. *Remote Sens.* **2016**, *8*, 48. [CrossRef]

46. Bandini, F.; Olesen, D.; Jakobsen, J.; Kittel, C.M.M.; Wang, S.; Garcia, M.; Bauer-Gottwein, P. Technical Note: Bathymetry Observations of Inland Water Bodies Using a Tethered Single-Beam Sonar Controlled by an Unmanned Aerial Vehicle. *Hydrol. Earth Syst. Sci.* **2018**, *22*, 4165–4181. [CrossRef]

47. Ruffell, A.; Lally, A.; Rocke, B. Dronar—Geoforensic Search Sonar from a Drone. *Forensic Sci.* **2021**, *1*, 202–212. [CrossRef]

48. Vélez-Nicolás, M.; García-López, S.; Barbero, L.; Ruiz-Ortiz, V.; Sánchez-Bellón, Á. Applications of Unmanned Aerial Systems (UASs) in Hydrology: A Review. *Remote Sens.* **2021**, *13*, 1359. [CrossRef]

49. Matrice 600 Pro-Product Information-DJI. Available online: https://www.dji.com/matrice600-pro/info#specs (accessed on 25 May 2022).

50. Alta X Specs-Freefly Systems. Available online: https://freeflysystems.com/alta-x/specs (accessed on 25 May 2022).

51. Metcalfe, B.; Thomas, B.; Treloar, A.; Rymansaib, Z.; Hunter, A.; Wilson, P. A Compact, Low-Cost Unmanned Surface Vehicle for Shallow Inshore Applications. In Proceedings of the 2017 Intelligent Systems Conference (IntelliSys), London, UK, 7–8 September 2017; pp. 961–968.

52. OPUS: The Online Positioning User Service, Process Your GNSS Data in the National Spatial Reference System. Available online: https://geodesy.noaa.gov/OPUS/ (accessed on 25 May 2022).

53. Yamasaki, S.; Tabusa, T.; Iwasaki, S.; Hiramatsu, M. Acoustic Water Bottom Investigation with a Remotely Operated Watercraft Survey System. *Prog. Earth Planet. Sci.* **2017**, *4*, 25. [CrossRef]

54. Elite-7 Ti2 US Inland, Active Imaging 3-in-1 | Lowrance USA. Available online: https://www.lowrance.com/lowrance/type/fishfinders-chartplotters/elite-7-ti2-us-inland-ai-3-in-1/#prl_specifications (accessed on 25 May 2022).
55. Halmai, Á.; Gradwohl-Valkay, A.; Czigány, S.; Ficsor, J.; Liptay, Z.Á.; Kiss, K.; Lóczy, D.; Pirkhoffer, E. Applicability of a Recreational-Grade Interferometric Sonar for the Bathymetric Survey and Monitoring of the Drava River. *ISPRS Int. J. Geo-Inf.* **2020**, *9*, 149. [CrossRef]

 *drones*

Article

Accuracy Assessment of Direct Georeferencing for Photogrammetric Applications Based on UAS-GNSS for High Andean Urban Environments

Rolando Salas López [1], Renzo E. Terrones Murga [1,2], Jhonsy O. Silva-López [1,*], Nilton B. Rojas-Briceño [1,3], Darwin Gómez Fernández [1], Manuel Oliva-Cruz [1] and Yuri Taddia [4]

[1] Instituto de Investigación para el Desarrollo Sustentable de Ceja de Selva, Universidad Nacional Toribio Rodríguez de Mendoza, Chachapoyas 01001, Peru

[2] Lidar Peru Sociedad Anónima Cerrada | Lidar Peru, Av. Alfredo Benavides Nro. 1944, Lima 15047, Peru

[3] Instituto de Investigación en Ingeniería Ambiental, Facultad de Ingeniería Civil y Ambiental, Universidad Nacional Toribio Rodríguez de Mendoza de Amazonas, Chachapoyas 01001, Peru

[4] Engineering Department, University of Ferrara, Via Saragat 1, 44122 Ferrara, Italy

* Correspondence: jhonsy.silva@untrm.edu.pe; Tel.: +51-998-769-936

Citation: Salas López, R.; Terrones Murga, R.E.; Silva-López, J.O.; Rojas-Briceño, N.B.; Gómez Fernández, D.; Oliva-Cruz, M.; Taddia, Y. Accuracy Assessment of Direct Georeferencing for Photogrammetric Applications Based on UAS-GNSS for High Andean Urban Environments. *Drones* **2022**, *6*, 388. https://doi.org/10.3390/drones6120388

Academic Editors: Arianna Pesci, Giordano Teza and Massimo Fabris

Received: 9 October 2022
Accepted: 28 November 2022
Published: 30 November 2022

Publisher's Note: MDPI stays neutral with regard to jurisdictional claims in published maps and institutional affiliations.

Abstract: Unmanned Aircraft Systems (UAS) are used in a variety of applications with the aim of mapping detailed surfaces from the air. Despite the high level of map automation achieved today, there are still challenges in the accuracy of georeferencing that can limit both the speed and the efficiency in mapping urban areas. However, the integration of topographic grade Global Navigation Satellite System (GNSS) receivers on UAS has improved this phase, leading to a reach of up to a centimeter-level accuracy. It is therefore necessary to adopt direct georeferencing (DG), real-time kinematic positioning (RTK)/post-processed kinematic (PPK) approaches in order to largely automate the photogrammetric flow. This work analyses the positional accuracy using Ground Control Points (GCP) and the repeatability and reproducibility of photogrammetric products (Digital Surface Model and ortho-mosaic) of a commercial multi-rotor system equipped with a GNSS receiver in an urban environment with a DG approach. It was demonstrated that DG is a viable solution for mapping urban areas. Indeed, PPK with at least 1 GCP considerably improves the RMSE (x: 0.039 m, y: 0.012 m, and z: 0.034 m), allowing for a reliable 1:500 scale urban mapping in less time when compared to conventional topographic surveys.

Keywords: UAS; SfM; RTK; PPK; direct georeferencing

1. Introduction

Originally, Unmanned Aircraft Systems (UAS) were developed mainly for military purposes and applications such as unmanned inspection, surveillance, reconnaissance, and mapping of hostile areas [1]. The first civil UAS experience in geomatics took place in the 1990s. Nowadays, UAS represent a common platform for photogrammetric data acquisition [2]. Accordingly, UAS is an alternative for mapping and to perform a spatial analysis of the territory [3,4]. This data collection technology offers similar quality and reliability results when compared to conventional topography, which requires higher costs for its application instead [5]. Therefore, the use of UAS is more economically profitable than other technologies such as the use of Global Navigation Satellite System (GNSS) receivers [6], Terrestrial Laser Scanning (TLS) [7], Light Detection and Ranging (LiDAR) [8] and conventional aerial photogrammetry [9].

Images acquired by UAS offer useful information for archaeological documentation [10], geological and geomorphological surveys and monitoring [11–13], urban modeling [14–17], emergency assessment [18], and engineering applications [19,20]. Based on the use of the structure-from-motion (SfM) algorithm [21], which could replace conventional surveying methods, centimeter accuracies can be obtained according to the scale of study [22,23]. In order to achieve such a level of accuracy in the mapping of urban areas using UAS, a high overlap between captured images is required [24,25]. Afterwards, detailed photogrammetric models (Point Cloud, Digital Elevation Model, Orthophoto) are obtained by indirect or direct georeferencing [26].

Indirect georeferencing (IG) of these models is based on extensive field deployment, survey deployment, and ground control point (GCP) reconnaissance. The latter are subsequently recognized on the images within the photogrammetric software in the lab/office [27–29]. On the other hand, direct georeferencing (DG) replaces GCPs with aerial control by GNSS-assisted Bundle Block Adjustment (BBA) using Automatic Aerial Triangulation (AAT), which is one of the most widely used techniques in aerial photogrammetry [30,31]. In AAT, the coordinates of the camera stations are measured by an integrated GNSS receiver at each exposure time and are introduced as constraints in the BBA adjustment [32,33], or they are used to calculate the transformation of the arbitrary SfM reference to the mapping system accurately [28].

However, the accuracy obtained is influenced by the type of data processing, and there are three common configurations for geotagging photographs in an SfM workflow. (i) On-board position calculation: photo positions are based on the location of the UAS and recorded as an Exchangeable Image File (Exif) data; (ii) Post Processed Kinematic (PPK): photo positions are calculated after flight based on records from the mobile (UAS) and a GNSS base station; and (iii) Real Time Kinematic (RTK): photo positions are calculated in real-time, with corrections sent to the mobile directly from the base station [34]. The base station can be local or in special cases, a correction can be sent from a co-commercial base station via Networked Transport of RTCM via Internet Protocol (NTRIP protocol) to the remote controller [22,35]. In this context, guaranteeing the recording and storing of coordinates in a known place for the PPK [36] and RTK [26] configuration depends largely on the location of the base station and mobile, which must share an appropriate geometrical configuration of satellites and carrier phase [37].

In DG, no GCPs collected with external GNSS are required to process aerial photographs, thus the workflow becomes less time-consuming and costly. Nevertheless, there is a need to mitigate UAS camera positioning errors for complex geometry scenarios, such as urban environments or linear mapping (urbanized corridors, runways, etc.) [38]. This can be done using a GNSS navigation/quality inertial navigation system (INS), thus obtaining a high accuracy, but at an increased cost for UAS [23,39]. In recent years, variations of such techniques have been developed to address the specific peculiarities of UAS-based aerial photogrammetry, including relative position/orientation control [2,40], raw observation, and multi-sensor adjustment [41]. However, the use of GCPs is important to establish adequate topographic control [41–43].

In that context, the Phantom 4 RTK UAS (DJI-P4RTK), having an integrated multi-frequency GNSS receiver, allows for the performance of either a PPK, RTK, or NTRIP DG approach [44,45]. In addition, P4RTK contains a non-metric camera, with limitations on identifying appropriate Interior Orientation Parameters (IOP) along with the package setting. However, this could be overcome by using at least one GCP or one chamber pre-calibration [37,42,46]. Furthermore, having a GNSS receiver allows for storing raw data that can be used for further evaluations. Accordingly, the UAS allows one to perform an indirect orientation using GCP, as well as the direct orientation of the measured image positions (external orientation—EO). Due to this, in this study, the combination of both (GCP and EO), known as integrated orientation was used, which has shown better accuracies [45].

Integrated guidance for UAS has been used both in coastal mapping [47] and in rural and urban mapping [48]. There is no current knowledge about the variants of the data capture mentioned in PPK and RTK [49], particularly for urban environments with complex physiography such as the city of Chachapoyas (narrow access roads, typical of colonial-era cities in Peru). Therefore, this research aims to analyze the accuracy of the photogrammetric models (DSM and ortho-mosaic) based on the accuracy standards set by the American Society for Photogrammetry and Remote Sensing (ASPRS) [50–52]. For this purpose, three types of flights are analyzed using the RTK and PPK approach acquired by the UAS, which are combined with a set of GCPs (1, 5, and 10) to create 27 photogrammetric projects. This study identifies the most suitable UAS data acquisition and processing method for mapping an urban area. It also has the potential for replicability in similar areas with rugged physiography and multiple colonial buildings, or other similar scenarios.

2. Materials and Methods

2.1. Study Area

The city of Chachapoyas (6°13′45.84″ S; 77°52′20.47″ W), which is the capital of the department of Amazonas, is located in the north-eastern Peruvian Andes at 2483 m above mean sea level (m.a.s.l.). The city is 1200 km far away from Lima, the capital of Peru (Figure 1). The climate varies from temperate to moderately rainy, with a cumulative annual rainfall of 777.8 mm and maximum and minimum temperatures (period 1960–1991) of 19.8 °C and 9.2 °C, respectively. It has a marked climatic seasonality with an alternating rainy season from November to April, and a dry season from May to October [53,54]. The urban area of the city of Chachapoyas covers 2239.35 ha, of which 15 ha belonging to the La Laguna neighborhood was mapped with six cadastral blocks, consisting of multiple colonial buildings and narrow, steeply sloping streets [54].

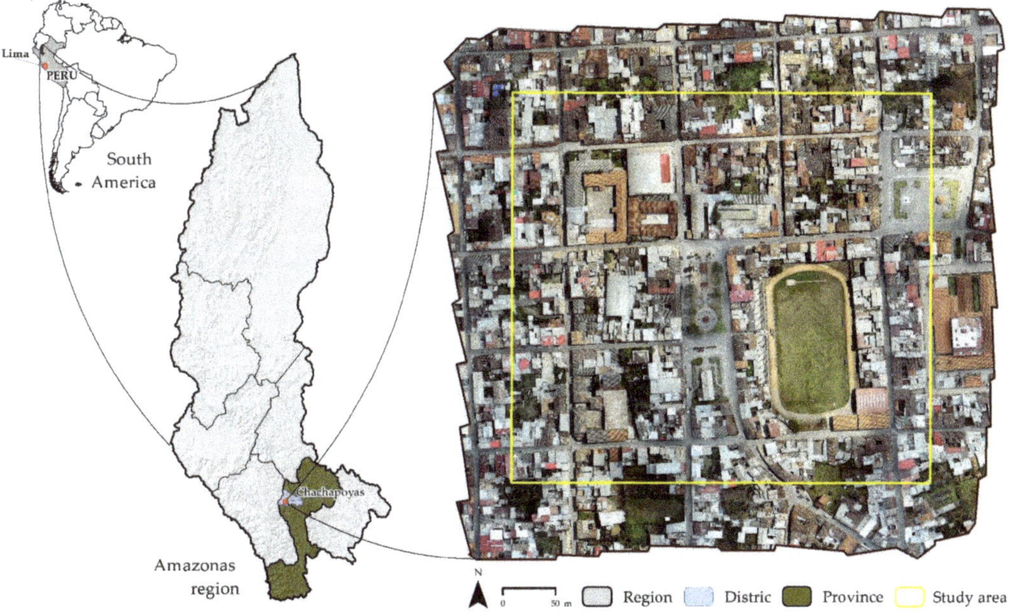

Figure 1. Location Map of the Surveyed Urban Area.

2.2. Data Acquisition

2.2.1. GNSS Survey

In this work, 10 GCPs and 4 Validation Points (VPs) were established. The GCPs and VPs were represented as square-shaped marks of 0.533 m in size (Figure 2a). The three-dimensional coordinates of these targets were collected with a Trimble R10 GNSS receiver operated in PPK mode (Figure 2b). The base station was established at a geodetic point of order C [55] with a record of 7 min and was calculated by a static survey, previously corrected with a geodetic pillar (AMA01) of the Peruvian National Geographic Institute (IGN) [55]. The metric coordinates of BM-01 (Bench Mark) were 18,2601.173 m E, 9,310,378.217 m N, and 2357.706 m.a.s.l. (Figure 2c). These horizontal coordinates refer to the Universal Transverse Mercator (UTM) Zone 18 South coordinate system, and the elevation refers to the MSL using the EGM08 geoid model. For PPK mode GCP measurements, these multi-frequency GNSS geodetic instruments have an accuracy specification set by the manufacturer of ± 3 mm + 0.5 ppm horizontal RMS and ± 5 mm + 1 ppm vertical RMS [56]. In particular, since the distance between the base station and the study area was approximately 260 m, the horizontal and vertical errors were 1 and 2 cm, respectively.

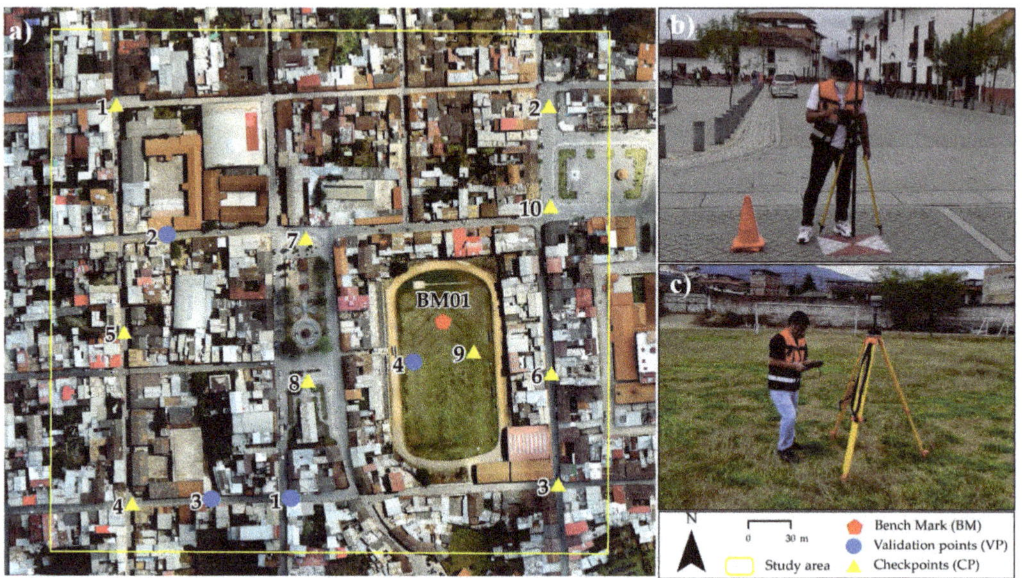

Figure 2. Photogrammetric Data Acquisition Process. (**a**) GCP distribution. (**b**) Target measurement. (**c**) Base station BM-01.

2.2.2. UAS Planning

Three types of flights were performed: grid (2D), double grid (3D), and terrain tracking, planned to use the GSRTK mobile application [45]. In addition, the altitude optimization option was used. This enabled the collection of oblique images ($-45°$) from the lower left corner to the center of the study area, except for the terrain-following flight type [45]. This "hybrid" data collection is useful for simulating the results of image acquisition in general [44]. These flights covered an approximate area of 15 ha with a Ground Sampling Distance (GSD) of 3.84 cm. A nadir position of the camera was established, except for the 3D flight ($-60°$) taking into account north-south flight lines, as well as an overlap of 80% longitudinally and 75% laterally, and an average flight height of 140 m. Each type of flight was executed using three positioning solutions (Table 1). This resulted in the execution of six photogrammetric flights over the study area. For RTK, three configurations were considered, in which CP (1, 3, and 5) were used for the type of 2D flight, 3D flight and ground tracking. Similarly, the same was applied to PPK1 and PPK2, resulting in a total of 27 photogrammetric projects.

Table 1. DJI Phantom 4 RTK UAS Flight and Positioning Configurations.

Flight Configuration ID [1]	Positioning Solution	Photogrammetric Projects According to N° CP		
		1	**3**	**5**
A	RTK (Refined position due to corrections sent by a GNSS base station in the field, the D-RTK 2 receiver placed at a point of known coordinates)	RTK_1, RTK_4, RTK_7.	RTK_2, RTK_5, RTK_8	RTK_3, RTK_6, RTK_9,
B	PPK1 (Refined position due to post-process corrections by a GNSS base station in cabinet, Trimble R10 receiver placed at a point of known coordinates)	PPK1_1, PPK1_4, PPK1_7.	PPK1_2, PPK1_5, PPK1_8.	PPK1_3, PPK1_6, PPK1_9.
	PPK2 (Refined position due to post-process corrections by a GNSS base station in a cabinet, the commercial receiver AMA01 established by IGN)	PPK2_1, PPK2_4, PPK2_7.	PPK2_2, PPK2_5, PPK2_8.	PPK2_3, PPK2_3, PPK2_9.

[1] Each configuration was run for the three types of flights: grid, double grid, and terrain tracking.

2.3. Positioning Configurations Adopted during Flight Tests

2.3.1. UAS-RTK Surveying

An RTK approach was performed (Figure 3a), with the specifications given by Teppati et al. [49]. The DJI GNSS base station DRTK2 [57] was used as a positioning solution, which transmits real-time corrections to improve the positioning accuracy of the UAS located at any point with known or unknown coordinates [34]. For this work, we adopted the solution that requires the use of a point of known coordinates (BM-01): the GNSS base station was placed in the central part of the study area (Figure 2c), enabling reconstruction of the orientation and the displacement of the camera concerning the APC (Antenna Phase Center) for the PPK approach.

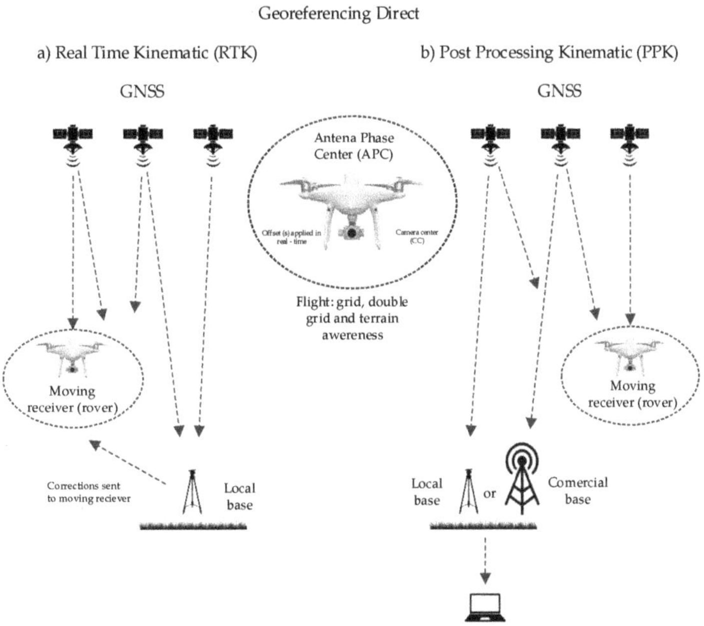

Figure 3. Direct Georeferencing. (**a**) PPK Approach and (**b**) RTK Approach of the UAS DJI-P4RTK, Adapted from [34,47].

2.3.2. UAS-PPK Surveying

For the PPK approach (Figure 3b), which is the correction of the UAS positions recorded by the GNSS receiver during flight and the estimation of the camera positions after the data acquisition phase, a post-process is required. Here, the Trimble R10 GNSS receiver with a known coordinate (BM-01) storing data in static mode at one second in the PPK1 configuration (Table 1) was used as a base. Similarly, the data from the AMA01 geodetic pillar (recorded every 5 s) was used in the PPK2 configuration. Consequently, in order to process the data, it is necessary to use either a self-built solution [47] or free third-party software [58], or alternatively commercial software [59]. For this work, PPK processing was performed using RED toolbox, which performed geotagging with fixed solutions for all images. RED toolbox, for P4RTK and other DJI products, automatically calculates the lever arm from the information stored in the UAS files, compared to the free RTKLIB software used by Teppati et al. [49].

2.4. Photogrammetric Processing of the Acquired Data

The photogrammetric processing flow (Figure 4) consists of matching waypoints from the UAS images. These images were automatically calibrated based on EXIF information for tie point generation. The calibrated images were georeferenced using an integrated orientation to generate a dense 3D point cloud, and to subsequently obtain the DSM and ortho-mosaic. Image georeferencing is the first step in obtaining photogrammetric products (ortho-mosaics, DSM, DTM, and contour lines) and represents the prerequisite for their metric exploitation [60].

2.5. Comparative Analysis

2.5.1. Accuracy Assessment

In order to perform the assessment of both horizontal (ortho-mosaic) and vertical (DSM) accuracy, the ASPRS guide [49] was used as a reference. For the altitude coordinate (Z), the location of the vertical control points should be surveyed on flat open terrain or with uniform slopes and slopes $\leq 10\%$. According to the guide, the vertical non-vegetation accuracy (NVA) at the 95% confidence level in unvegetated terrain is approximated by multiplying the vertical accuracy class accuracy values (or RMSEz) by 1.96.

The first method to assess accuracy is the root mean square of the quadratic differences between the reconstructed model and the surveyed GCP coordinates, known as the root mean square error (RMSE) [61]. The horizontal (x and y) and vertical (z) positional accuracies were determined for the photogrammetric projects from the GCP coordinates, specifically with the validation points (VPs) within Agisoft Metashape [60].

The RMSE values for the X, Y, Z, and r components are estimated as shown in Equations (1)–(5). Where the sub-indices oi and GNSSi of X, Y, and Z represent the coordinates estimated by the ortho-mosaic and measured by the GNSS receiver, respectively.

$$\mathrm{RMSE}_{\mathrm{X}} = \pm \sqrt{\frac{\sum_{i=1}^{n}\left(X_{\mathrm{oi}} - X_{\mathrm{GNSSi}}\right)^2}{n}} \tag{1}$$

$$\mathrm{RMSE}_{\mathrm{Y}} = \pm \sqrt{\frac{\sum_{i=1}^{n}\left(Y_{\mathrm{oi}} - Y_{\mathrm{GNSSi}}\right)^2}{n}} \tag{2}$$

$$RMSE_{\mathrm{Z}} = \pm \sqrt{\frac{\sum_{i=1}^{n}\left(Z_{\mathrm{oi}} - Z_{\mathrm{GNSSi}}\right)^2}{n}} \tag{3}$$

$$RMSE_{\mathrm{XY}} = \pm \sqrt{\mathrm{RMSE}_{\mathrm{X}}^2 + \mathrm{RMSE}_{\mathrm{Y}}^2} \tag{4}$$

$$RMSE_{r} = \pm \sqrt{\mathrm{RMSE}_{\mathrm{X}}^2 + \mathrm{RMSE}_{\mathrm{Y}}^2 + \mathrm{RMSE}_{\mathrm{z}}^2} \tag{5}$$

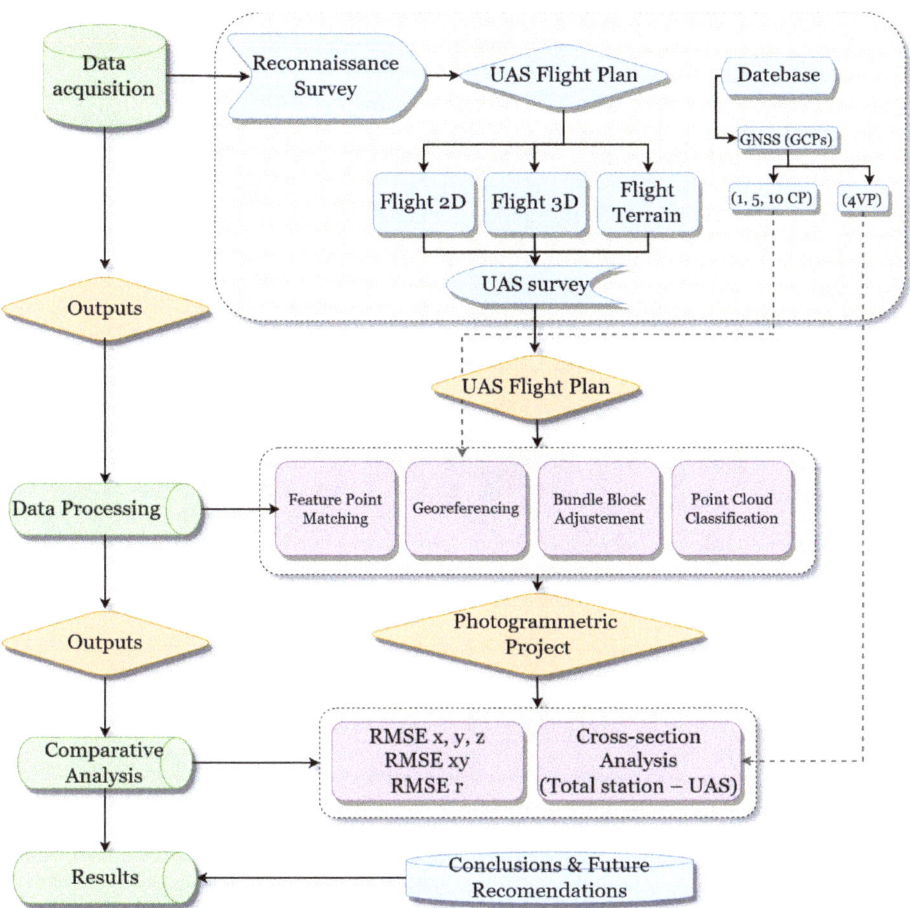

Figure 4. Methodological Flowchart of UAS Photogrammetric Processing for the Evaluation of Direct Georeferencing Accuracy for UAS-GNSS-based Photogrammetric Applications for Urban Environments.

2.5.2. Cross-Section Analysis

In urban mapping, assessing height differences through cross-sectional profiles is an important issue [62,63]. For this reason, a profile was extracted from the DSMs, to be compared with the cross-sectional profile surveyed in the field by a total station (Trimble M5 1').

The DSM resolution of the 27 photogrammetric projects ranged from 0.06 to 0.105 m. Thereafter, all DSMs were re-sampled using the nearest neighbor technique to a resolution of 0.105 m, to make them compatible and consistent for the comparative analysis of transverse profiles. To generate the cross-sectional profile, a 40 m line was drawn on the ortho-mosaic in the central part of the study area; the altimetric data was hence collected in the field using the total station with a step of 0.10 m, monitoring the changes in altitude.

Notably, the cross-section has been selected in the region where the 2D, 3D, and Terrain flight plans overlap to compare the 27 DSMs generated in this work, with the data collected from the cross-section obtained by the total station during the survey operations.

3. Results

3.1. Geometric Accuracy of Aerial Photographic Mosaics

3.1.1. RTK Accuracy

RMSE was calculated based on the VP (04) for the RTK photogrammetric projects (Figure 5). Using the real-time correction of the GNSS receiver base DRTK2 linked to a known point BM-01 (Figure 2), different solutions were tested according to the type of images acquired in the field (2D, 3D, and terrain) and the number of GCPs used for the block orientation (1, 5 and 10), obtaining 9 UAS photogrammetric projects.

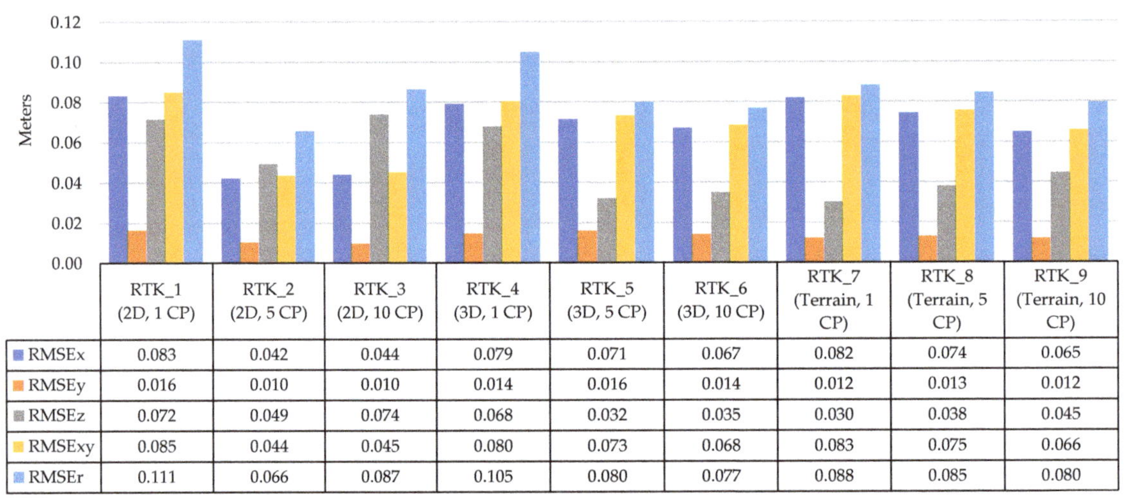

	RTK_1 (2D, 1 CP)	RTK_2 (2D, 5 CP)	RTK_3 (2D, 10 CP)	RTK_4 (3D, 1 CP)	RTK_5 (3D, 5 CP)	RTK_6 (3D, 10 CP)	RTK_7 (Terrain, 1 CP)	RTK_8 (Terrain, 5 CP)	RTK_9 (Terrain, 10 CP)
RMSEx	0.083	0.042	0.044	0.079	0.071	0.067	0.082	0.074	0.065
RMSEy	0.016	0.010	0.010	0.014	0.016	0.014	0.012	0.013	0.012
RMSEz	0.072	0.049	0.074	0.068	0.032	0.035	0.030	0.038	0.045
RMSExy	0.085	0.044	0.045	0.080	0.073	0.068	0.083	0.075	0.066
RMSEr	0.111	0.066	0.087	0.105	0.080	0.077	0.088	0.085	0.080

Figure 5. RMSE of the VPs for Different RTK Configurations.

All of the 9 configurations present an RMSEr between 0.066 m and 0.111 m computed through the VPs. The RTK4 configuration has the highest RMSEr value, whereas the RTK2 configuration shows the lowest value (0.06 m). Indeed, using 5 GCPs reduces the precession by 0.05 m compared to RTK_1 with 10 GCPs, reducing the fieldwork by surveying GCPs. According to ASPRS, the RTK_2 configuration with a GSD of 3.84 cm/px belongs to Class II, which allows one to perform mapping at a scale of 1/500.

The VPs distributed within the urban area have an RMSEz value ranging from 0.030 m to 0.074 m, with the RTK_7 Terrain mode configuration with 1 GCP having the best accuracy, meaning that the non-vegetated vertical accuracy (NVA) values are equal to RMSEz (58.8 mm). This dataset was tested to meet ASPRS positional accuracy standards for a vertical accuracy class of 10 cm. The actual accuracy of the NVA was found to be 7 cm, which is equivalent to +/− 15 cm at a 95% confidence level. This is considered acceptable for urban mapping applications.

3.1.2. PPK1 and PPK2 Accuracy

Both the results of the PPK1 and PPK2 approaches consist of 9 photogrammetric projects, from PPK1_1 to PPK1_9 (Figure 6a) and from PPK2_1 to PPK2_9 (Figure 6b), respectively. Figure 6 shows the RMSEr of the VPs for the different configurations, the error for the x, y, z, xy components, and the total error.

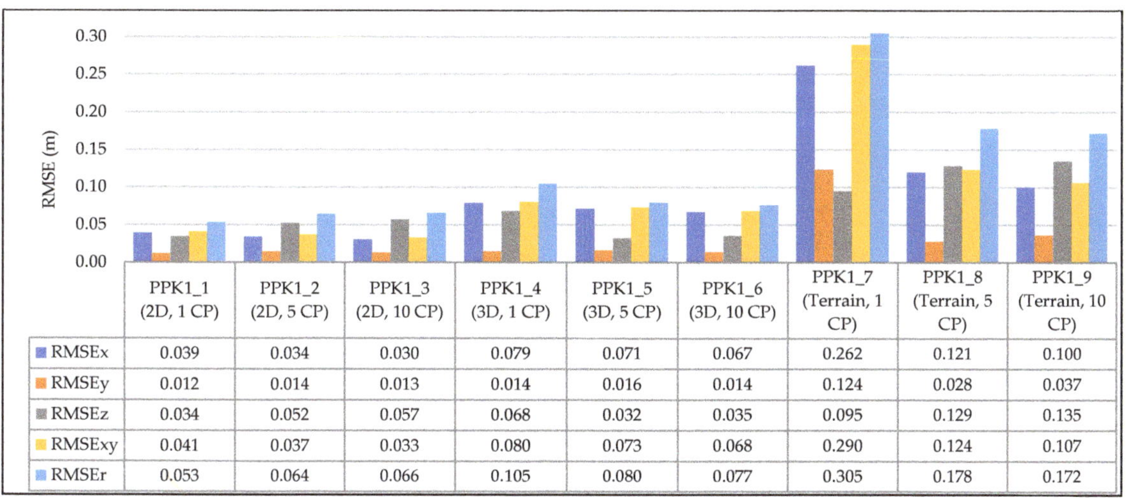

	PPK1_1 (2D, 1 CP)	PPK1_2 (2D, 5 CP)	PPK1_3 (2D, 10 CP)	PPK1_4 (3D, 1 CP)	PPK1_5 (3D, 5 CP)	PPK1_6 (3D, 10 CP)	PPK1_7 (Terrain, 1 CP)	PPK1_8 (Terrain, 5 CP)	PPK1_9 (Terrain, 10 CP)
RMSEx	0.039	0.034	0.030	0.079	0.071	0.067	0.262	0.121	0.100
RMSEy	0.012	0.014	0.013	0.014	0.016	0.014	0.124	0.028	0.037
RMSEz	0.034	0.052	0.057	0.068	0.032	0.035	0.095	0.129	0.135
RMSExy	0.041	0.037	0.033	0.080	0.073	0.068	0.290	0.124	0.107
RMSEr	0.053	0.064	0.066	0.105	0.080	0.077	0.305	0.178	0.172

(a)

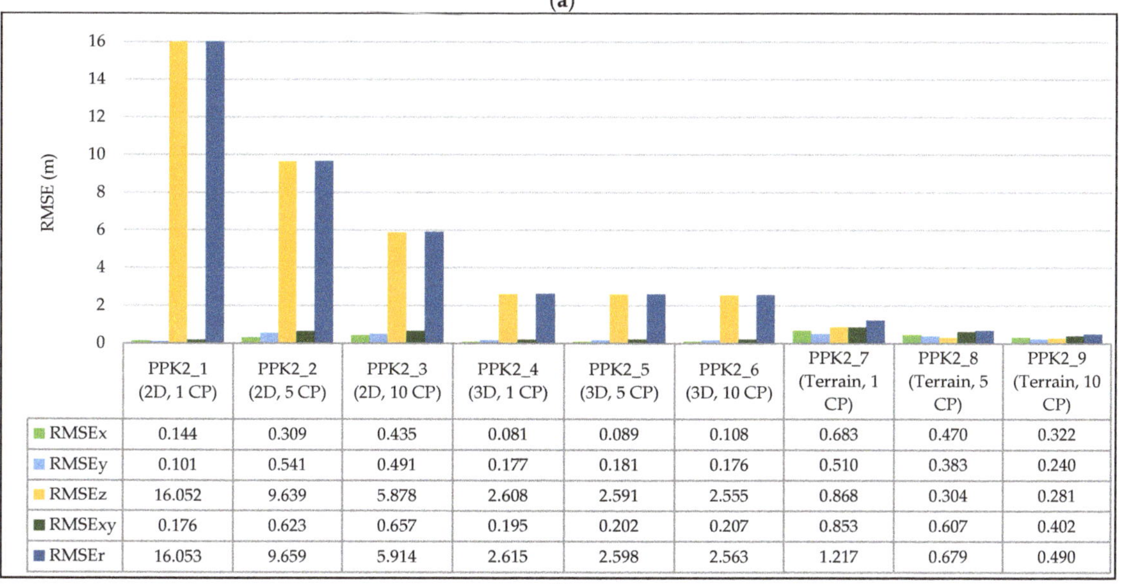

	PPK2_1 (2D, 1 CP)	PPK2_2 (2D, 5 CP)	PPK2_3 (2D, 10 CP)	PPK2_4 (3D, 1 CP)	PPK2_5 (3D, 5 CP)	PPK2_6 (3D, 10 CP)	PPK2_7 (Terrain, 1 CP)	PPK2_8 (Terrain, 5 CP)	PPK2_9 (Terrain, 10 CP)
RMSEx	0.144	0.309	0.435	0.081	0.089	0.108	0.683	0.470	0.322
RMSEy	0.101	0.541	0.491	0.177	0.181	0.176	0.510	0.383	0.240
RMSEz	16.052	9.639	5.878	2.608	2.591	2.555	0.868	0.304	0.281
RMSExy	0.176	0.623	0.657	0.195	0.202	0.207	0.853	0.607	0.402
RMSEr	16.053	9.659	5.914	2.615	2.598	2.563	1.217	0.679	0.490

(b)

Figure 6. PPK1 and PPK2 Accuracy (**a**) RMSE of VP for the PPK1 Approach (**b**) RMSE of VP for the PPK2 Approach.

The best RMSEr of the PPK1 approach was found for the PPK1_1 configuration (2D flight with 01 GCPs), with an overall accuracy of 0.053 m. For the PPK2 approach, the best RMSEr was obtained using the PPK2_9 configuration (0.490 m), which consisted of a ground flight mode with 10 GCPs. The latter is a high value when compared to the RTK and PPK1 approaches.

Similarly, the VP within the urban area have an RMSEz value of 0.034 (PPK1_1) m and 0.281 (PPK2_9) m. Here, the PPK1_1 GCP configuration obtained the best accuracy results, which means that the NVA = RMSEz × 1.96 = 63 mm. This dataset was tested to meet ASPRS positional accuracy standards for a vertical accuracy class of 10 cm. The actual accuracy of the NVA was found to be 5 cm, which is equivalent to $+/-$ 10 cm at a 95% confidence level, and is therefore considered acceptable for urban cadastral applications.

3.2. Cross-Section

Regarding the cross-section of the DSM (Figure 7), it can be noted that, in the RTK approach (Figure 7a), on average, all the configurations present small variations of 0.05 m between the profiles extracted from the model and the ones surveyed by the total station (TS). On the contrary, the PPK approach (Figure 7b) presents similarities in altitude in all of the profiles except for PPK_8 and PPK_9, which have a difference of 0.10 m in the central part of the graph.

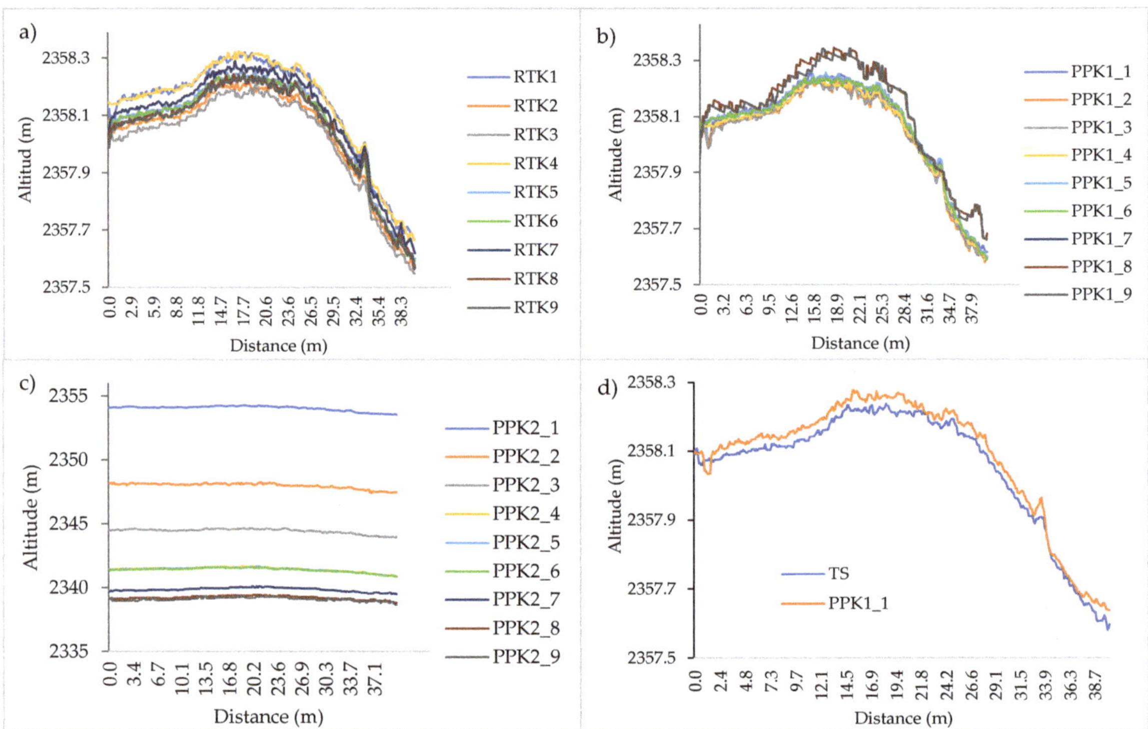

Figure 7. Comparative Analysis of the TS Cross-section. (**a**) Cross-section of the RTK Approach. (**b**) Cross-section of the PPK1 Approach. (**c**) Cross-section of the PPK2. PPK1 Approach and (**d**) The Most Accurate Cross-section of the Approaches to the ST.

In the case of the PPK approach (Figure 7c), this presents a very high-altitude bias versus the TS plot of approximately 2–15 m along the cross-section. On the other hand, the configurations that integrated some oblique images in the nadiral dataset (PPK_1, PPK_5, PPK_6, RTK5, RTK6) produced profiles that generally presented less altitude bias, considering also a vertical level of uncertainty of a few centimeters.

Figure 7d shows the most accurate terrain profile concerning the data collected by the total station. Overall, the PPK_1 2D flight mode configuration with 1 GCP seems to have the highest accuracy.

4. Discussion

For the 27 photogrammetric projects analyzed in this study, using RTK, PPK1, and PPK2 approaches, a 3D accuracy of 0.066 m, 0.053 m, and 0.49 m was obtained, respectively. Such values conform to the APRSS [50] mapping standard of a nominal map scale of 1:500. For all solutions, an integrated orientation approach (EO + GCPs) was adopted, since this approach presents better results, as discussed by Przybilla [46]. However, to adopt the best solution for a direct georeferencing (GI) approach, specific guidelines must be followed, such as collecting GCPs in situ for block adjustment (at least 01) and flying with a minimum overlap of 80% in urban areas and within a range of 80 to 120 m in altitude above ground level [36].

In regards to the flight planning, when covering the area to be mapped with two flights or more, these should be connected to a base station by RTK or NTRIP, as it reinforces the acquisition geometry improving the estimation of internal parameters of the IOP. This was not the case for the PPK2 approach, which had an offset between flight plans, decreasing the overlap of the photo capture and resulting in a considerable increase of the RMSEr 0.437 m to PPK1 (0.53 m). This is due to the desynchronization of the collected data [23]. In Peru, the IGN's commercial GNSS bases are in the process of being modernized so that they can collect static data at 1Hz [64], which would significantly reduce the RMSRr of photogrammetric projects.

In addition, for the PPK1 and PPK2 approach in the ground-based mode, it has a high RMSE (33 times more) because a global DEM (90 m) was used to map the area, which resulted in a risky flight and low overlap of photos. Therefore, to perform a flight with proper terrain tracking, a high-resolution DEM with a topographic precision is necessary [65].

The RTK approach presents difficulties because there is a need to have at least one known coordinate point in the area of interest, as well as a real-time signal connection at all times from the DRTK2–P4RTK to obtain optimal results. The signal in urban areas tends to be lost, causing a poor photograph geotagging accuracy due to signal structures and interferences by communication antennas [46]. Therefore, DJI implemented the RTK FIX option, which allows keeping the connection fixed for up to 10 min [45]. That is, the longer P4RTK and D-RKT2 remain unconnected (10 min max.), the more the RTK signal drops.

This caused the RTK approach using the GNSS D-RTK2 GNSS base, specifically the RTK_2 configuration, to have a slightly higher RMSEr (0.066 m) than the PPK1_1 approach (0.053 m). Besides, the accuracy of the reference point coordinates can directly influence the accuracy of the coordinates stored in the geotag of the acquired images [61].

Practical experience also showed that the PPK approach [34,36,46,47,49], specifically PPK1, is preferable to RTK technology [22,46,62,66] in an environment where frequent occlusion of satellite signals occurs due to the movement of the drone or its surrounding communication signals (e.g., radio signals, telephone lines, television, etc.). Although there is the option of using the PPK approach with the D-RTK2 GNSS receiver, it is not yet available as an official DJI solution to perform such post-processing of data derived from static mode acquisitions [45]. This makes it necessary to use a third-party GNSS base, increasing the costs of the photogrammetric survey. Furthermore, it is worth mentioning that there is the RTK approach via NTRIP using a local or commercial GNSS base receiver. The latter approach has been used in previous studies [34,59], but Peru currently lacks the NTRIP service, hence we have not used it for this study.

It was also demonstrated that the cross-sectional profiles of the PPK1 approach best match the profile obtained using the total station. This is of the utmost importance for the remodeling, construction, and design of roads in urban environments, leading to the convenience of this comparison [47,66].

Overall, the PPK1 approach led to the best 3D accuracy of the block orientation phase of the block using at least one GCP. It was demonstrated that using data derived from a GNSS receiver allows for a faster acquisition in the field, as well as lowering requirements in terms of deployment of instrumentation. For large-scale mapping purposes, it is strongly recommended to use a GNSS receiver within the surveyed area.

For future research and in similar study area conditions, it is recommended to perform camera calibration a priori, in order to evaluate whether the accuracy of the photogrammetric model is improved or not. This is because the DJI UAS cameras cannot be considered metric and stable cameras [46]. It is also recommended to use PPK approaches in NTRIP mode on a commercial basis. In addition, geo-tagging processing should be done with the use of free and paid software on a comparative basis.

5. Conclusions

Direct georeferencing was shown to be a viable solution for mapping urban areas with multiple colonial buildings and narrow streets. It was noted that the PPK1 approach is better than PPK2 and RTK. PPK1 errors are between 0.053 to 0.305 m, unlike PPK2 ranging from 0.49 to 16.053 m, the latter being about 33 times greater than PPK1. This was due to the temporary desynchronization of data stored in AMA01 (5Hz) and P4RTK (1Hz). In the case of the PPK1 (PPK1_1 to PPK1_6) approach, this method is slightly better than the RTK approach, presenting RMSEr 0.053 to 0.105 m and 0.066 to 0.111, respectively. This is because inside a city there is often an interference caused by telecommunications networks, which generate positioning errors in some cameras due to a signal loss.

In addition, if we consider the PPK1_1 approach in a 2D flight with at least 1 GCP, this kind of configuration significantly improves the RMSEz when compared to the use of 5 or 10 GCPs. The precision also improves horizontally, obtaining an RMSEr of 0.053 m, which allows for the development of urban cartography on a scale of 1:500.

Photogrammetry application is possible using UAS and applying the DG technique, in particular the PPK1 approach. Therefore, in the case of Peru, it is not recommended to use the PPK2 approach due to the desynchronization of the collected data, despite the use of different GCPs. Additionally, signage such as GCP recognition requires considerable effort in cartography with UAS, especially in complex terrain such as cities with multiple colonial buildings in which it is not possible to conduct an RTK study of GNSS. In this sense, we recommend a more in-depth analysis of the new UAS configurations with the GD approach to use GCPs to a lesser extent.

Author Contributions: Conceptualization, R.S.L., R.E.T.M. and N.B.R.-B.; Methodology, R.E.T.M. and J.O.S.-L.; Software, R.E.T.M. and J.O.S.-L.; Validation, R.E.T.M. and J.O.S.-L.; Formal analysis, R.E.T.M., J.O.S.-L., N.B.R.-B. and Y.T.; Investigation, R.S.L., J.O.S.-L. and D.G.F.; Resources, R.E.T.M. and M.O.-C.; Data curation, R.E.T.M., J.O.S.-L., D.G.F. and Y.T.; Writing—original draft, R.E.T.M. and J.O.S.-L.; Writing—review & editing, R.E.T.M. and N.B.R.-B.; Visualization, R.S.L. and D.G.F.; Supervision, R.S.L., N.B.R.-B., M.O.-C. and Y.T.; Project administration, R.S.L. and M.O.-C.; Funding acquisition, R.E.T.M. and N.B.R.-B. All authors have read and agreed to the published version of the manuscript.

Funding: This work was carried out with the support of the Public Investment Project GEOMÁTICA (CUI N° 2255626), executed by the Instituto de Investigación para el Desarrollo Sustentable de Ceja de Selva (INDES-CES) of the Universidad Nacional Toribio Rodríguez de Mendoza de Amazonas (UNTRM).

Data Availability Statement: The data used to support the findings of this study are available from the corresponding author upon request.

Acknowledgments: The authors acknowledge and appreciate the support of the Instituto de Investigación para el Desarrollo Sustentable de Ceja de Selva (INDES-CES) of the Universidad Nacional Toribio Rodríguez de Mendoza de Amazonas (UNTRM). We deeply thank Nilton Atalaya Marin and Jhon A. Zabaleta Santisteban for their technical and logistical assistance.

Conflicts of Interest: The authors declare no conflict of interest.

References

1. Koslowski, R.; Schulzke, M. Drones along Borders: Border Security UAVs in the United States and the European Union. *Int. Stud. Perspect.* **2018**, *19*, 305–324. [CrossRef]
2. Blázquez, M.; Colomina, I. Relative INS/GNSS aerial control in integrated sensor orientation: Models and performance. *ISPRS J. Photogramm. Remote Sens.* **2012**, *67*, 120–133. [CrossRef]
3. Kerle, N.; Nex, F.; Gerke, M.; Duarte, D.; Vetrivel, A. UAV-Based Structural Damage Mapping: A Review. *ISPRS Int. J. Geo-Inf.* **2019**, *9*, 14. [CrossRef]
4. Jiang, S.; Jiang, C.; Jiang, W. Efficient structure from motion for large-scale UAV images: A review and a comparison of SfM tools. *ISPRS J. Photogramm. Remote Sens.* **2020**, *167*, 230–251. [CrossRef]
5. Grubesic, T.H.; Nelson, J.R. *UAVs and Urban Spatial Analysis*; Springer: Cham, Switzerland, 2020.
6. Mozas-Calvache, A.T.; Pérez-García, J.L. Analysis and Comparison of Lines Obtained from GNSS and UAV for Large-Scale Maps. *J. Surv. Eng.* **2017**, *143*, 04016028. [CrossRef]
7. Roberts, J.; Koeser, A.; Abd-Elrahman, A.; Wilkinson, B.; Hansen, G.; Landry, S.; Perez, A. Mobile Terrestrial Photogrammetry for Street Tree Mapping and Measurements. *Forests* **2019**, *10*, 701. [CrossRef]
8. Xu, S.; Vosselman, G.; Elberink, S.O. Multiple-entity based classification of airborne laser scanning data in urban areas. *ISPRS J. Photogramm. Remote Sens.* **2014**, *88*, 1–15. [CrossRef]
9. Pepe, M.; Fregonese, L.; Scaioni, M. Planning airborne photogrammetry and remote-sensing missions with modern platforms and sensors. *Eur. J. Remote Sens.* **2018**, *51*, 412–435. [CrossRef]
10. Jones, C.A.; Church, E. Photogrammetry is for everyone: Structure-from-motion software user experiences in archaeology. *J. Archaeol. Sci. Rep.* **2020**, *30*, 102261. [CrossRef]
11. Vasuki, Y.; Holden, E.-J.; Kovesi, P.; Micklethwaite, S. Semi-automatic mapping of geological Structures using UAV-based photogrammetric data: An image analysis approach. *Comput. Geosci.* **2014**, *69*, 22–32. [CrossRef]
12. Taddia, Y.; Corbau, C.; Zambello, E.; Russo, V.; Simeoni, U.; Russo, P.; Pellegrinelli, A. UAVs to Assess the Evolution of Embryo Dunes. In Proceedings of the International Conference on Unmanned Aerial Vehicles in Geomatics, Bonn, Germany, 23 August 2017; Volume 42, pp. 363–369.
13. Taddia, Y.; Pellegrinelli, A.; Corbau, C.; Franchi, G.; Staver, L.; Stevenson, J.; Nardin, W. High-Resolution Monitoring of Tidal Systems Using UAV: A Case Study on Poplar Island, MD (USA). *Remote Sens.* **2021**, *13*, 1364. [CrossRef]
14. Gaitani, N.; Burud, I.; Thiis, T.; Santamouris, M. High-resolution spectral mapping of urban thermal properties with Unmanned Aerial Vehicles. *Build. Environ.* **2017**, *121*, 215–224. [CrossRef]
15. Tokarczyk, P.; Leitao, J.P.; Rieckermann, J.; Schindler, K.; Blumensaat, F. High-quality observation of surface imperviousness for urban runoff modelling using UAV imagery. *Hydrol. Earth Syst. Sci.* **2015**, *19*, 4215–4228. [CrossRef]
16. Salvo, G.; Caruso, L.; Scordo, A. Urban Traffic Analysis through an UAV. *Procedia Soc. Behav. Sci.* **2014**, *111*, 1083–1091. [CrossRef]
17. Zhang, M.; Rao, Y.; Pu, J.; Luo, X.; Wang, Q. Multi-Data UAV Images for Large Scale Reconstruction of Buildings. In Proceedings of the Multi Media Modeling 26th International Conference, MMM 2020, Daejeon, Republic of Korea, 5–8 January 2020; Springer: Cham, Switzerland, 2020; Volume 11962, pp. 254–266.
18. Nex, F.; Remondino, F. UAV for 3D mapping applications: A review. *Appl. Geomat.* **2014**, *6*, 1–15. [CrossRef]
19. Zarco-Tejada, P.J.; Diaz-Varela, R.; Angileri, V.; Loudjani, P. Tree height quantification using very high resolution imagery acquired from an unmanned aerial vehicle (UAV) and automatic 3D photo-reconstruction methods. *Eur. J. Agron.* **2014**, *55*, 89–99. [CrossRef]
20. Casapia, X.T.; Falen, L.; Bartholomeus, H.; Cárdenas, R.; Flores, G.; Herold, M.; Coronado, E.N.H.; Baker, T.R. Identifying and Quantifying the Abundance of Economically Important Palms in Tropical Moist Forest Using UAV Imagery. *Remote Sens.* **2020**, *12*, 9. [CrossRef]
21. Westoby, M.; Brasington, J.; Glasser, N.F.; Hambrey, M.J.; Reynolds, J.M. 'Structure-from-Motion' photogrammetry: A low-cost, effective tool for geoscience applications. *Geomorphology* **2012**, *179*, 300–314. [CrossRef]
22. Kalacska, M.; Lucanus, O.; Arroyo-Mora, J.; Laliberté, E.; Elmer, K.; Leblanc, G.; Groves, A. Accuracy of 3D Landscape Reconstruction without Ground Control Points Using Different UAS Platforms. *Drones* **2020**, *4*, 13. [CrossRef]
23. Cledat, E.; Jospin, L.; Cucci, D.; Skaloud, J. Mapping quality prediction for RTK/PPK-equipped micro-drones operating in complex natural environment. *ISPRS J. Photogramm. Remote Sens.* **2020**, *167*, 24–38. [CrossRef]
24. Trujillo, M.M.; Darrah, M.; Speransky, K.; DeRoos, B.; Wathen, M. Optimized flight path for 3D mapping of an area with structures using a multirotor. In Proceedings of the 2016 International Conference on Unmanned Aircraft Systems (ICUAS), Arlington, VA, USA, 7–10 June 2016. [CrossRef]
25. Backes, D.; Schumann, G.; Teferele, F.N.; Boehm, J. Towards a High-Resolution Drone-Based 3D Mapping Dataset to Optimise Flood Hazard Modelling. In Proceedings of the ISPRS Geospatial Week 2019, Enschede, The Netherland, 10–14 June 2019; Volume 42, pp. 181–187.
26. Gabrlik, P. The Use of Direct Georeferencing in Aerial Photogrammetry with Micro UAV. *IFAC-Pap.* **2015**, *48*, 380–385. [CrossRef]
27. Agüera-Vega, F.; Carvajal-Ramírez, F.; Martínez-Carricondo, P. Assessment of photogrammetric mapping accuracy based on variation ground control points number using unmanned aerial vehicle. *Measurement* **2017**, *98*, 221–227. [CrossRef]

28. Martínez-Carricondo, P.; Agüera-Vega, F.; Carvajal-Ramírez, F.; Mesas-Carrascosa, F.J.; García-Ferrer, A.; Pérez-Porras, F.-J. Assessment of UAV-photogrammetric mapping accuracy based on variation of ground control points. *Int. J. Appl. Earth Obs. Geoinf.* **2018**, *72*, 1–10. [CrossRef]

29. Sanz-Ablanedo, E.; Chandler, J.H.; Rodríguez-Pérez, J.R.; Ordóñez, C. Accuracy of unmanned aerial vehicle (UAV) and SfM photogrammetry survey as a function of the number and location of ground control points used. *Remote Sens.* **2018**, *10*, 1606. [CrossRef]

30. Heipke, C.; Jacobsen, K.; Wegmann, H.; Andersen, Ø.; Nilsen, B. Test Goals and Test Set up for the OEEPE Test. In *Integrated Sensor Orientation*; OEEPE Official Publication: Amsterdam, The Netherlands, 2002.

31. Bilker, M.; Honkavaara, E.; Jaakkola, J. GSPS Supported Aerial Triangulation Using Untargeted Ground Control. *Int. Arch. Photogramm. Remote Sens.* **1998**, *32*, 2–9.

32. Ip, A.; El-Sheimy, N.; Mostafa, M. Performance Analysis of Integrated Sensor Orientation. *Photogramm. Eng. Remote Sens.* **2007**, *73*, 89–97. [CrossRef]

33. Cramer, M.; Stallmann, D.; Haala, N. Direct Georeferencing Using GPS/Inertial Exterior Orientations for Photogrammetric. *Int. Arch. Photogramm. Remote Sens.* **2000**, *33*, 198–205.

34. Losè, L.T.; Chiabrando, F.; Tonolo, F.G. Boosting the Timeliness of UAV Large Scale Mapping. Direct Georeferencing Approaches: Operational Strategies and Best Practices. *ISPRS Int. J. Geo-Inf.* **2020**, *9*, 578. [CrossRef]

35. Xiang, T.-Z.; Xia, G.-S.; Zhang, L. Mini-unmanned aerial vehicle-based remote sensing: Techniques, applications, and prospects. *IEEE Geosci. Remote Sens. Mag.* **2019**, *7*, 29–63. [CrossRef]

36. Zhang, H.; Aldana-Jague, E.; Clapuyt, F.; Wilken, F.; Vanacker, V.; Van Oost, K. Evaluating the potential of post-processing kinematic (PPK) georeferencing for UAV-based structure- from-motion (SfM) photogrammetry and surface change detection. *Earth Surf. Dyn.* **2019**, *7*, 807–827. [CrossRef]

37. Benassi, F.; Dall'Asta, E.; Diotri, F.; Forlani, G.; Morra Di Cella, U.; Roncella, R.; Santise, M. Testing Accuracy and Repeatability of UAV Blocks Oriented with GNSS-Supported Aerial Triangulation. *Remote Sens.* **2017**, *9*, 172. [CrossRef]

38. Rehak, M.; Skaloud, J. FIXED-WING Micro Aerial Vehicle for Accurate Corridor Mapping. In Proceedings of the International Conference on Unmanned Aerial Vehicles in Geomatics, Toronto, ON, Canada, 30 August–2 September 2015; Volume 2, pp. 23–31.

39. Stöcker, C.; Nex, F.; Koeva, M.; Gerke, M. Quality Assessment of Combined IMU/GNSS Data for Direct Georeferencing in the Context of UAV-Based Mapping. In Proceedings of the International Conference on Unmanned Aerial Vehicles in Geomatics, Bonn, Germany, 4–7 September 2017; Volume 42, pp. 355–361.

40. Rehak, M.; Mabillard, R.; Skaloud, J. A Micro Aerial Vehicle with Precise Position and Attitude Sensors. *Photogramm. -Fernerkund. -Geoinf.* **2014**, *4*, 239–251. [CrossRef]

41. Cucci, D.A.; Rehak, M.; Skaloud, J. Bundle adjustment with raw inertial observations in UAV applications. *ISPRS J. Photogramm. Remote Sens.* **2017**, *130*, 1–12. [CrossRef]

42. Rabah, M.; Basiouny, M.; Ghanem, E.; Elhadary, A. Using RTK and VRS in direct geo-referencing of the UAV imagery. *NRIAG J. Astron. Geophys.* **2018**, *7*, 220–226. [CrossRef]

43. Hugenholtz, C.; Brown, O.; Walker, J.; Barchyn, T.; Nesbit, P.; Kucharczyk, M.; Myshak, S. Spatial Accuracy of UAV-Derived Orthoimagery and Topography: Comparing Photogrammetric Models Processed with Direct Geo-Referencing and Ground Control Points. *Geomatica* **2016**, *70*, 21–30. [CrossRef]

44. Forlani, G.; Diotri, F.; Morra Di Cella, U.; Roncella, R. UAV Block Georeferencing and Control by ON-BOARD GNSS Data. In Proceedings of the XXIV ISPRS Congress, Nice, France, 31 August–2 September 2020; Volume 43, pp. 9–16.

45. DJI Phantom 4 RTK, User Manual v2.4. Available online: https://www.dji.com/downloads/products/phantom-4-rtk (accessed on 3 May 2022).

46. Przybilla, H.-J.; Bäumker, M.; Luhmann, T.; Hastedt, H.; Eilers, M. Interaction between direct georeferencing, control point configuration and camera self-calibration for rtk-based uav photogrammetry. In Proceedings of the XXIV ISPRS Congress, Nice, France, 31 August–2 September 2020; pp. 485–492.

47. Taddia, Y.; Stecchi, F.; Pellegrinelli, A. Coastal Mapping Using DJI Phantom 4 RTK in Post-Processing Kinematic Mode. *Drones* **2020**, *4*, 9. [CrossRef]

48. Štroner, M.; Urban, R.; Reindl, T.; Seidl, J.; Brouček, J. Evaluation of the Georeferencing Accuracy of a Photogrammetric Model Using a Quadrocopter with Onboard GNSS RTK. *Sensors* **2020**, *20*, 2318. [CrossRef]

49. Losè, L.T.; Chiabrando, F.; Tonolo, F.G. Are measured ground control points still required in uav based large scale mapping? Assessing the positional accuracy of an RTK multi-rotor platform. In Proceedings of the XXIV ISPRS Congress, Nice, France, 31 August–2 September 2020; pp. 507–514.

50. American Society for Photogrammetryand Remote Sensing (ASPRS). ASPRS Positional Accuracy Standards for Digital Geospatial Data. *Photogramm. Eng. Remote Sens.* **2015**, *81*, A1–A26. [CrossRef]

51. Whitehead, K.; Hugenholtz, C.H. Applying ASPRS Accuracy Standards to Surveys from Small Unmanned Aircraft Systems (UAS). *Photogramm. Eng. Remote Sens.* **2015**, *81*, 787–793. [CrossRef]

52. Agüera-Vega, F.; Carvajal-Ramírez, F.; Martínez-Carricondo, P. Accuracy of Digital Surface Models and Orthophotos Derived from Unmanned Aerial Vehicle Photogrammetry. *J. Surv. Eng.* **2017**, *143*, 04016025. [CrossRef]

53. Rscón, J.; Angeles, W.G.; Oliva, M.; Huatangari, L.Q.; Grurbillon, M.A.B. Determinación de Las Épocas Lluviosas y Secas En La Ciudadde Chachapoyas Para El Periodo de 2014–2018. *Rev. Climatol.* **2020**, *20*, 15–28.

54. Municipalidad Provincial de Chachapoyas (MPCH). *Plan de Desarrollo Urbano de La Ciudad de Chachapoyas*; Scribd: Chschapoyas, Peru, 2013.

55. Instituto Geográfico Nacional (IGN). *Norma Técnica Geodésica: Especificaciones Técnicas Para Posicionamiento Geodésico Estático Relativo Con Receptores Del Sistema Satelital de Navegación Global*; IGN: Lima, Peru, 2015.

56. TRIMBLE. *Trimble R10 GNSS Receiver User Guide*; IGN: Lima, Peru, 2014.

57. DJI. *D-RTK 2 High Precision GNSS Mobile Station Release Notes*; DJI: Shenzhen, China, 2021.

58. Takasu, T.; Yasuda, A. Development of the Low-Cost RTK-GPS Receiver with an Open Source Program Package RTKLIB. In *International Symposium on GPS/GNSS*; International Convention Center Jeju Korea: Seogwipo-si, Korea, 2009; Volume 1, pp. 1–6.

59. REDcatch. *REDtoolbox v2.77 User Manual*; REDcatch: Fulpmes, Austria; pp. 1–29.

60. Agisoft Metashape User Manual, Standard Edition, Version 1.7. Available online: https://www.agisoft.com/downloads/user-manuals/ (accessed on 3 May 2021).

61. Congalton, R.G. Thematic and Positional Accuracy Assessment of Digital Remotely Sensed Data. In Proceedings of the Seventh Annual Forest Inventory and Analysis Symposium, Portland, ME, USA, 3–6 October 2005; pp. 149–154.

62. Taddia, Y.; Stecchi, F.; Pellegrinelli, A. Using Dji Phantom 4 Rtk Drone for Topographic Mapping of Coastal Areas. In Proceedings of the ISPRS Geospatial Week 2019, Enschede, The Netherlands, 10–14 June 2019; Volume 42, pp. 625–630.

63. Tenedório, J.A.; Estanqueiro, R.; Lima, A.M.; Marques, J. Remote Sensing from Unmanned Aerial Vehicles for 3D Urban Modelling: Case Study of Loulé, Portugal. In *Back to the Sense of the City: International Monograph Book*; Centre de Política de Sòl i Valoracions: Loulé, Portugal, 2016.

64. Instituto Geográfico Nacional. *Diario el Peruano Resolución Jefatural No. 149-2022_IGN_DIG_SDPG*; Normas y Documentos Legales; Gobierno Del Perú: Lima, Peru, 2022.

65. Trajkovski, K.K.; Grigillo, D.; Petrovič, D. Optimization of UAV Flight Missions in Steep Terrain. *Remote Sens.* **2020**, *12*, 1293. [CrossRef]

66. Forlani, G.; Dall'Asta, E.; Diotri, F.; di Cella, U.M.; Roncella, R.; Santise, M. Quality Assessment of DSMs Produced from UAV Flights Georeferenced with On-Board RTK Positioning. *Remote Sens.* **2018**, *10*, 311. [CrossRef]

Article

Expeditious Low-Cost SfM Photogrammetry and a TLS Survey for the Structural Analysis of Illasi Castle (Italy)

Massimo Fabris * , **Pietro Fontana Granotto and Michele Monego**

Department of Civil, Environmental and Architectural Engineering, University of Padua, 35131 Padua, Italy
* Correspondence: massimo.fabris@unipd.it

Abstract: The structural analysis of degraded historical buildings requires an adequate 3D model of the object. Structure from motion (SfM) photogrammetry and laser scanning geomatics techniques can satisfy this request by providing geometrically affordable data. The accuracy and resolution depend on the instruments and procedures used to extract the 3D models. This work focused on a 3D survey of Illasi Castle, a strongly degraded historical building located in northern Italy, aimed at structural analysis in the prevision of a static recovery. A low-cost drone, a single-lens reflex (SLR) camera, and a smartphone were used in the survey. From each acquired dataset, using the integration between the images acquired by the drone and the SLR camera, a 3D model of the building was extracted by means of the SfM technique. The data were compared with high-precision and high-resolution terrestrial laser scanning (TLS) acquisitions to evaluate the accuracy and performance of the fast and low-cost SfM approach. The results showed a standard deviation value for the point cloud comparisons in the order of 2–3 cm for the best solution (integrating drone and SLR images) and 4–7 cm using smartphone images. Finally, the integration of the best SfM model of the external walls and the TLS model of the internal portion of the building was used in finite element (FE) analysis to provide a safety assessment of the structure.

Keywords: SfM photogrammetry; TLS measurements; smartphone acquisitions; 3D models; Illasi Castle; 3D comparisons

Citation: Fabris, M.; Fontana Granotto, P.; Monego, M. Expeditious Low-Cost SfM Photogrammetry and a TLS Survey for the Structural Analysis of Illasi Castle (Italy). *Drones* **2023**, *7*, 101. https://doi.org/10.3390/drones7020101

Academic Editor: Geert Verhoeven

Received: 11 January 2023
Revised: 24 January 2023
Accepted: 28 January 2023
Published: 1 February 2023

1. Introduction

Three-dimensional survey techniques, such as structure from motion (SfM) photogrammetry and laser scanning, with both terrestrial- and drone-based approaches are widely used in different fields (architectural, archaeological, cultural heritage, and structural) to reconstruct the geometric characteristics of objects, with the aims of documentation, analysis, preservation, maintenance, structural analysis, and restoration [1–9]. These techniques provide sparse and/or dense point clouds that can be used to enhance details [10]. Their use together allows for the improvement of coverage completeness, enabling the modeling of objects and complex cultural heritage buildings [11]; moreover, their integration with classical topography and the GNSS (Global Navigation Satellite System) system allows for the measurement of control networks and reference targets and the georeferencing of data in local, national, and/or international systems [12–15].

The structural analysis of historical buildings requires a detailed knowledge of their construction, where in situ investigations are carried out together with numerical analysis [16,17]. In particular, in the conservation and restoration of historical buildings, their unique characteristics must be examined, including structure type, the materials employed and their source, construction methods and enlargement, reconstruction and restoration interventions over time, types of use, etc. [18]. In order to prevent the possible damage/collapse of cultural heritage, the availability of complete and updated 3D models simplifies the planning and intervention phases. For these reasons, low-cost, expeditious, and non-destructive methods can be of great help in performing 3D surveys [19]. In many cases, a simplified model of the building is acceptable for use in seismic analysis by means of the pushover approach [20]. In this context, the use of low-cost sensors, including consumer-level smartphones, which are increasingly used in 3D model reconstruction, proved to be very useful [21,22]. Data obtained from 3D surveys (usually dense point clouds) are not directly suitable for the generation of a numerical model; generally, several operations are necessary [15]. In fact, the process of converting the acquired 3D geometric data, which provides a realistic model of the building, into a convenient format for finite element (FE) analysis software is crucial [23]. For this reason, several studies on different cases and situations are available in the literature. At present, FE analysis is used in most studies and simulation applications [24]. This procedure requires the generation of a solid model of the structure, built using basic 3D shapes, such as parallelepipeds, cones, cylinders, spheres, wedges, and toroids. The combination of these forms allows the generation of complex solids (Boolean operations) where the surfaces, volumes, and other geometric properties are known.

Many authors used terrestrial laser scanning (TLS) and SfM photogrammetric techniques, separately or integrated, to extract 3D models for the structural analysis of buildings and infrastructure. Hinks et al. [25] proposed an automatic and fast method to transform the point cloud acquired using the laser scanning technique in a solid model for FE analysis. The method requires better shape detection and the separation of the real openings and the occlusions in the laser scanning data; in many cases, integration with the SfM technique can better improve this separation. Barazzetti et al. [26] used a semiautomatic procedure to obtain a building information model (BIM) that preserved the TLS point cloud complexity of the building through the parametrization of NURBS (non-uniform rational B-splines) surfaces for a structural simulation based on FE analysis; this approach allowed the authors to take advantage of TLS surveys, which are characterized by a higher level of detail and precision than the SfM technique. Castellazzi et al. [27] created an FE model from a laser scanner survey using a semiautomatic method based on slide extraction for the creation of the discretized geometry of the studied building. The proposed method was applied to a structure with irregular geometry that could be easily acquired with dense TLS data, while using the SfM approach may not have been effective. Sánchez-Aparicio et al. [28] investigated the combined use of photogrammetric techniques (DIC and SfM) and geometrical (NURBS and Hausdorff distance) and FE strategies to assess the origin of the damage and give an evaluation of the current stability of the San Lorenzo Church (Zamora, Spain), a historical building. However, the results obtained even for a simple structure (the dome) suggested the need to restrict this strategy to constructions with large deformations, due to the SfM model's accuracy of several millimeters. Tucci et al. [23] combined TLS surveys, deviation analysis (DA), and FE numerical modeling for a brick minaret located in Aksaray (Turkey). They extracted the different geometric parameters of the structure based on a detailed point cloud 3D mesh model and introduced a method for the direct transfer of the high accuracy of TLS based on a 3D model to FE structural analysis. From the FE analysis of this structure, variation in the maximum base shear ranging between 0.07 and 0.18 of the total weight of the minaret was observed, denoting the importance of the precise characterization of minaret geometry that can be obtained only from dense TLS data. Bassier et al. [29] presented a semi-automated approach to create accurate models of complex heritage buildings for the purpose of structural analysis. A complex mesh of the structure was created using TLS and photogrammetry; the authors used photogrammetry only for the extraction of crack information from images, which is troublesome in TLS point clouds due to the sparsity of laser scan data. For this reason, they integrated TLS and photogrammetry only partially. Rissolo et al. [30] conducted a comprehensive digital documentation of the exterior and interior surfaces and spaces of Satunsat (Mexico) by means of TLS and SfM photogrammetry in order to perform a structural health assessment and an evaluation of the varying masonry techniques employed in its construction. The authors fused the point clouds derived from the two techniques to provide the highest possible radiometric fidelity of the structures (from the SfM); however, they did not carry out a critical analysis of the quality of the extracted datasets. Bruno et al. [31] integrated TLS and drone- and terrestrial-based photogrammetry for the geometric analysis of a 19th-century bridge in Italy to better understand the historical evolution of its structural disorders and its present conservation status, and as a reliable base to define future interventions. The point clouds obtained from the two approaches were compared using only GCPs (ground control points) and merged; the authors did not provide indications on the degree of accuracy for each dataset. Selvaggi et al. [7] explored the potential contribution of TLS and classical topography to structural engineering by investigating the vertical structure of the San Luzi Church bell tower (Switzerland). Due to the position of the TLS stations, some parts of the building (in particular the roof of the tower, characterized by very sparse data) were not modeled because of missing data. Moreover, many holes were identified in the building, especially inside, due to narrow spaces; in these cases, the more flexible SfM acquisitions could help to reduce the occlusions. Fang et al. [32] integrated TLS and FE to assess the structural deformation of a historic brick masonry building, the Beamless Hall at Linggu Temple in Nanjing, China. They deduced that its asymmetrical layout under self-weight was likely the main reason for its structural deformation. This analysis was possible due to the high-resolution and high-precision data provided by TLS, which is particularly useful in the investigation of complex structures. Yang and Xu [33] used TLS and photogrammetric data to monitor composite tunnel structures and discussed segmentation deformation, cracks, and water seepage. The authors reported that deformation analysis based on TLS can effectively verify the location of water seepage and cracks, and photogrammetry technology can clearly identify and quantify the dimension of the damage. In this context, they highlighted the advantages derived from the integration of the two techniques. Many other studies integrating TLS and SfM photogrammetry for structural analysis are available in the literature.

In general, the above-mentioned works are characterized by (i) 3D data acquired by means of high-resolution and high-precision sensors; (ii) the use of expensive instruments; and (iii) little information on the time required for the survey. For these reasons, this work aimed to analyze 3D photogrammetric models extracted using expeditious survey procedures and low-cost sensors in the field of archaeology/cultural heritage, focusing on structural analysis for restoration activities. In surveys, SfM photogrammetry allows for the collection of a large amount of data in a short time, providing better coverage of the object by integrating ground- and drone-based acquisitions when compared to the expensive TLS technique. Despite the high-accuracy and high-resolution data provided by the latter approach, it requires more time for the complete survey and 3D coverage of an object.

By comparing the fast SfM models with the high-precision, high-resolution, high-cost, and time-consuming TLS data, an estimation of accuracies can be performed. In the case of poor geometric quality from the 3D photogrammetric reconstruction (low accuracy), integration with TLS data is necessary to ensure the reliability of the structural analysis; however, this integration should be limited only to critical areas where the SfM approach provides inaccurate surfaces.

In this context, low-cost drones and consumer-level cameras (including a smartphone sensor) were applied in the survey of Illasi Castle, a medieval building located in northern Italy (Figure 1) for which future restoration activities are planned. To highlight the advantages and limitations of the 3D SfM models extracted using fast procedures, the data were (i) georeferenced using control points (CPs); (ii) validated using checkpoints (ChPs); and (iii) compared with a 3D reference model extracted from a TLS survey using the Leica ScanStation P20, a high-resolution and high-precision instrument, to evaluate the final accuracies.

Figure 2 shows a flowchart of the methodology applied in this work. By means of different techniques and sensors, 3D models were extracted, georeferenced (using CPs), validated (using ChPs), and compared with TLS models as a reference.

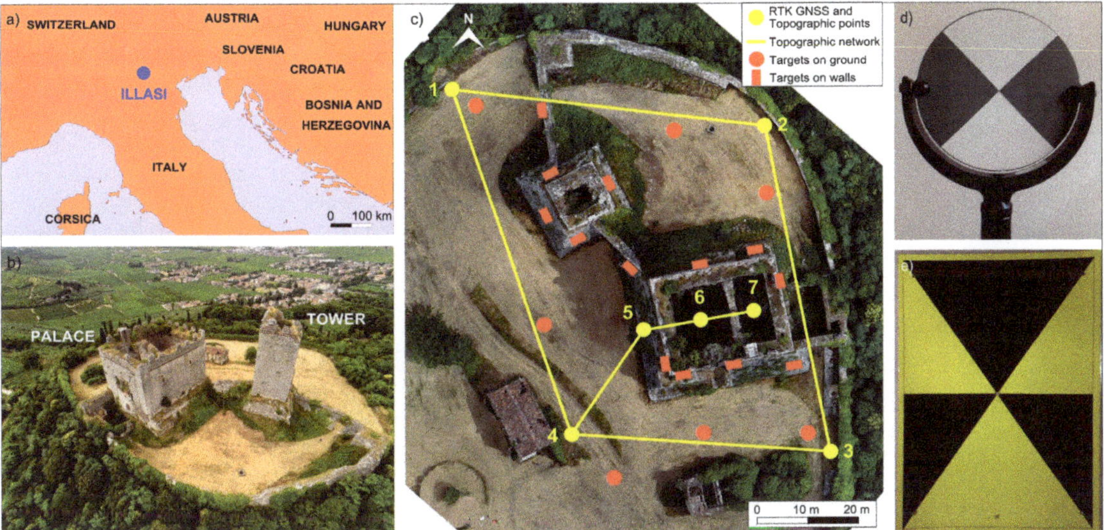

Figure 1. (**a**) Location of Illasi Castle in northern Italy. (**b**) Aerial view of the surveyed palace and tower along with the surrounding area. (**c**) Map of the surveyed area showing the RTK GNSS and topographic points (stations 1–7) and the topographic network with the location of the ground and wall targets (both control points (CPs) and checkpoints (ChPs)). Targets used for TLS (**d**) and SfM (**e**).

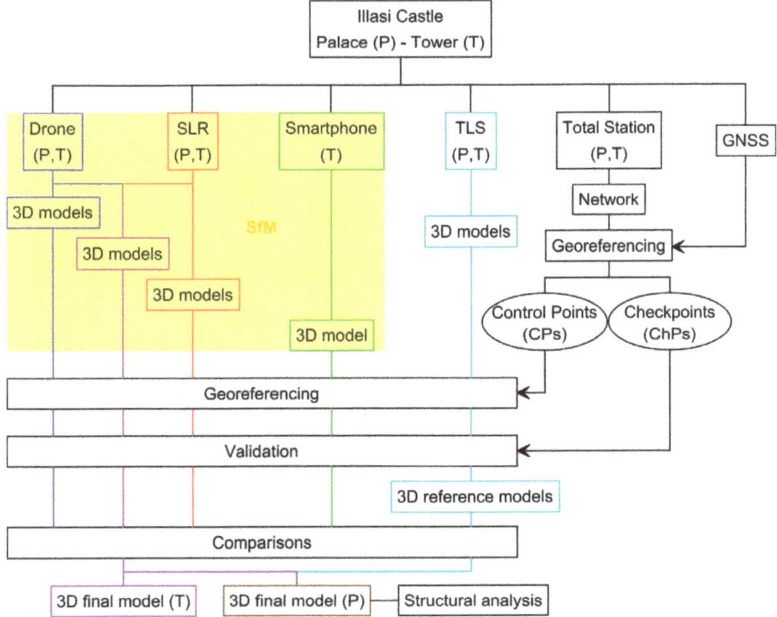

Figure 2. Flowchart of the methodology proposed in this work.

Based on the results of the comparisons, the final 3D models used in the structural analysis were by obtained integrating the SfM data with TLS data only where strictly necessary, due to the high cost and time-consuming nature of the latter approach.

2. Illasi Castle

The structure under investigation is composed of two main buildings, the palace and the tower, which were built simultaneously at the beginning of the 13th century.

The castle is enclosed by an almost circular wall with a radius of about 50 m; the elevated walls constitute what remains of the structures, while the floors and the roof are no longer present.

The palace walls are approximately 27 m high. The four perimeter walls form a rectangular area of about 19.6 × 24 m, where the walls are between 2.7 and 2.8 m thick. The internal space of the building is divided into two unequal rooms by a 1.2 m thick wall, running parallel to the perimeter walls in the north–south direction, with an arch in the center.

From historical documents [34] and the most recent studies based on archaeological excavations performed in 2003–2008 [35,36], there appears to be a basement in the building, which is now covered by the ground. The floor, made up of detrital material, exhibits an irregular level, on average just under the level of the entrance door.

The tower walls are about 32 m high in the highest area. The plan of the tower has a square shape of about 10.3 × 10.3 m with a wall thickness of approximately 3.2 m. On the walls of both the palace and the tower, there are various openings at different heights (i.e., doors and windows), likely added in different periods.

Beginning in the mid-17th century, the castle was gradually abandoned, starting the process of deterioration. At present, due to the general state of decay and frequent localized collapses of material, access to the interior of the two main buildings is limited. Furthermore, the castle is protected by the Archaeological Superintendence of the Veneto Region, part of the Italian Ministry of Culture.

Currently, the structures are overgrown with vegetation, which was not removed before the survey both to lower costs and avoid possible collapses of the walls (Figure 3).

Figure 3. Characteristics of the tower and palace of Illasi Castle highlighting the complexity of the structures and showing the unremoved vegetation (1–5 show the portions enclosed by the colored polylines).

3. Materials and Methods

3.1. Surveys

3.1.1. RTK GNSS and Topographic Measurements of the Reference Network

A reference network is required to co-register data acquired with different techniques to the same reference system.

Here, seven intervisible wooden pickets in the ground, located inside and outside the castle, were measured with a Leica Viva GS 15 GNSS receiver using the RTK approach. In particular, 2 points inside the palace were used to acquire the visible portions of the 2 rooms separated by an arch structure; they were connected with the external points of the network by the entrance door.

Subsequently, 20 photogrammetric targets (both CPs and ChPs, Figure 1e) were positioned, 7 on the ground and 13 on the palace and tower walls (uniformly distributed and located only on accessible surfaces).

The locations of the reference points were designed for visibility and acquisition from the ground, using the SLR and smartphone cameras, and from the air, using the drone (Figure 1). All the targets and points in the network were measured in a local reference system using a Leica TCR1201 total station.

In the second phase, the coordinates of the network points and targets were transformed in the UTM 32 zone cartographic reference system using GNSS data.

3.1.2. 3D Photogrammetric Survey by Drone

The study area and the castle walls were surveyed using a low-cost drone (Parrot Anafi) equipped with a camera, which had a 1/2,4″ CMOS image sensor, a diagonal of 7.83 mm, a resolution of 21 MP, and a focal length of 4 mm (an equivalent focal length of 23 mm). The survey was performed by combining images from different flights. The flights were conducted in manual mode to better cover all the surfaces of the buildings, which are characterized by different depth plans and geometries. A total of 5 flights were executed, each taking 10–12 min, with an estimated overlap of at least 60% between subsequent images and a mean GSD (ground sample distance) of 16 mm/pixel.

The inclination of the camera was almost nadiral for the general flight from above. For the flights dedicated to the walls, two directions were used: perpendicular to the surfaces and slightly inclined.

A general flight of the study area was executed at a relative altitude between 45 and 60 m, acquiring 152 images for the survey of the top of the hill and the walls together with the internal vertical walls of the palace, which had a very complex acquisition geometry. This was due to the limited spaces inside the building that, even in the absence of the roof, did not allow for the easy movement of the drone inside the structure. Subsequently, a flight for each vertical external wall of the palace and tower was carried out at a distance of about 10 m, acquiring 258 images in total. In all cases, to achieve the best accuracy, the relations between the geometry of acquisition, coverage, overlap, and the number and distribution of targets were set according to works published previously by other authors [37,38].

3.1.3. Ground-Based Photogrammetric Acquisitions

A total of 254 images of the structures with high overlap were also acquired from the ground using a single-lens reflex (SLR) Canon EOS 5Ds camera (50.6 MP) with a full-frame 36 × 24 mm CMOS sensor, a diagonal of 43.27 mm, and a focal length of 35 mm. The survey was performed using a camera–object distance of about 5–10 m, which provided a wall pixel size of about 1.5 mm. The images were acquired from different positions around the buildings, with different inclinations of the camera, and from different perspectives.

Using the same approach and acquisition points as for the SLR camera, a survey of the tower was performed using a low-cost and commonly available device: a Redmi Note 7 smartphone equipped with a 48 MP camera and with a 1/2″ sensor that provides 8000 × 6000-pixel images. A total of 83 images were obtained (only for the tower).

3.1.4. TLS Acquisitions

A TLS survey was performed to acquire the high-resolution and high-precision data necessary to evaluate the accuracies of the 3D models generated with the SfM technique using images acquired by low-cost sensors.

A time-of-flight (TOF) Leica ScanStation P20 laser scanner was used. This instrument can acquire the coordinates of the visible surfaces in a local reference system, covering 360° in the horizontal plane and 305° in the vertical plane; it is characterized by an accuracy of 2 mm for distance and 8″ for angular measurements. The instrument was stationed at points in the reference network to acquire 6 point clouds using a resolution of 3 mm at 10 m for the 4 external scans, which were taken further away from the structure, and 6 mm at 10 m for the internal scans of the palace. It should be noted that a point cloud was not acquired from station 5 because it did not add significant information with respect to the time required for execution (Figure 1). Each scan was performed by adequately overlapping it with the previous, and at least 2 TLS targets were acquired (Figure 1d).

This relatively quick procedure was possible because of the use of "known backsight" measurements. The profile of acquisition employed here allowed us to optimize the time taken for scans, target measurements, and alignments by using the topographic network, whose points correspond to the station points and target positions. This allowed us to accurately locate, measure, and verify the positions of the scanning stations and improve and control the accuracy of the alignment and co-registration processes in the UTM 32 zone system.

Overall, using the different techniques described above, the complete survey of the study area and the castle required one working day.

3.2. 3D Data Processing

Photogrammetric 3D point clouds of the palace and the tower were generated via the SfM technique using the Agisoft Metashape software version 1.8.4 and the available reference targets. In detail, the total available points—7 ground targets and 13 wall targets (Figure 1)—were subdivided into 14 CPs (5 ground and 9 wall targets), used in model computation, and 6 ChPs (2 ground and 4 wall targets), which were not included in the processing (Figure 4).

For the photo alignment procedure, an automatic camera calibration tool was used. Subsequently, 4 different 3D models were generated using (i) the aerial-based images acquired with the drone, (ii) the ground-based images acquired with the SLR camera, (iii) the ground-based images of the tower acquired with the consumer-level smartphone camera, and (iv) the integration between the images obtained from the SLR and drone cameras. The last approach was representative of the best photogrammetric solution, integrating ground-based SLR acquisitions with images from a low-cost drone for surveying the higher portions of the buildings.

The processing parameters in the Agisoft Metashape software were set by selecting "high accuracy" for the extraction of the tie points (sparse cloud), "high quality" and "mild filtering" for the extraction of the dense cloud, and "dense cloud" and "high face count" for the extraction of the 3D mesh. The processing times on a workstation with good performance are listed in Table 1 for the different operations and datasets. It should be noted that the areas under investigation around the buildings covered by aerial and terrestrial acquisitions were very different in terms of extension, and the smartphone survey was limited to the tower. For these reasons, the models have different dimensions and the processing times had to be evaluated separately.

Figure 4. Separation of the available targets into 14 CPs (5 on the ground and 9 on the walls) and 6 ChPs (2 on the ground and 4 on the walls).

Table 1. Processing times related to each step of the Agisoft Metashape workflow separated for each dataset (d, h, and m indicate days, hours, and minutes, respectively). A workstation with good performance was used for the processing.

Processing Time	Operation	Drone (h, m)	SLR (h, m)	Smartphone (m)	Drone + SLR (d, h, m)
Tie points (sparse cloud)	Matching:	0, 31	0, 51	3	0, 1, 16
	Alignment:	0, 4	0, 14	2	0, 0, 23
Dense cloud	Depth map generation:	4, 57	1, 38	14	0, 5, 34
	Dense cloud generation:	19, 16	2, 48	15	2, 17, 10
3D model (mesh)	Reconstruction:	0, 19	0, 44	22	0, 1, 2
	Texturing:	0, 4	0, 36	6	0, 1, 7

The processing of the dense point clouds was performed by computing the confidence values for each dataset. The results highlighted the lower reliability of point clouds corresponding to vegetated areas, some edges (especially those acquired from a complex perspective), partially shaded portions, and areas with low visibility. On the other hand, the main parts of the buildings and the ground had high confidence values, as expected.

In order to understand the potentialities and limits of a commonly available low-cost sensor in the geometric reconstruction of a historic building, the 3D model of the tower obtained from the smartphone images was extracted for a direct comparison with the point cloud provided by the SLR camera.

Moreover, the images acquired by the drone were used to extract the point cloud of the whole study area (i.e., the top of the hill that hosts the castle and surroundings).

The processing of the TLS point clouds was performed using the Cyclone software version 9.3.1, and the final 3D model was generated starting from the coordinates of the reference topographic network (the stationing of the TLS measurements) and the TLS targets.

After the generation of the different 3D global models, validation was performed for each extracted product by comparing the coordinates of the measured ChPs with those obtained by the processing.

In this context, the availability of CPs allows for the georeferencing of the data, which provides advantages for future restoration projects (here, the FE analysis was devoted to restoration), such as for the castle under investigation, while also helping to preserve the architectural and historical value of the site by providing the only complete and up-to-date survey of the buildings. Similarly, the availability of ChPs allows for an estimation of the absolute accuracy that otherwise would not be possible. Moreover, the comparison between point clouds, used to evaluate the mean distance of the two models, is possible only if the data are co-registered. In the absence of CPs, co-registration can be performed by aligning the SfM point cloud with the TLS model; however, this operation inevitably influences the results of the comparisons.

Finally, the 3D point cloud models of the castle were compared using the CloudCompare software version 2.12 alpha [39] by applying the "M3C2 distance" plugin (multiscale model to model cloud comparison (M3C2)), which is useful for determining robust signed distances directly between two aligned point clouds.

In detail, this tool allows for the calculation of the distance between two point clouds, setting one as the reference with a set of core points (using the full or resampled resolution data). In most cases, the core points are a sub-sampled version, although the computations are performed on the original full dataset. This option speeds up processing and takes into account the fact that computation results are generally required at a lower and more uniform spatial resolution [40].

To evaluate the performances of the low-cost sensors in the photogrammetric surveys, the point cloud obtained from the TLS measurements was used as a reference due to the relatively high resolution, reliability, and precision of the scanning data. Five comparisons were performed between the different datasets: (i) TLS and drone, (ii) TLS and SLR, (iii) TLS and integrated drone/SLR, (iv) TLS and smartphone, and (v) smartphone and SLR. Some specific areas of interest on the tower and castle walls were considered for more detailed distance calculations to better understand the performances of the surveys in relation to the type of surface and limitations in the geometry of the acquisitions.

4. Results and Discussion

4.1. Processing of the Point Clouds

Applying the SfM technique by means of the Agisoft Metashape software, the centers of the images acquired by the drone (Figure 5a,b) and SLR (Figure 5c,d) cameras were calculated (blue flags represent the image plane).

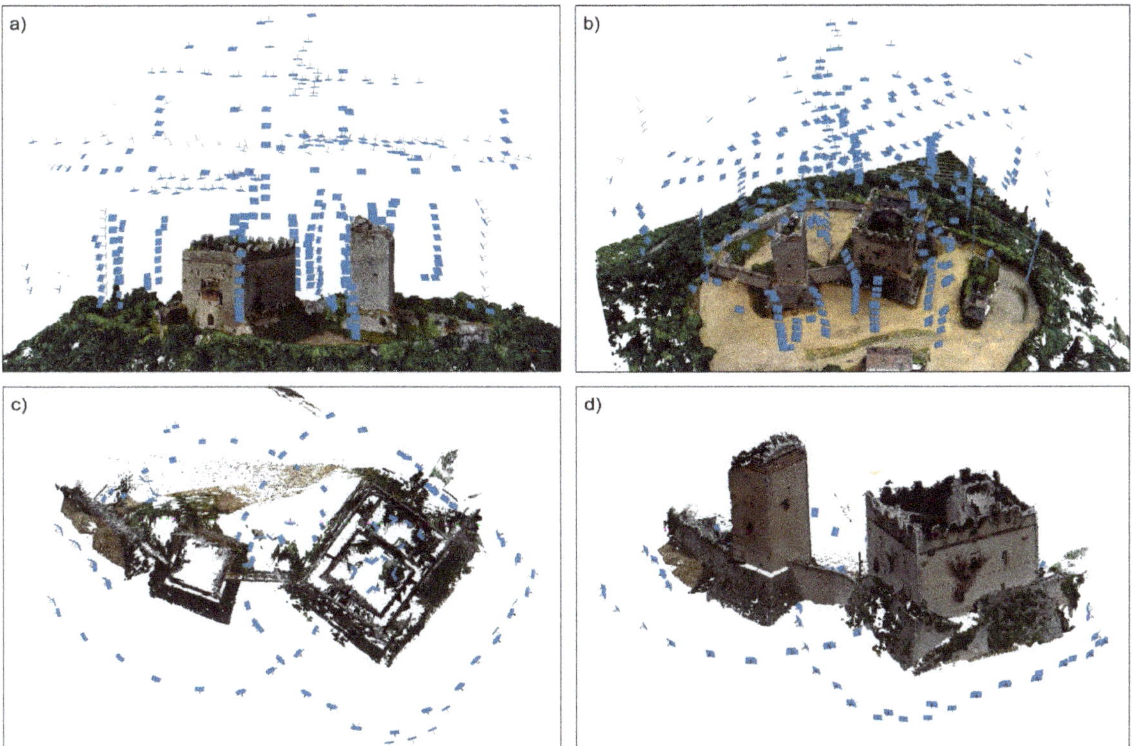

Figure 5. Center of image acquisitions (blue flags represent the image plane); (**a**,**b**) center of images acquired by the drone camera; and (**c**,**d**) center of images acquired by the SLR camera.

Four 3D models (both point cloud and triangular mesh models) of the castle were generated (i) using four hundred and ten images acquired with the drone and fourteen CPs (Figure 6a); (ii) using two hundred and ninety-four images acquired with the SLR camera and nine CPs (Figure 6c); (iii) using eighty-three images acquired with the smartphone camera and five CPs only for the tower (Figure 6e); and (iv) using seven hundred and four images from the drone and SLR camera together with fourteen CPs (Figure 6g). The same detail of the tower extracted using these four models is shown in Figure 6b,d,f,h, respectively.

These results highlight the differences in the four final 3D models, both in terms of the capability of detail representation (data resolution) and photographic quality. In this context, the best result with respect to completeness and cleanliness was provided by the combination of images from the SLR and drone cameras, while both the drone and smartphone models were of lower quality.

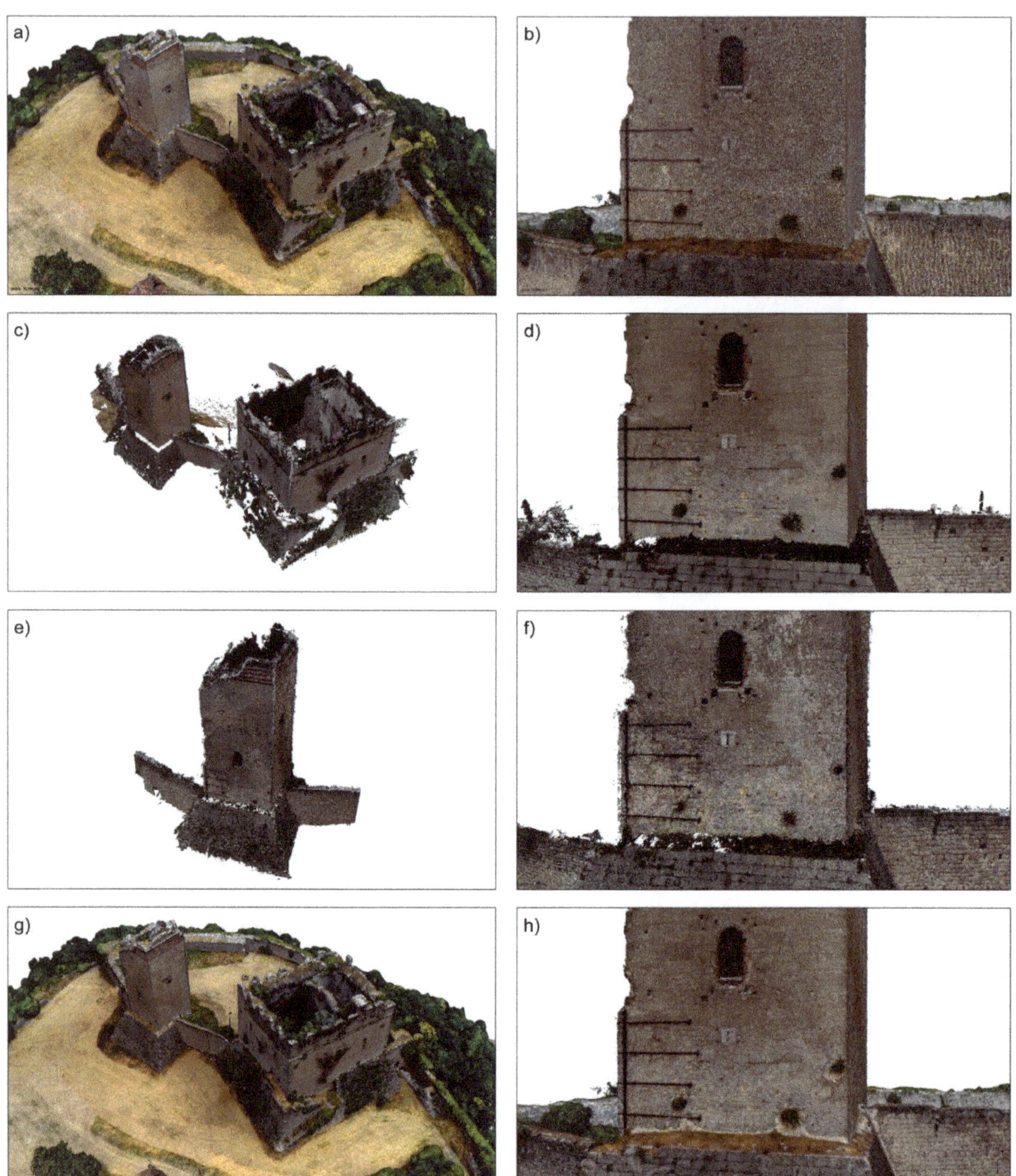

Figure 6. Three-dimensional models obtained via SfM processing of images acquired by the drone (**a**), SLR camera (**c**), and smartphone (**e**), and by the integration between drone and SLR imagery (**g**). A common detailed area of interest of each model is shown in (**b**,**d**,**f**,**h**) for the drone, SLR, smartphone, and drone together with SLR images, respectively.

The geometric accuracy of the models was validated automatically using the Agisoft Metashape software by means of the CPs and ChPs. Table 2 provides the root-mean-square error (RMSE) of the 3D comparison for each model. Taking into account the ChPs, the best result was obtained by the model extracted using images from the SLR camera: however, data derived from the low-cost drone provided a similar RMSE value. Similar accuracies were reported by Capolupo [41], who compared the results of SfM photogrammetry with GCPs and ChPs for the All Saints' Monastery of Cuti (Italy), and Sanseverino et al. [9], who analyzed a structure similar to that under investigation here (Calabrian fortress "Castle of Charles V"). However, in these cases, the surveys were not performed using low-cost sensors and expeditious approaches, highlighting the good results obtained in this work.

Table 2. The number of CPs used in processing, available ChPs (see Figure 4), and comparisons between the 3D coordinates and the four extracted 3D photogrammetric SfM models in terms of RMSE.

3D Model	N. CPs	N. ChPs	RMSE (cm)	
			CPs	ChPs
Drone	14	6	1.4	1.1
SLR	9	4	0.5	1.0
Drone + SLR	14	6	1.1	1.2
Smartphone	5	-	1.9	-

The TLS point clouds were aligned with the Cyclone software using the known coordinates of the six stations and thirteen targets. The co-registration of the scans had an average error of 3 mm, according to the results obtained in similar studies [19,26,42].

Figure 7 shows the obtained final 3D point cloud for the external walls of the tower and the palace, the internal rooms and architectural elements of the palace, and the surrounding area, representing the top of the hill where the structure is located up to the boundary walls.

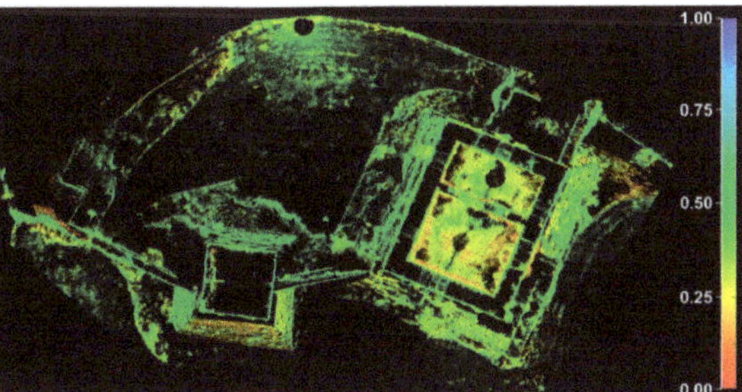

Figure 7. Different points of view of the Illasi Castle 3D model obtained from the alignment of the six TLS point clouds. Colors represent point reflectivity in the range of 0–1. On the right, the two internal rooms of the palace are visible.

4.2. Comparisons between Point Clouds

To evaluate the accuracies of the extracted low-cost photogrammetric point clouds, each of the four SfM 3D models was compared with the TLS reference model. The final TLS model was used as a reference due to the high accuracy and high resolution of the data acquired with the TLS instrument.

No filter was applied to the SfM point clouds in order to directly use all the data obtained from the processing; this choice was made because the noise produced in the data and in the low confidence points, which were different between the four datasets, influenced the comparisons and highlighted specific problems for each type of sensor. The comparisons were performed using the M3C2 distance computation plugin in CloudCompare software.

Figure 8 shows the results of the comparisons between the 3D models—TLS and drone (a), TLS and SLR (b), and TLS and drone + SLR (c)—in terms of distances between the compared point clouds. The colors, which indicate 3D distances, show maximum differences in the order of 20 cm.

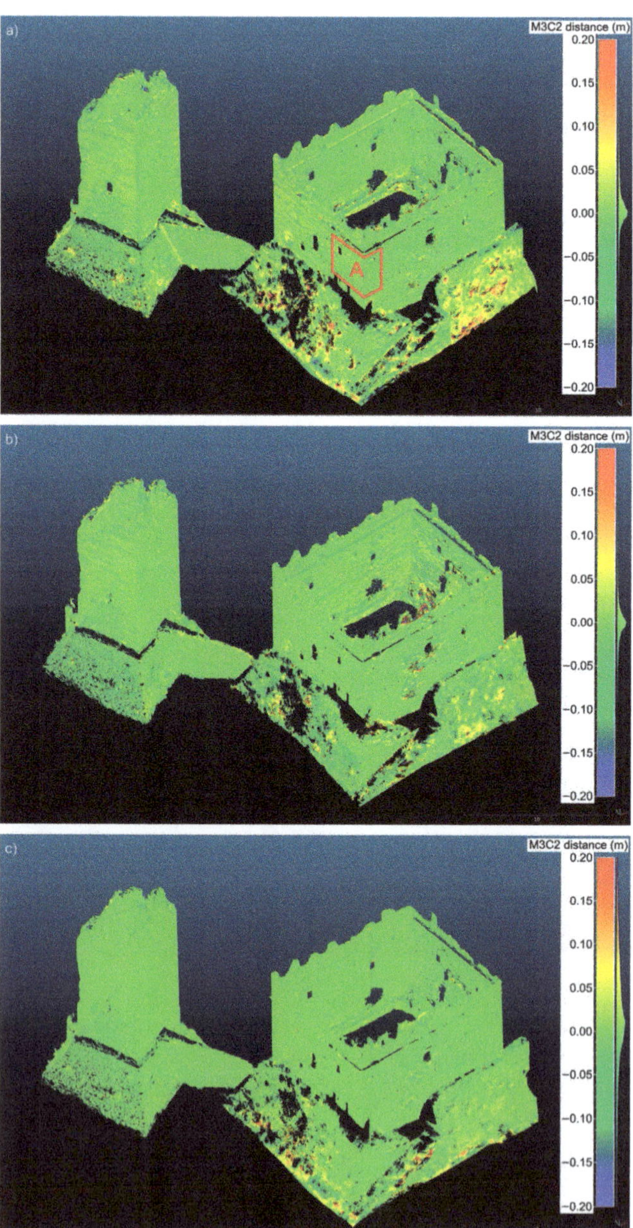

Figure 8. Comparisons between point clouds using the TLS model as reference. TLS and drone (**a**), TLS and SLR (**b**), and TLS and drone + SLR (**c**). Colors represent the 3D distances between models.

The values that could be considered as outliers were mainly located in the lower part of the palace walls and the internal portions, where the vegetation (see Figures 1, 3 and 6) limited the reconstruction of the model, created shadow areas and, above all, represented a surface that was subject to changes during the survey.

The comparisons involving the smartphone model (Figure 9, TLS and smartphone (a) and SLR and smartphone (b)), while presenting the same maximum differences included in the 20 cm range, showed more surfaces that were characterized by high distance values. This was likely due to the lower quality of the smartphone camera. In fact, (i) in the areas of the tower with more differences between the point clouds, the vegetation coverage was irrelevant (see Figures 1, 3 and 6), and (ii) in the same areas, the comparisons between the TLS, SLR, and drone 3D models provided good results (Figure 8). A detailed investigation highlighted that the high values were located in the upper part of the tower, over the walls connecting the tower and the palace and the tower and the boundary walls, and in the inclined walls of the lower portion, where the axis of the camera was far from being perpendicular to the surfaces. These considerations suggest a greater difficulty of the low-cost sensor in the 3D reconstruction of the structures acquired in non-optimal geometric conditions (foreshortened walls, elements in strong perspective, a complex field of view of the surfaces, etc.).

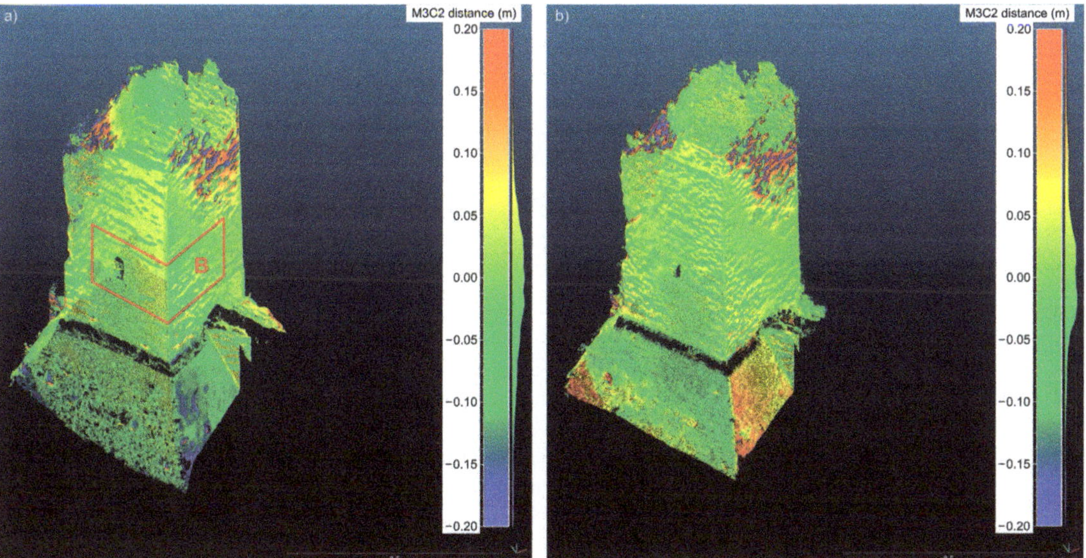

Figure 9. Comparisons between point clouds of the tower. TLS and smartphone (**a**) and SLR and smartphone (**b**). Colors represent the 3D distances between models.

To better analyze the differences between point clouds, three areas were investigated in detail: (i) portion A of the palace (Figure 8); (ii) portion B of the tower (Figure 9); and (iii) the internal surfaces of the palace. The first two cases were considered as two elements of the walls, a corner and an opening, to generalize the study (the analyzed portion of the tower also had good conditions of acquisition for the smartphone).

Table 3 shows the results of the comparisons in terms of mean distance and standard deviation for the analysis of all available data. High standard deviation values, from 9.6 cm to 11.9 cm, were obtained using all the generated 3D data (including the portions with vegetation); these values were drastically reduced after taking into account only areas A and B (i.e., walls without vegetation). In the last case, the best results were obtained when comparing the TLS and SLR models in area A and the TLS and Drone + SLR models in area

B; in addition, the values of the comparison between TLS and SfM were in agreement with those obtained by Monego et al. [43], who compared TLS and SfM 3D models of a rocky spire. Moreover, similar values were obtained by Moyano et al. [44], who validated SfM photogrammetry for heritage applications by analyzing and comparing the point density and the 3D mesh geometry obtained using the TLS technique for a 15th-century façade of Casa de Pilatos in Seville (Spain), a building comparable to the one under investigation. The comparison between smartphone and SLR, although not necessary given the results of the TLS and SLR, and TLS and smartphone comparisons, better clarified the limits, in terms of accuracies, of the SfM technique with the sensors used here.

Table 3. Mean distances and standard deviation values of the comparisons between the TLS model (as a reference) and the obtained SfM point clouds considering all the available data, area A of Figure 8 (palace), area B of Figure 9 (tower), and the internal portion of the palace.

	Mean Distance (cm)	**Standard Deviation (cm)**
Available Data		
TLS and Drone	0.9	10.1
TLS and SLR	−0.5	9.6
TLS and Drone + SLR	0.6	9.6
TLS and Smartphone	−1.5	10.9
Smartphone and SLR	1.5	11.9
Area A		
TLS and Drone	0.2	2.4
TLS and SLR	−0.3	1.8
TLS and Drone + SLR	0.1	2.2
Area B		
TLS and Drone	0.3	2.4
TLS and SLR	0.2	2.7
TLS and Drone + SLR	0.4	2.3
TLS and Smartphone	−0.1	4.7
Smartphone and SLR	−0.1	7.1
Internal Palace		
TLS and Drone + SLR	1.1	7.7

It should be noted that, with respect to the smartphone 3D model (area B), the standard deviation of the comparison with the TLS final point cloud was decreased by 51% compared with the best SfM solution (Drone + SLR). Moreover, a standard deviation of 7.1 cm was obtained when comparing the smartphone and Drone + SLR 3D models. Although these results are noteworthy, only one specific smartphone was used in this work; in future studies, other smartphone sensors will be tested in order to better analyze their performances.

In the internal part of the palace, the comparison with TLS data was performed considering only the SfM 3D model obtained by integrating the images acquired with the drone and SLR cameras. This was because data from the drone were characterized as distorted images due to the impossibility of flying inside the rooms. Similarly, the images acquired from the ground of the upper part of the internal walls with the SLR camera had complex geometry. The integration between the two datasets was carried out to improve the final 3D SfM model. The obtained data were compared with the TLS model, and the differences between the point clouds generated a high standard deviation value (7.7 cm). The reasons for this could be (i) the explained complex photogrammetric geometric acquisitions and/or (ii) the presence of vegetation on the walls that the photogrammetric technique had more difficulty penetrating (see above) compared to the TLS method. For these reasons, the final 3D model of the palace was created via the integration of two contributions: (i) the SfM data derived from the integration of the images acquired with the drone and SLR cameras for the external surfaces due to better coverage of the shadow areas when compared to the TLS acquisitions (there were few TLS stations, Figure 1), and (ii) the TLS point cloud for the internal portion due to the low accuracy of the SfM results (Figure 10). In this case, to provide a reliable 3D model for the FE analysis, the integration of the low-cost and expeditious SfM technique with the high-cost TLS data was necessary. On one side, in general, if the FE model of a substantial structure, such as the one investigated in this work, is not very sensitive to uncertainties of a few centimeters, which are typical of the SfM technique, the results will still be acceptable; on the other hand, the complex conditions of image acquisition and the presence of significant localized vegetation in the internal portion of the palace generated artifacts in the geometry of the 3D photogrammetric model that, in many cases, could influence the structural analysis by changing the results of the FE model. In this context, the ability of the TLS technique to better penetrate the vegetation made it easier to eliminate these artifacts in the subsequent editing operations. For these reasons, the use of the TLS technique—only for critical areas and where it was strictly necessary—slightly extended the time of the survey.

Figure 10. Final 3D model of the palace integrating the SfM results for the external surfaces (drone + SLR, in photorealistic representation) with TLS data for the internal portions (in reflectivity representation, where colors highlight the points' reflectivity in the range of 0–1).

4.3. Structural Analysis

The geometric 3D survey of the surfaces combined with investigations that allowed for the analysis of the characteristics of the materials and construction of the structural elements made it possible to define the resistant system of the palace. From this, once the actions on the building were defined, local and global seismic analyses were carried out on the basis of Italian legislation. For local mechanisms, rigid body kinematics analyses were performed. For global mechanisms, with the aid of an FE model, modal and pushover analyses (static non-linear) were carried out in the four main directions.

The FE model was created starting from the 3D survey. For this type of degraded historical building, the simplified 3D model was acceptable [45]. The geometric volume was created using the Leica Cyclone 3DR and Autodesk AutoCAD 2020 software version 23.1.47.0. Plans and sections were automatically extracted from the 3D model by imposing an interdistance of 1 m using the former software: a reference plane and the offset value were fixed in order to progressively intersect the point cloud along a specific direction. This creates a set of planar sections in the form of polylines, computed by interpolation between the points of the cloud and the reference plane. Subsequently, using the latter software, polylines were imported and edited to remove noise and vegetation. Points in the range of significant cut planes were used as a reference to generate, through solid modeling procedures, basic 3D shapes; from 3D shapes, the complex geometry of the building was then generated with Boolean operations. This was possible because the FE analysis of massive structures, such as the one investigated in this study, generally does not require detailed models; a few centimeters of 3D model uncertainty do not lead to significant variations in the FE model. The solid model obtained, which contains information on all geometric properties (vertices, edges, faces, and volumes) was imported into Midas FEA NX software version 1.2.0.

Starting from the geometry, the 3D FE discretization was carried out; the properties of the materials were attributed to the finite elements (separately for each of the wall types identified in the survey phase for the different portions of the walls [24,46]). Moreover, the gravitational loads for the pushover analysis were applied (Figure 11). The obtained results for one of the analyzed directions (in particular, along the X-axis) in terms of main strain and main stresses are shown in Figure 12.

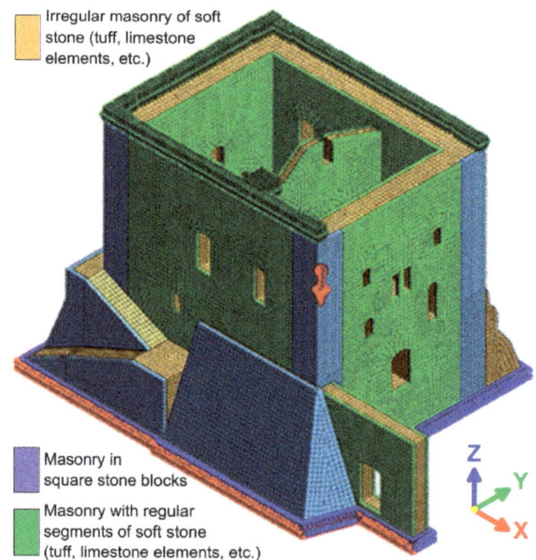

Figure 11. FE model of the palace, where each wall type is represented by a different color (Midas FEA NX software).

These results were confirmed by applying the same structural analysis procedure to the 3D model extracted using only the high-cost and high-precision TLS data; no significant differences were obtained in the FE analysis with respect to the expeditious and low-cost 3D model.

On the basis of the analyses of the local and global collapse mechanisms, a safety assessment of the structure was carried out. The controls on the mechanisms were performed in terms of the relationship between seismic capacity and seismic acceleration demand. For the unsatisfied controls, an increase in safety level must be obtained with global or local interventions.

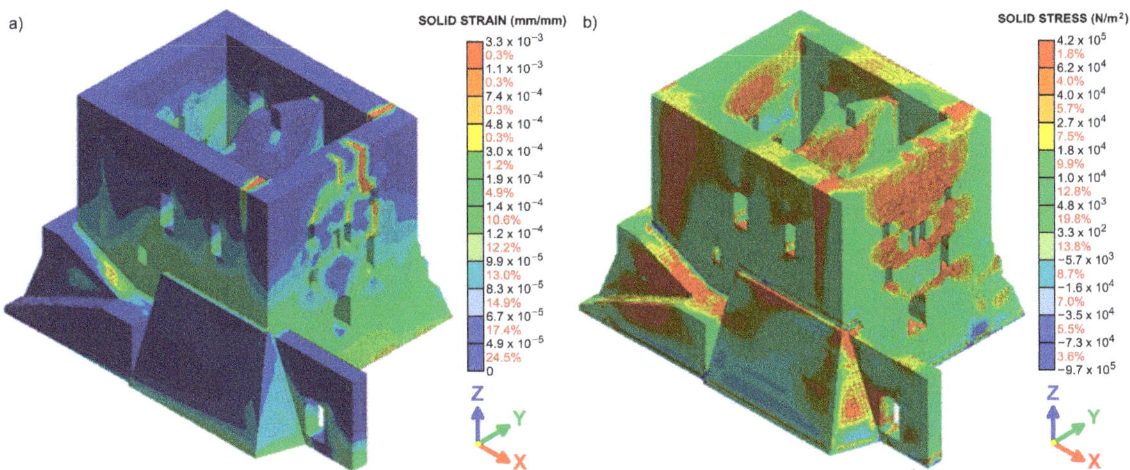

Figure 12. Contours of main strains (**a**) and main stress (**b**) from the pushover analysis for one of the analyzed directions (along the X-axis). The red numbers, in each interval, indicate the percentage of finite elements that assumed values of solid strains and solid stress within the range of values in black.

5. Conclusions

In this work, low-cost sensors and expeditious surveys were applied to generate reliable 3D models. Illasi Castle, a degraded historical building in Italy, was surveyed for structural analysis. Since very high accuracies were not required for this type of object, a low-cost drone, an SLR camera, and a smartphone were used to acquire images, and 3D models were extracted by applying SfM photogrammetry. Together with the images, a TLS survey was performed to validate the SfM models and integrate the 3D data where unsatisfactory results were obtained. All data were acquired in a single working day.

SfM photogrammetry using low-cost sensors can generate geometrically reliable point clouds. In the application considered here, an RMSE of up to 2–3 cm could be achieved. In this context, the use of a consumer-level smartphone (sensors that are increasingly used in SfM) demonstrated adequate performance considering the purposes of the survey, with a standard deviation of up to 4–7 cm when compared to the TLS and standard SfM data. In particular, taking into account the smartphone sensor used in this work, performances were strongly influenced by the geometric acquisitions, with standard deviation values that degraded rapidly when moving away from optimal conditions. It should be noted that further investigations are still required. Future studies should consider the use of different smartphones, conditions of acquisition, objects (surfaces), inclinations, fields of view, perspectives, camera–object distances, etc.

In the case study presented here, ground-based TLS acquisitions generated a shadow area in the upper portion of the walls. In this case, the use of a drone equipped with a laser scanner sensor increases the costs of the survey, while the high accuracies obtained using this technique in many cases are not required.

From the results of the structural analysis, comparable values for the relationship between the capacity and demand of the local and global mechanisms affecting the same portion of the structure were observed. Finally, a correlation between the crack pattern of the structures and the concentrations of main strains and stresses in the FE model was observed from the global analyses.

Author Contributions: Conceptualization, M.F.; methodology, M.F. and M.M.; software, M.F., P.F.G. and M.M.; validation, M.F., P.F.G. and M.M.; formal analysis, M.F. and M.M.; investigation, M.F. and M.M.; resources, M.F.; data curation, M.F.; writing—original draft preparation, M.F.; writing—review and editing, M.F., P.F.G. and M.M.; visualization, M.F. and M.M.; supervision, M.F.; project administration, M.F. All authors have read and agreed to the published version of the manuscript.

Funding: This research received no external funding.

Data Availability Statement: Not applicable.

Acknowledgments: The authors would like to thank the following Institutions: the Laboratory of Geomatics and Surveying and the Department of Civil, Environmental, and Architectural Engineering staff of the University of Padova.

Conflicts of Interest: The authors declare no conflict of interest.

References

1. Grussenmeyer, P.; Landes, T.; Voegtle, T.; Ringle, K. Comparison methods of terrestrial laser scanning, photogrammetry and tacheometry data for recording of cultural heritage buildings. In Proceedings of the International Archives of the Photogrammetry, Remote Sensing and Spatial Information Sciences, Beijing, China, 3–11 July 2008; pp. 213–218.
2. Fabris, M.; Boatto, G.; Achilli, V. 3D Laser scanning surveys in the modelling of cultural heritage. In *Recent Advances in Non-Destructive Inspection*; Meola, C., Ed.; Nova Science Publishers: New York, NY, USA, 2011; pp. 1–31.
3. Fabris, M.; Achilli, V.; Artese, G.; Bragagnolo, D.; Menin, A. High resolution survey of Phaistos Palace (Crete) by TLS and terrestrial photogrammetry. In Proceedings of the International Archives of the Photogrammetry, Remote Sensing and Spatial Information Sciences, Melbourne, Australia, 25 August–1 September 2012; Volume XXXIX-B5, pp. 81–86. [CrossRef]
4. Moropoulou, A.; Labropoulos, K.; Delegou, E.T.; Karoglou, M.; Bakolas, A. Non-destructive techniques as a tool for the protection of built cultural heritage. *Constr. Build. Mater.* **2013**, *48*, 1222–1239. [CrossRef]
5. Santagati, C.; Lo Turco, M. From structure from motion to historical building information modeling: Populating a semantic-aware library of architectural elements. *J. Electron. Imaging* **2017**, *26*, 011007. [CrossRef]
6. Bartoš, K.; Pukanská, K.; Repáň, P.; Kseňak, L.; Sabová, J. Modelling the Surface of Racing Vessel's Hull by Laser Scanning and Digital Photogrammetry. *Remote Sens.* **2019**, *11*, 1526. [CrossRef]
7. Selvaggi, I.; Bitelli, G.; Serantoni, E.; Wieser, A. Point cloud dataset and FEM for a complex geometry: The San Luzi bell tower case study. In Proceedings of the GEORES 2019—2nd International Conference of Geomatics and Restoration, Milan, Italy, 8–10 May 2019; Volume XLII-2/W11, pp. 1047–1052. [CrossRef]
8. Plata, A.R.M.d.l.; Franco, P.A.C.; Franco, J.C.; Gibello Bravo, V. Protocol Development for Point Clouds, Triangulated Meshes and Parametric Model Acquisition and Integration in an HBIM Workflow for Change Control and Management in a UNESCO's World Heritage Site. *Sensors* **2021**, *21*, 1083. [CrossRef] [PubMed]
9. Sanseverino, A.; Messina, B.; Limongiello, M.; Guida, C.G. An HBIM Methodology for the Accurate and Georeferenced Reconstruction of Urban Contexts Surveyed by UAV: The Case of the Castle of Charles V. *Remote Sens.* **2022**, *14*, 3688. [CrossRef]
10. Berrett, B.E.; Vernon, C.A.; Beckstrand, H.; Pollei, M.; Markert, K.; Franke, K.W.; Hedengren, J.D. Large-Scale Reality Modeling of a University Campus Using Combined UAV and Terrestrial Photogrammetry for Historical Preservation and Practical Use. *Drones* **2021**, *5*, 136. [CrossRef]
11. Xu, Z.; Wu, L.; Shen, Y.; Li, F.; Wang, Q.; Wang, R. Tridimensional Reconstruction Applied to Cultural Heritage with the Use of Camera-Equipped UAV and Terrestrial Laser Scanner. *Remote Sens.* **2014**, *6*, 10413–10434. [CrossRef]
12. Bitelli, G.; Dellapasqua, M.; Girelli, V.; Sanchini, E.; Tini, M. 3D geomatics techniques for an integrated approach to Cultural Heritage knowledge: The case of San Michele in Acerboli's church in Santarcangelo di Romagna. In Proceedings of the International Archives of the Photogrammetry, Remote Sensing and Spatial Information Sciences, Florence, Italy, 22–24 May 2017; Volume XLII-5/W1, pp. 291–296. [CrossRef]
13. Monego, M.; Menin, A.; Fabris, M.; Achilli, V. 3D survey of Sarno Baths (Pompeii) by integrated geomatic methodologies. *J. Cult. Herit.* **2019**, *40*, 240–246. [CrossRef]
14. Chen, X.; Achilli, V.; Fabris, M.; Menin, A.; Monego, M.; Tessari, G.; Floris, M. Combining Sentinel-1 Interferometry and Ground-Based Geomatics Techniques for Monitoring Buildings Affected by Mass Movements. *Remote Sens.* **2021**, *13*, 452. [CrossRef]
15. Betti, M.; Bonora, V.; Galano, L.; Pellis, E.; Tucci, G.; Vignoli, A. An Integrated Geometric and Material Survey for the Conservation of Heritage Masonry Structures. *Heritage* **2021**, *4*, 585–611. [CrossRef]
16. Bartoli, G.; Betti, M.; Facchini, L.; Orlando, M. Non-destructive characterization of stone columns by dynamic test: Application to the lower colonnade of the Dome of the Siena Cathedral. *Eng. Struct.* **2012**, *45*, 519–535. [CrossRef]
17. Lysandrou, V.; Agapiou, A. Comparison of documentation techniques for the restoration and rehabilitation of cultural heritage monuments: The example of Pyrgos 'Troulli' medieval tower in Cyprus. In Proceedings of the Third International Euro-Mediterranean Conference, EuroMed 2010, Lemessos, Cyprus, 8–13 November 2010.

18. Solla, M.; Gonçalves, L.M.S.; Gonçalves, G.; Francisco, C.; Puente, I.; Providência, P.; Gaspar, F.; Rodrigues, H. A Building Information Modeling Approach to Integrate Geomatic Data for the Documentation and Preservation of Cultural Heritage. *Remote Sens.* **2020**, *12*, 4028. [CrossRef]

19. Pieraccini, M.; Dei, D.; Betti, M.; Bartoli, G.; Tucci, G.; Guardini, N. Dynamic identification of historic masonry towers through an expeditious and no-contact approach: Application to the "Torre del Mangia" in Siena (Italy). *J. Cult. Herit.* **2014**, *15*, 275–282. [CrossRef]

20. Xu, W.; Neumann, I. Finite Element Analysis based on A Parametric Model by Approximating Point Clouds. *Remote Sens.* **2020**, *12*, 518. [CrossRef]

21. Ding, Y.; Zheng, X.; Zhou, Y.; Xiong, H.; Gong, J. Low-Cost and Efficient Indoor 3D Reconstruction through Annotated Hierarchical Structure-from-Motion. *Remote Sens.* **2019**, *11*, 58. [CrossRef]

22. Shih, N.-J.; Wu, Y.-C. AR-Based 3D Virtual Reconstruction of Brick Details. *Remote Sens.* **2022**, *14*, 748. [CrossRef]

23. Tucci, G.; Bartoli, G.; Betti, M.; Bonora, V.; Korumaz, M.; Korumaz, A.G. Advanced procedure for documenting and assessment of Cultural Heritage: From Laser Scanning to Finite Element. *IOP Conf. Ser. Mater. Sci. Eng.* **2018**, *364*, 012085. [CrossRef]

24. Ghiassi, B.; Vermeltfoort, A.; Lourenço, P. Masonry Mechanical Properties. In *Numerical Modelling of Masonry and Historical Structures—From Theory to Application*, 1st ed.; Ghiassi, B., Milani, G., Eds.; Woodhead Publishing: Sawston, UK; Elsevier: Cambridge, UK, 2019; pp. 239–261.

25. Hinks, T.; Carr, H.; Truong-Hong, L.; Laefer, D.F. Point Cloud Data Conversion into Solid Models via Point-Based Voxelization. *J. Surv. Eng.* **2013**, *139*, 72–83. [CrossRef]

26. Barazzetti, L.; Banfi, F.; Brumana, R.; Gusmeroli, G.; Oreni, D.; Previtali, M.; Schiantarelli, G. BIM from laser clouds and finite element analysis: Combining structural analysis and geometric complexity. In Proceedings of the International Archives of the Photogrammetry, Remote Sensing and Spatial Information Sciences, Ávila, Spain, 25–27 February 2015; Volume XL-5/W4, pp. 345–350. [CrossRef]

27. Castellazzi, G.; D'Altri, A.M.; Bitelli, G.; Selvaggi, I.; Lambertini, A. From Laser Scanning to Finite Element Analysis of Complex Buildings by Using a Semi-Automatic Procedure. *Sensors* **2015**, *15*, 18360–18380. [CrossRef]

28. Sánchez-Aparicio, L.J.; Villarino, A.; García-Gago, J.; González-Aguilera, D. Photogrammetric, Geometrical, and Numerical Strategies to Evaluate Initial and Current Conditions in Historical Constructions: A Test Case in the Church of San Lorenzo (Zamora, Spain). *Remote Sens.* **2016**, *8*, 60. [CrossRef]

29. Bassier, M.; Hardy, G.; Bejarano-Urrego, L.; Drougkas, A.; Verstrynge, E.; Van Balen, K.; Vergauwen, M. Semi-automated Creation of Accurate FE Meshes of Heritage Masonry Walls from Point Cloud Data. In *Structural Analysis of Historical Constructions: An Interdisciplinary Approach*; RILEM Bookseries, 18; Springer International Publishing: Cham, Switzerland, 2019; pp. 305–314. [CrossRef]

30. Rissolo, D.; Hess, M.R.; Huchim Herrera, J.; Lo, E.; Petrovic, V.; Amador, F.E.; Kuester, F. Comprehensive digital documentation and preliminary structural assessment of Satunsat: A unique Maya architectural labyrinth at Oxkintok, Yucatan, Mexico. In Proceedings of the International Archives of the Photogrammetry, Remote Sensing and Spatial Information Sciences, Ávila, Spain, 1–5 September 2019; Volume XLII-2/W15, pp. 989–992. [CrossRef]

31. Bruno, N.; Coïsson, E.; Diotri, F.; Ferrari, L.; Mikolajewska, S.; Morra di Cella, U.; Roncella, R.; Zerbi, A. History, Geometry, Structure: Interdisciplinary analysis of a historical bridge. In Proceedings of the International Archives of the Photogrammetry, Remote Sensing and Spatial Information Sciences, Milan, Italy, 8–10 May 2019; Volume XLII-2/W11, pp. 317–323. [CrossRef]

32. Fang, J.W.; Sun, Z.; Zhang, Y.R. TLS-FEM integrated structural deformation analysis on the Beamless Hall at Nanjing, China. In Proceedings of the International Archives of the Photogrammetry, Remote Sensing and Spatial Information Sciences, Beijing, China, 28 August–1 September 2021; Volume XLVI-M-1-2021, pp. 215–220. [CrossRef]

33. Yang, H.; Xu, X. Structure monitoring and deformation analysis of tunnel structure. *Compos. Struct.* **2021**, *276*, 114565. [CrossRef]

34. State Archive of Verona. Private Archive of Pompei Family, (ASVr, Pompei), *section XXVI*, 68, *section XXVII*, 71, *section XXVII*, 73.

35. Saggioro, F.; Varanini, G.M. Il castello di Illasi. Storia e archeologia. In *Archaeologica*; Bretschneider, G., Ed.; università degli studi di padova: Rome, Italy, 2009; Volume 151, p. 300. ISBN 9788876892370.

36. Saggioro, F.; Mancassola, N. Il castello di Illasi (VR): Dati archeologici sull'insediamento medioevale. In *Paesaggi, Comunità, Villaggi Medievali*; Fondazione Centro Italiano di Studi sull'Alto Medioevo: Spoleto (Perugia), Italy, 2012; pp. 639–644. ISBN 9788879883474.

37. Sanz-Ablanedo, E.; Chandler, J.H.; Rodríguez-Pérez, J.R.; Ordóñez, C. Accuracy of Unmanned Aerial Vehicle (UAV) and SfM Photogrammetry Survey as a Function of the Number and Location of Ground Control Points Used. *Remote Sens.* **2018**, *10*, 1606. [CrossRef]

38. Ferrer-González, E.; Agüera-Vega, F.; Carvajal-Ramírez, F.; Martínez-Carricondo, P. UAV Photogrammetry Accuracy Assessment for Corridor Mapping Based on the Number and Distribution of Ground Control Points. *Remote Sens.* **2020**, *12*, 2447. [CrossRef]

39. CloudCompare (Version 2.12 Alpha). Available online: http://www.cloudcompare.org/ (accessed on 15 July 2022).

40. Lague, D.; Brodu, N.; Leroux, J. Accurate 3D comparison of complex topography with terrestrial laser scanner: Application to the Rangitikei canyon (N-Z). *ISPRS J. Photogramm. Remote Sens.* **2013**, *82*, 10–26. [CrossRef]

41. Capolupo, A. Accuracy Assessment of Cultural Heritage Models Extracting 3D Point Cloud Geometric Features with RPAS SfM-MVS and TLS Techniques. *Drones* **2021**, *5*, 145. [CrossRef]

42. Monego, M.; Fabris, M.; Menin, A.; Achilli, V. 3-D Survey applied to industrial archaeology by tls methodology. In Proceedings of the International Archives of the Photogrammetry, Remote Sensing and Spatial Information Sciences, Florence, Italy, 22–24 May 2017; Volume XLII-5/W1, pp. 449–455. [CrossRef]
43. Monego, M.; Achilli, V.; Fabris, M.; Menin, A. 3-D Survey of Rocky Structures: The Dolomitic Spire of the Gusela del Vescovà. *Commun. Comput. Inf. Sci.* **2020**, *1246*, 211–228. [CrossRef]
44. Moyano, J.; Nieto-Julián, J.E.; Bienvenido-Huertas, D.; Marín-García, D. Validation of Close-Range Photogrammetry for Architectural and Archaeological Heritage: Analysis of Point Density and 3D Mesh Geometry. *Remote Sens.* **2020**, *12*, 3571. [CrossRef]
45. Quattrini, R.; Clementi, F.; Lucidi, A.; Giannetti, S.; Santoni, A. From TLS to FE analysis: Points cloud exploitation for structural behaviour definition. The San Ciriaco's Bell Tower. In Proceedings of the International Archives of the Photogrammetry, Remote Sensing and Spatial Information Sciences, Ávila, Spain, 1–5 September 2019; Volume XLII-2/W15, pp. 957–964. [CrossRef]
46. De Villiers, W.I. Computational and Experimental Modelling of Masonry Walling towards Performance-Based Standardisation of Alternative Masonry Units for Low-Income Housing. Dissertation Presented for the Degree of Doctor of Philosophy in Engineering in the Faculty of Engineering, at Stellenbosch University, Stellenbosch, South Africa. 2019. Available online: https://core.ac.uk/download/pdf/268883067.pdf (accessed on 6 September 2022).

MDPI

St. Alban-Anlage 66

4052 Basel

Switzerland

www.mdpi.com

Drones Editorial Office

E-mail: drones@mdpi.com

www.mdpi.com/journal/drones